The ASCENT *of* BIRDS

The ASCENT *of* BIRDS

How Modern Science
is Revealing their Story

JOHN REILLY

Published by Pelagic Publishing
www.pelagicpublishing.com
PO Box 874, Exeter EX3 9BR, UK

The Ascent of Birds: How Modern Science is Revealing their Story

ISBN 978-1-78427-203-6 (Pbk)
ISBN 978-1-78427-170-1 (ePub)
ISBN 978-1-78427-171-8 (PDF)

Originally published in 2018
This paperback edition published in 2019

British Library Cataloguing in Publication Data
A catalogue record for this book is available from the British Library.

Also by John Reilly:
Greetings from Spitsbergen: Tourists at the Eternal Ice 1827–1914
Spitsbergen's Early Postcards: an Annotated Catalogue

Cover image: paintings by Jon Fjeldså, Natural History Museum of Denmark; phylogeny by Guojie Zhang, Avian Phylogenomics Consortium

FSC
www.fsc.org
MIX
Paper from responsible sources
FSC® C114687

To Jon Fjeldså for the idea to start, and Nick Davies and Gareth Dyke for the encouragement to finish

'The Creation is never over. It had a beginning but has no ending. Creation is always busy making new scenes, new things, and new Worlds.'

Immanuel Kant (1724–1804)

'The time will come I believe, though I shall not live to see it, when we shall have fairly true genealogical trees of each great kingdom of nature.'

Charles Darwin (1809–1882)

'You can know the name of that bird in all the languages of the world, but when you're finished, you'll know absolutely nothing whatever about the bird. You'll only know about humans in different places and what they call the bird. So let's look at the bird and see what it is doing – that's what counts.'

Richard Feynman (1918–1988)

Contents

Acknowledgements

The idea for this book was the result of the chance input of two individuals: Nik Borrow, who first raised the issue of picathartes speciation, and Jon Fjeldså, who kindly provided the answer. In their different ways, both were responsible for kick-starting this three-year project – and for that I am especially grateful.

I would also like to thank the many professional guides whose field skills enabled me to observe first-hand the leading characters of this book, and in so doing opened my eyes to the bewildering variation and complexity of the world's avifauna. I am also indebted to those scientists who replied to my pestering emails, forwarded relevant papers, provided helpful suggestions, and allowed me to use their photographs and artwork. I have listed all of the above in alphabetical order: Peter Alfrey, Craig Benkman, Charles Bishop, Nik Borrow, Leslie Christidis, Stefan Christmann, Julia Clarke, Joel Cracraft, Willie de Vries, Jared Diamond, Jon Fjeldså, Ewan Fordyce, Nicole Fuller, Robert Furness, Peter Garrity, Chris Gaskin, Peter and Rosemary Grant, Lucy Hawkes, Jason Horn, Des Hume, Walter Jetz, Daniel Ksepka, Tim Laman, Markus Lilje, Kevin McCracken, Andrew Meade, John Megahan, Jin Meng, Steve Mills, Ian Montgomery, Pete Morris, Rolf Nussbaumer, János Oláh, Nuno Oliveira, Mark Pagel, Tony Parker, Eduardo Patrial, Lars Petersson, Richard Prum, Petra Rank, Forrest Rowland, Edwin Scholes, Graham Scott, Victor Soria-Carrasco, Chubzang Tangbi, Shaun Templeton, Gavin Thomas, Daniel Thomas, Andreas Trepte, Mark van Beirs, Katrina van Grouw, Dylan van Winkel, Chris Venditti and Christopher C. Witt. I am especially indebted, however, to Gerald Mayr, who kindly checked the manuscript for errors relating to palaeontology, and whose encyclopaedic knowledge and attention to detail has saved me many a blush.

Acknowledgement of all those who helped, of course, does not imply their endorsement of the views expressed in this book; I alone am answerable for them. Similarly, any errors of fact or omissions are entirely my responsibility.

Thanks must also go to Nick Davies and Gareth Dyke for their confidence in the project, and to Frank Ryan for his gentle editorial guidance. I would like to thank my son Mark for his patience in producing the many drafts of the figures. I am indebted to the skills of copy-editor Hugh Brazier, whose eagle eyes have spotted more grammatical and factual errors than I would care to admit. I am also grateful to Nigel Massen and his team at Pelagic Publishing, all of whom have helped transform my words and pictures into the book you now hold in your hands.

Last of all, but by no means least, I would like to thank Janette, my wife and birding companion, who convinced me that the writing of this book was not beyond me, and who provided the support needed during the times of self-doubt.

Illustrations

FIGURES

PLATES

28. King of Saxony Bird-of-Paradise
29. Trumpet Manucode: male trachea
30. Greater Blue-eared Starling; Superb Starling
31. Common Blackbird
32. Italian Sparrow; House Sparrow; Spanish Sparrow
33. Zebra Finch
34. Louisiade White-eye
35 Cassia Crossbill
36. Darwin's finches: Grey Warbler-Finch; Common Cactus Finch; Large Ground Finch
37. Masked Flowerpiercer

Timeline

Years before present	Major events
4.5 billion	Formation of Earth
3.8 billion	Emergence of life on Earth
231–66 million	Age of dinosaurs
150 million	Archaeopteryx
90–70 million	Ratites evolve into separate lineages
90 million	Shorebirds
80–40 million	New Zealand wrens (Acanthisittidae)
72.8 million	Oilbird
71.4 million	Suboscine–oscine divergence
66.5 miilion	*Vegavis iaai*
66 million	K–Pg mass extinction event
64.2 million	Suboscine divergence into New World and Old World lineages
62 million	Oldest penguin fossil (*Waimanu manneringi*)
60 millon	Emergence of owls
58.5–53.8 million	Storm petrels
58.0 million	Diversification of crown lineage of parrots
55.6 million	Diversification of pittas and broadbills
52.5 million	Sapayoa splits from broadbills
52.3 million	Albatrosses (Diomedeidae)
50.6 million	Pan-Apodiformes fossil *Eosypselus rowei*
48 million	Hummingbirds split from their sister group, the swifts and treeswifts
47 million	Picathartes; rockjumpers; Rail-babbler
45 million	Core corvoids and transitory oscines emerged from the proto-Papuan archipelago
41.2 million	Origin of Asity lineage
41 million	South America separates from Antarctica
35 million	Australia separates from Antarctica
28 million	Common ancestor of ducks, geese and swans.

24 million	Origins of the birds-of-paradise (Paradisaeidae)
23.5 million	Coscoroba Swan; Cape Barren Goose
23 million	Diversification of pittas
22 million	Common Hummingbird ancestor reaches South America
22 million	Oxpeckers (Buphagidae)
21 million	Mockingbirds (Mimidae)
20 million	Old World Sparrows (*Passer*)
	Common ancestor of geese and swans
17.5 million	Genus *Corvus*
16.5 million	Estrildid finches
16.2 million	Origin of African pittas
16 million	Starlings and mynas (Sturnidae)
10–15 million	Marine flooding of Amazon
9 million	Origin of Australian pittas
	Divergence of crossbills (genus *Loxia*)
5–3 million	Formation of Panamanian isthmus
4.5 million	Bar-headed Goose
3 million	Taita and Abyssinian Thrushes
2 million	White-eyes (genus *Zosterops*)
1.4 million	Steamer ducks
600,000	Hispaniolan Crossbill
100,000	Madeiran and Monteiro's Storm Petrels
12,000–10,000	Agricultural revolution
8,000	Italian Sparrow
6,000	Cassia Crossbill
600	Extinction of the moas
356	Extinction of the Dodo
18	Extinction (in the wild) of Spix's Macaw

Geological Ages

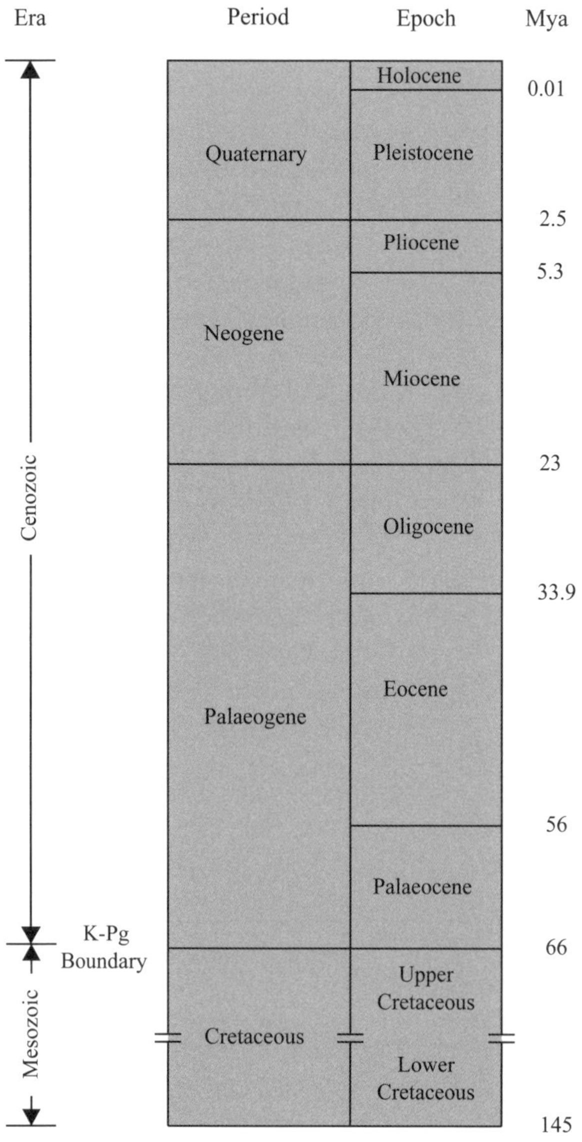

Prologue

EVOLUTION OF AN IDEA

It was the treasury of forest sounds that I remember most, that and the heat. The low buzz of tiresome bees about the head, drawn by rivulets of sweat; the scuttling of unknown animals in the dense undergrowth and the alarm call of a startled bird; all recollections spawned by the enhanced acuity of eager anticipation. While, sadly, the seemingly ubiquitous whirr of a distant chainsaw sullied the oppressive air.

Our small party had left the remote Ghanaian village of Bonkro several hours before, with its throng of inquisitive schoolchildren and equally bemused adults: scattered adobe huts and palm-leaf shelters nestling among fields of corn and cocoa. In single file, we inched our way up the trail's sinuous windings, following our local guide higher and higher towards the summit ridge with its overhanging cliffs. For it is here, amid west Africa's Upper Guinean rainforest, that one of the world's most sought-after and enigmatic birds is to be found, the elusive Yellow-headed Picathartes, or White-necked Rockfowl.

After a strenuous final push, we emerged from the dappled forest light and found ourselves beneath a small rocky outcrop where the Picathartes' cup-shaped nests were clearly visible, plastered against the shaded cliff. However, true to form, the birds were not at home. It was mid-afternoon, and they were out foraging deep in the surrounding forest, searching for their wide range of invertebrate prey. While this feeding strategy renders them difficult to observe during the day, a few individuals typically return to their nesting site to roost. As a result, we settled down on several makeshift wooden benches and awaited their arrival.

Not a word was spoken as we sat, each immersed in a frisson of expectancy and anxiety, oblivious to the cooling air and the lengthening shadows. Suddenly, the briefest of movements caught my eye, a vague suggestion, and I sensed the group's alertness. Moments later, a ghost-like form topped a distant rock, only to vanish instantaneously. We collectively held our breaths, unwilling to accept that the long-awaited encounter could be so transient. Seconds became minutes, and then, as if with a flick of a conjurer's cloak, two birds appeared, nonchalantly preening on the nearest of rocks, seemingly oblivious to our presence.

The Picathartes is one of the most extraordinary birds on the planet, with its bright yellow featherless head that resembles a falconer's hood with interstices for jet-black eyes, nape and beak. The bird's pure-white underbelly and silvery-grey legs add to the spectral illusion (Plate 1A).

All too soon, our time was up, and the performance was over. The birds hopped and bounced in turn towards the cliff's darkest recesses, hesitated momentarily, and then disappeared for good. Dusk had fallen when we began our return to the village, and already the forest's night players were out in force, buzzing and thrumming about our heads. In Ghana, darkness conquers twilight fast, and our descent became an obstacle course of arboreal snares – gnarled tree roots, intertwined lianas and low-slung gossamer-coated branches – in a world defined by the throw of our head-torches. Except, that is, for the evanescent fireflies dancing drunkenly in the still air and the occasional whirring night bird.

Once we had gained the lower slopes and the difficulties had eased, I became lost in thought, reliving the day's events and reflecting on one of the most wondrous of avian encounters. The intense experience negated the relegation of the Yellow-headed Picathartes to a perfunctory 'tick' on my world list. It deserved so much more, and as we neared the village I determined to learn as much as I could about this iconic bird. However, it soon became apparent that the information available from both field guides and fellow birders was insufficient to answer a query that had begun to dominate my thoughts. Why are there two species of picathartes? While both birds are restricted to the Guinea–Congolian rainforest of west Africa, the Yellow-headed Picathartes is found in fragmented populations from Guinea to Ghana while the Grey-necked Picathartes lives further east in Cameroon and Gabon (Plate 1B).[1] Intriguingly, there is no geographical overlap between the two species, despite them having almost identical habitat requirements, plumage patterns and biology. How could such a situation have arisen?

Picathartes are undoubtedly avian oddities, exhibiting a smörgåsbord of biological characteristics. Indeed, to quote the mischievous musings of the ornithologist William Serle, writing in the 1950s, 'they are bald like vultures, lay eggs like crows, build nests like swallows and have the cranial bones of a starling.'[2] Not surprisingly, taxonomists have struggled to classify picathartes, and the birds have been assigned variously to the starling, babbler and crow families. Today they have been accorded a family of their own, the Picathartidae.[3] But why are the picathartes so different as to warrant separate family status? Furthermore, recent molecular studies have revealed the picathartes' closest relatives, or sister taxa, to be the rockjumpers of South Africa and the forest floor-dwelling Rail-babbler of Malaysia, Borneo and Sumatra. Such unexpected kinships must surely hold significant evolutionary clues, but what could they be?

These puzzling questions, posed during a state of post-picathartes euphoria, continued to intrigue and demand answers on my return. While convincing explanations seemed elusive, I was ultimately beguiled by the 'forest refugia' theory, a hypothesis in which climatic and vegetative upheavals during the Pleistocene's ice ages are thought to have led to fragmentation of previously continuous forest habitats. Was it possible that such events had resulted in the early picathartes population being split, leading to the evolution of the two extant species? I was aware that Jürgen Haffer, a German geologist working in Colombia, had first proposed this theory to account for the distribution of some Amazonian birds.[4] Seeking clarification, I sent an outline of my tentative thoughts to several experts for comment.

Fortunately, one of my emails was kindly forwarded to Professor Jon Fjeldså, curator of the Natural History Museum of Denmark, an authority on the evolution, biogeography and taxonomy of passerines, or 'perching birds', an order that includes the picathartes. Not only did he tactfully indicate that my conjecture was incorrect (his recent molecular studies have shown that the divergence occurred much earlier than the Pleistocene), but he also provided me with a summary of his latest ideas on the birds' early history. It appears that the picathartes, along with rockjumpers and the Rail-babbler, are members of an ancient lineage within the songbirds, or oscines, which colonised Africa from Australasia, independently of other songbirds. Once in Africa, the founder population spread out to give rise to the three families that ornithologists recognise today.[5] The fact that these radiations occurred in the distant past, approximately 35 million years ago, may account for the picathartes' unique set of features that has so challenged taxonomists. Furthermore, Fjeldså revealed that after the separation of Antarctica from South America and Australia, the resultant cold austral current was able to reach the west coast of Africa, causing a large savanna gap to develop in the region's rainforest. This area, the so-called 'Dahomey Gap', divided the ancestral picathartes population and enabled the two species to evolve. Fjeldså also emphasised that geological processes are dynamic and ongoing. At the time of the picathartes split, approximately 6 million years ago, the equator would have crossed over Guinea and Sierra Leone and the warm equatorial current would have hit west Africa, enabling the maximal effects of the cold austral current further south. Since then, as continental drift drags Africa continually northwards, the equatorial current has had an increasing influence, leading to milder conditions and a gradual narrowing of the Dahomey Gap. The subsequent expansion in the area's rainforest has probably enlarged the picathartes' range so that the two species could eventually come into contact – if it were not for the deforestation caused by humans.

Fjeldså's email was a game-changer. It was not the individual factors that he had highlighted to explain the picathartes' evolution – continental drift, changing ocean currents, and phylogenetics (the evolutionary relationships among species) – that had the most profound effect. It was less subtle than that. Rather, it was the realisation that every one of the world's 10,000 bird species, from tinamous to tanagers, has a unique evolutionary history that stretches back to the age of the dinosaurs.

Inspired by Fjeldså's complex and intriguing story of the Picathartidae, I determined to find out what was known about the evolution of the rest of the world's species. As it turned out, this was not as straightforward as I had expected. Surprisingly, no accessible account has yet been published. The closest offering, *Where Song Began* by Tim Low, focuses on the emergence of Australia's earliest birds.[6] Furthermore, most information lies buried in a plethora of scholarly journals, often with the important message hidden by a fog of jargon, acronyms and arcane statistical methodology. Each speciality, whether it is molecular biology, genetics, palaeontology, geology, phylogenetics or bioinformatics, has a unique vocabulary that is only understood by a small band of cognoscenti. As a result, awareness of this fascinating aspect of ornithology is mainly restricted to those who work in the field. To compound the problem, the increase in scientific knowledge is relentless and shows no signs of slowing. In 2010, whole-genome sequences were available for only three species, but by 2014 the work of the Avian Phylogenomic Consortium (200 scientists from 80 institutions in 20 countries) had increased this number to 48. Indeed, so much information was generated by the Consortium that their findings were reported simultaneously in 28 papers, eight of which were published back-to-back in a special issue of *Science*.[7] Plans are already under way to generate complete genomic data for all avian species over the next five years, with the aim of finally resolving one of the biggest challenges in systematic biology, namely the establishment of the evolutionary relationships of every modern bird.[8]

It was against this backdrop that the idea of writing the present book took root, with the wish to share the latest scientific findings with a wider audience. However, my nascent project faced a formidable challenge – how best to organise the wealth of available information. To try and squeeze over 100 million years of avian evolution into a single volume is a tall order. Indeed, the sheer complexity of the data and the number of species involved seemed to preclude a logical and comprehensive approach. According to the latest update from the International Ornithological Union, there are 10,694 extant species, classified in 40 orders, 238 families, and 2297 genera (as well as a bewildering number of subspecies).[9] A format was required that would give the story some degree of structure rather than allowing it to become an unwieldy catalogue or an encyclopaedic tome. With flagging spirits and a desk overrun with paper, I was fortunate to re-read Richard Dawkins' book *The Ancestor's Tale.*[10] For hidden among this magnum opus are several 'species' tales', each relating to an evolutionary process that is germane to the topic as a whole. Dawkins' format offered a glimmer of hope, and I have unashamedly adopted this approach. As a result, *The Ascent of Birds* is divided into 27 chapters, or stories, that collectively encompass the evolution of all modern birds or Neornithes. Even so, I have had to be painfully selective, relying on the filter of my own idiosyncratic choices – and, as a result, the responsibility for any omission, or commission, lies with me alone.

I have not dwelt on the earliest history of birds, one that extends from the theropod dinosaurs to the common ancestor of all modern species, as this topic has been explored comprehensively in several recent publications.[11] Instead, the present narrative commences with the ratites (a diverse group of flightless birds that includes the rheas and ostriches), which lie at the base of the avian phylogenetic tree, and then proceeds chronologically towards the terminal branches or offshoots that give rise to the finches and tanagers. For consistency, I have followed the phylogeny constructed by Jetz and colleagues: one based on all the available genetic sequences and scientific publications up to 2012.[12] Recent modifications to this schema are only highlighted if they have a significant impact on our story. A summary is available online, based on OneZoom software – a program that allows the interrogation of a wealth of data by laying it out in ever smaller bubbles using a fractal structure, together with a zooming interface, so that the computer never runs out of space.[13] As a result, data relating to nearly every known species is accommodated – from the Ostrich to the Grey-bellied Flowerpiercer. I would encourage readers to explore this valuable resource (www.onezoom.org/EDGE_birds.htm), as it is not only fun but it makes the avian phylogenetic tree immediately comprehensible in a way that only years of study could have done in the past.

Although each chapter is spearheaded by a named bird, the narrative will often digress to emphasise themes and biological processes that have a wider relevance to the evolution of not just birds, but all fauna and flora. Our chronicle opens with *The Tinamou's Story*, one that explains the presence of flightless ratites in South America, Africa and Australasia, and in so doing dispels the cherished role of continental drift as an explanation for their biogeography. It also introduces the concept of neoteny, an evolutionary trick that enabled dinosaurs to become birds and humans to conquer the planet. *The Vegavis's Story* explores the evidence for a Cretaceous origin of modern birds and how they were able to survive the asteroid collision that saw the demise of the dinosaurs and up to three-quarters of all plant and animal species on Earth. In *The Waterfowl's Story*, our attention switches to sex: why is it that so few avian species have retained the ancestral copulatory organ, or, to put it another way, why do most birds exhibit the paradoxical phenomenon of penis loss, despite requiring internal fertilisation? *The Hoatzin's Story* reveals unexpected oceanic rafting from Africa to South America: a stranger-than-fiction means of dispersal that is now thought to account for the presence of other South American vertebrates, including worm lizards, caviomorph rodents and monkeys.

The latest ideas to explain the emergence of new species are also explored. *The Manakin's Story*, for example, reveals how South America's extraordinarily rich avifauna has been moulded by past geologic, oceanographic and climatic changes, while *The Storm Petrel's Story* examines how species inhabiting the same geographical area can evolve into new taxa. The thorny issue of what constitutes a species is covered in *The Albatross's Story*, while *The Penguin's Story* highlights the effects of environment on phenotype: in the case of the Emperor Penguin, the harshest environment on Earth.

Recent genomic advances have provided scientists with novel approaches to explore the distant past and have revealed many unexpected journeys. They include the unique overland dispersal of an Old World suboscine from Asia to South America (*The Sapayoa's Story*) and the Blackbird's sweepstake dispersals across the Atlantic (*The Thrush's Story*).

Additional vignettes will update more familiar concepts underpinning speciation: sexual selection (birds-of-paradise), extended phenotypes (bowerbirds), hybridisation (sparrows) and 'great speciators' (white-eyes). Finally, we will explore the raft of recent publications that have added to our knowledge of the evolution of cognitive skills (crows), plumage colouration (starlings) and birdsong (Zebra Finch).

I have tried to keep this book as concise and focused as possible, in part by eschewing any unnecessary terminology. Where this is unavoidable, the scientific term is explained in the text when first encountered and, as a failsafe, included in a glossary at the end.

I am also aware that comprehensiveness and readability do not make for comfortable bedfellows. Rather than obfuscate an evolving storyline by inclusiveness, I have resorted to consensus science and highlighted the prevailing view. Of course, the adoption of such an approach can have limitations, as current doctrines may well be found wanting and dumped tomorrow in favour of alternative scenarios. Indeed, while I was completing the chapter on the emergence of passerines (see *The New Zealand Wren's Story*), a number of research groups published results suggesting that the date for their origin may be more recent than previously thought. Although I have stressed the conventional view, readers should be aware that if the latest findings are confirmed, then the geographical origin of perching birds will need to be reassessed. Fortunately, these recent studies have little impact on our main storyline, one that explores the various dispersal routes and speciation mechanisms that underpin their existence.

In *The Ascent of Birds*, I invite you to accompany me on an exciting odyssey, one that hopefully will open your eyes to the wonders of avian evolution and its resultant diversity. If, like me, you find that this little-known story enhances your enjoyment and appreciation of the natural world, then I will have succeeded in my aim.

PART ONE

The Non-Passerines

CHAPTER 1

The Tinamou's Story

DEATH OF A PARADIGM

The wintry gusts failed to dampen the great man's enthusiasm for delivery. Thomas Henry Huxley, aptly nicknamed 'Darwin's Bulldog', was on top form that evening, and he knew it. Standing behind the Royal Institution's desk, he fixed his 'hawk-like' eyes on the assembled socialites and gauged their reaction to his latest ideas. In the raked amphitheatre's gaslight, Huxley cut an assured figure, dressed in a subfusc frock coat, bow-tie and pince-nez, topped off with a shock of dark hair and matching sideburns – manna for the caricaturists of the day. Nevertheless, his confident mien and evangelical oratory masked a decade of struggle with Darwin's theory and the genealogical approach to classification. But not now, for Huxley had become convinced that his friend's radical views offered the best explanation for the emergence of new species. Taxa were not static but evolved from one to another over time, although how long this required remained unclear. Simply put, all organisms are descended from a single common ancestor and therefore related to one another. Armed with his newly acquired conviction, Huxley proceeded to captivate his audience with the pinnacle of his palaeontological studies. His lecture was a *tour de force*, one that contained the most astonishing of pedigrees – a reptilian ancestry for birds. In doing so, the self-taught biologist became the first to propose that the ancient flightless birds, of which the kiwis and rheas are but 'scanty modern heirs', had evolved from dinosaurs.[1]

The catalyst for Huxley's proposal was a visit six months earlier, in October 1867, to Oxford University Museum. Here, among the collection of precious relics, he noticed a fossil that had been incorrectly identified as part of the shoulder girdle of a *Megalosaurus*. Instead, Huxley realised that it was part of the dinosaur's upper pelvis or ilium: a bone that struck him as 'so bird-like as to be astonishing'. His belief that dinosaurs and birds are closely related was further bolstered after he had the foresight to reconstruct the British Museum's *Iguanodon* as a biped rather than a quadruped. Despite the specimen's size, approximately 9 metres tall, Huxley noted that there was a 'considerable touch of a bird about the pelvis and legs!' Later, he would go much further:

> If the whole hindquarters from the ilium to the toes, of a half-hatched chicken could be suddenly enlarged, ossified, and fossilised as they are, they would furnish us with the last step of the transition between Birds and Reptiles, for there would be nothing in their characters to prevent us from referring them to the Dinosauria.[2]

During the last 10 years, a wealth of well-preserved feathered dinosaurs and bird fossils from China has helped confirm Huxley's belief that birds are the descendants of dinosaurs. Indeed, it is not hard to imagine his smug, self-satisfied look were he able to learn of the insights that these and other astounding finds have provided, not just in the linkage of dinosaurs to birds but also concerning the origins of feathers and powered flight.[3] Huxley's prescience, however, was not the result of sudden genius, but rather the outcome of many years of dogged and gritty work, involving the detailed study of thousands of avian bones. Such painstaking anatomical comparisons led him to another crucial insight, one that provides the foundation stone for *The Ascent of Birds* – the flightless ratites are the most primitive of all modern birds.

Based on the structure of the breastbone, Huxley was able to classify modern birds into two groups, or superorders: the Ratitae, which he considered to be closest to the non-avian dinosaurs, and the Carinatae, which included all other birds. Indeed, his confidence in the basal position of ratites is evident from a letter he wrote to the German biologist and polymath, Ernst Haeckel: 'I am engaged [in] a revision of the Dinosauria, with an eye to the "Descendenz Theorie." The road from Reptiles to Birds is by way of Dinosauria to the Ratitae.'[4]

The term ratite (derived from the Latin *ratis*, meaning 'raft') refers to the shape of the sternum or breastbone – one that is flat because it lacks the median ridge, or keel, needed to anchor the strong flight muscles of flying species. As a result, all ratites are flightless, since their feeble vestigial wings cannot lift their heavy bodies off the ground. Four extant families make up the ratites. The largest are the ostriches (Struthionidae), which live in the savannas and Sahel of Africa. Slightly smaller are the rheas (Rheidae), which are native to the pampas and Chaco forests of South America, and the cassowaries (Casuariidae), the most colourful of all the ratites, which live in the tropical forests of New Guinea and northeastern Australia. They are the most dangerous, and when surprised or cornered can attack with razor-sharp talons. Until recently the Emu, which inhabits inland and coastal regions of Australia, was assigned its own family (Dromaiidae), but it has now been moved to join the cassowaries as a member of the Casuariidae. The most unusual ratites are the nocturnal kiwis (Apterygidae), species endemic to New Zealand that nest in burrows and locate their invertebrate prey using a highly developed sense of smell.

Several extinct families also belong to the ratites, including the elephant birds (Aepyornithidae) and the moas (Dinornithidae). The flightless elephant birds were huge species, reaching 2–3 metres in height and weighing up to half a tonne. They were widespread on the island of Madagascar up until the thirteenth century, when habitat loss and hunting pressure probably led to their extinction 400 years later. Elephant birds are believed to have been the inspiration for the fabled *roc* of Sinbad fame, a giant eagle-like bird that reportedly was capable of carrying off and devouring full-grown elephants (hence its name). The species' eggs were just as impressive, with a circumference of nearly a metre and a capacity of 9 litres – the equivalent of 200 chicken eggs. However, it is the moas of New Zealand that have captured our imagination more than any other ratite, ever since the anatomist Richard Owen famously deduced their existence from a single fragment of bone (Plate 2). Sadly, these herbivores met the same fate as the elephant birds, although this time at the hands of Polynesian settlers who colonised the islands during the thirteenth century. Indeed, by the time Europeans had reached New Zealand, the moas were reduced to 'mere bone and egg fragments in hunting middens'.[5] As we will discuss in *The Buzzard's Story*, not long after the moas were eliminated, the largest known bird of prey, the Haast's Eagle, was also lost. It is now clear that the Māoris had disrupted a food chain, and once the eagle's primary food source – the moas – had disappeared the raptor was unable to thrive and rapidly became extinct.

Huxley identified another skeletal commonality among ratites, a primitive, reptilian-looking palate.[6] In general, ratites possess a more complex, stronger and less flexible palate than the light-boned, flexible forms found in all other birds. Anatomists use the term palaeognathous to describe such a primitive palate and, as a result, Huxley's two avian superorders are now known as Palaeognathae ('ancient jaws') and Neognathae ('new jaws'). Of all the bird families that Huxley studied, only one caused him a taxonomic headache: the South American tinamous (family Tinamidae) (Plate 3). For these medium-sized ground-dwelling birds seemed to defy his neat taxonomic dichotomy. Tinamous possess a sternal keel, with associated wing muscles. Indeed, as birders know only too well, tinamous, although reluctant to fly, can suddenly rise when disturbed and, with loud, frantic wingbeats, disappear quickly from view. Flight distances are always short, since they possess small hearts in relation to their body size. Tinamous, therefore, are not ratites. However, they do possess a primitive palate, indicating that they are palaeognaths. So where should tinamous be placed – with the palaeognaths, or among the neognaths, or in a division all by themselves? In the end, Huxley opted to group them with the neognaths, as he believed that they were most closely related to the ground-feeding Galliformes, a group that includes grouse and turkeys. Today, as we will highlight, tinamous are firmly placed within the Palaeognathae, although their precise phylogenetic position has been the subject of intense debate.

A role for neoteny

In contrast, Huxley's conclusion that palaeognaths are the most primitive of modern birds has stood the test of time, although the idea has not been without its critics. A popular counterview was that palaeognaths are not primitive birds but only appear so because of the retention of juvenile features into adulthood, a process called neoteny or paedomorphism. Domesticated animals, for example, are thought to be neotenous versions of their wild counterparts. Compared to wolves, dogs retain many anatomical and behavioural features that are characteristic of puppies: floppy ears, large eyes, as well as playfulness and bouts of affection. The novel idea that neoteny might account for the primitive features of palaeognaths was promulgated by the respected zoologist and comparative embryologist Sir Gavin de Beer. In the 1950s, de Beer argued that ratites had evolved from neognaths and that their downy plumage, unfused cranial bones and primitive palates are merely retained juvenile features.[7] Although this hypothesis has not been supported by recent studies, the role of neoteny in avian evolution turns out to be far more profound than de Beer could ever have imagined. For it is now apparent that neoteny enabled the rapid evolution of all modern birds and facilitated their subsequent global success.

In 2012, a group from Harvard University, headed by Bhart-Anjan Bhullar and Arkhat Abzhanov, concluded that the evolution of modern birds occurred through a neotenous change in the development of dinosaurs.[8] With the use of sophisticated x-ray technology, the team scanned juvenile and adult fossilised skulls from non-avian theropod dinosaurs and ancient birds, as well as the skulls of modern birds. Crucially, the researchers had access to fossilised dinosaur eggs that contained developing embryos. After highlighting various 'landmarks' on each scan, Bhullar and colleagues were able to track how the skulls had evolved over millions of years. The results were a surprise. It turns out that all dinosaurs, even those that are most closely related to modern birds, underwent dramatic maturational changes in their skull structure. In contrast, the skulls of modern birds remain similar to their juvenile forms throughout life. The researchers concluded that birds evolved from dinosaurs by a block in maturation so that they retained the large brain, big eyes and short face of infantile dinosaurs. Paedomorphism, therefore, enabled modern birds to become smaller and to reach sexual maturity far more rapidly – in as few as 12 weeks in some species – and this process opened up new opportunities for evolutionary experimentation. Being small would have had the added advantages of requiring less food and being less susceptible to predation. Maybe it is not so surprising that the only dinosaurs to have survived the dramatic mass extinction event at the Cretaceous–Palaeogene (K–Pg) boundary were the small neotenous ones (see *The Vegavis's Story*).[9] According to Abzhanov, 'what is interesting about this research is the way it illustrates evolution as a developmental phenomenon. By changing the developmental biology in early species, nature has produced the modern bird – an entirely new creature.'[10]

Neoteny, therefore, provides life with an efficient and rapid evolutionary route, one that works by taking something already available and modifying it, rather than having to develop a whole new set of genetic instructions. Maturation block may also explain why humans are so radically different from their nearest cousins, the chimpanzees and bonobos. The retention of the primate's juvenile features, including hair distribution, large brain and flat face, may have enabled humans to evolve more rapidly, despite sharing most of the same genes.[11] Indeed, the process appears to have been so dramatic that some scientists refer to our species as the 'neotenous clan of apes'.[12]

The neotenous switch to a juvenile-like skull shape in dinosaurs allowed the potential of modern birds to be unleashed, as the development of a larger skull-to-body ratio enabled the formation of bigger and more complex brains. More neuronal connections would have allowed early birds to evolve innovative and flexible behaviours to help compensate for any environmental changes, as well as increasing their ability to colonise novel ecological niches. It is ironic, therefore, that, while de Beer's ideas have been side-lined, neoteny is now seen as a crucial step in the evolution of all modern birds, enabling them to become one of the most successful groups of organisms on the planet.

In theory, the role of neoteny in avian evolution could be confirmed by tweaking the relevant genes in developing embryos and seeing if it resulted in a reversion to a dinosaurian phenotype. Indeed, in 2015, Bhullar, now working at the University of Yale, together with Abzhanov reported the results of a study that suggested that such an approach might be possible. By using small-molecule inhibitors that downregulate protein essential for beak formation, the researchers were able to induce chicken embryos to express a snout and palate similar to those of the small *Velociraptor*-like dinosaurs.[13] Once the genes controlling neoteny have been identified, there is no reason why similar experiments could not be undertaken.

We have discussed the role of neoteny long enough. Let us move on and consider how the entire group of palaeognaths – ratites and tinamous – evolved, and how their unique biogeography can be explained. As I will reveal, the tinamous provided the key that unlocked the group's evolutionary past, although the story is a labyrinthine one, beset with many twists and turns.

The unravelling of vicariance

From the beginning, scientists were uncertain whether palaeognaths (tinamous and ratites) are merely a hodgepodge of unrelated forms that have followed a parallel line of evolution with multiple ancestral origins (termed 'polyphyly') or whether they form a natural group, with a single common ancestor (termed 'monophyly'). You will recall that Huxley identified tinamous as palaeognaths, although he placed them with the Neognathae: a taxonomy that implied a separate ancestry for ratites and tinamous. He did, however, admit that tinamous were 'the most struthious [ostrich-like] of all carinate birds.'[6] His arch rival, and nemesis, Richard Owen, went much further, suggesting an independent origin for most ratites, since he believed that ostriches were allied to bustards and that the moas and kiwis were close to megapodes. Support for the so-called 'polyphyly hypothesis' was still evident in the 1980s when such palaeontological heavyweights as Alan Feduccia and Storrs Olson believed that different ratites could have variously evolved from ducks, geese, cranes and even ibises. Protagonists of the multiple-ancestor theory offered two explanations: either the group's characteristic anatomy had evolved independently by convergent evolution as adaptations for a flightless cursorial lifestyle, or their key features were due to neotenous changes, as argued by de Beer. The key question, therefore, was whether various neognath families could have given rise to the different ratites, given enough time?

The simple answer is no. During the last 10 years, it has become clear that all ratites possess anatomical features not present in any other bird. Furthermore, a large number of molecular studies have consistently shown that ratites and tinamous are more closely related to each other than they are to any other bird family. Palaeognaths, therefore, are monophyletic, and any biogeographical explanation needs to take this into account.

A further controversy, however, proved more intractable: the precise phylogenetic relationships between the various palaeognaths, extant and extinct. Until the end of the twentieth century, the flightless ratites were thought to be monophyletic, with the whole group being sister to the flying tinamous (sister groups are lineages that are each other's closest evolutionary relatives). Such a conclusion, however, implied that the ratites' common ancestor must have been flightless – an idea that troubled biogeographers, since a non-flying ancestor could not have crossed the wide ocean barriers to reach the various continents and islands in the southern hemisphere where they are found today.

Then, biogeographers were offered an unexpected lifeline. In 1962, Harry Hess, a Princeton geologist, published what would turn out to be one of the most influential papers in modern science.[14] Succinctly entitled 'History of ocean basins', it revealed a mechanism for how landmasses could drift apart; it was a seminal publication that laid the foundations for the unifying theory of plate tectonics. Hess's elegantly compelling concept was understandably seized upon by biogeographers, including the influential American ornithologist and palaeontologist, Joel Cracraft. In 1974, Cracraft proposed that the flightless ancestor of the ratites, which had already split off from the flying tinamous, roamed widely across Gondwana during the late Cretaceous. Gondwana is the name given by geologists to the southern supercontinent which included present-day Antarctica, South America, Africa, Madagascar, India, Australia and New Zealand. As this vast southern landmass broke up, between 130 and 50 million years ago, the early flightless ratites would have been split into several isolated populations, a process termed vicariance. Over time, the ancestral groups were then postulated to have drifted on fragments of the Earth's crust to reach their current locations, before giving rise to the various species we recognise today.[15] Despite several nagging inconsistencies, the convenient serendipity of continental drift proved irresistible, and the ratite story remained a paradigm of vicariance biogeography. Indeed, the idea was regularly rehashed by journalists and popular science writers. Even the renowned biologist Richard Dawkins devoted many pages to continental drift and ratite evolution in the first edition of his popular book *The Ancestor's Tale*, published in 2004.[16]

However, around the turn of the twenty-first century, what had seemed a robust model started to crumble. The first cracks resulted from studies that analysed a greater number of genes and which gave a clearer resolution of ratite relationships. The resultant phylogenies, coupled with better molecular dating, revealed a poor agreement between the branching order of the ratite tree and the sequence of Gondwana break-up. Madagascar, for example, has been an island for at least 88 million years, a landmass that separated from Gondwana well before the arrival of the elephant bird. Similarly, the kiwi lineage reached New Zealand millions of years after the islands' separation from Gondwana.

It was an international study in 2008, however, led by John Harshman at the Field Museum of Natural History in Chicago, that shook the model's foundations.[17] The American team analysed genetic material from 18 species of palaeognath, including all the living ratites and four tinamou taxa. Instead of relying on just one or two regions of DNA, they sequenced 20 stretches that were widely dispersed throughout the genome. Contrary to the prevailing view, Harshman showed that it is the ostrich and not the tinamou that is the sister group to all the other ratites. Indeed, the flying tinamous lie deeply embedded within the ratites' family tree (Figure 1.1). The implications of Harshman's study are profound: the common ancestor of the palaeognaths must have been a flying bird, capable of long-distance dispersal, which crossed the southern oceans to reach the different continents. It was only after their arrival that the various ratite lineages – ostrich, rhea, kiwi, cassowary, elephant bird and moa – lost the ability to fly. The group's characteristic anatomy, therefore, was the result of convergent evolution, in which

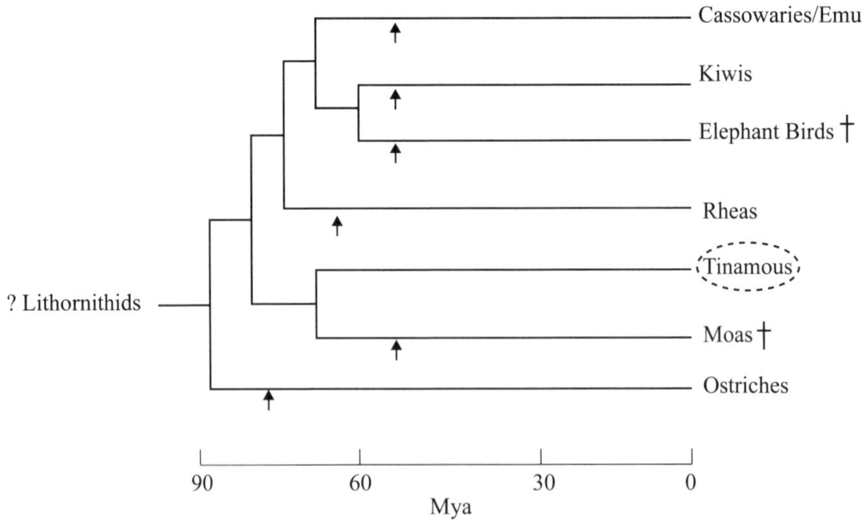

Figure 1.1 Phylogeny of palaeognaths, showing that the flying tinamous (dotted circle) lie deeply within the ratites (a paraphyletic group): a finding that implies a loss of flight on at least six separate occasions (arrows). Mya = millions of years ago. † = extinct species. Adapted from Baker *et al.* (2014),[18] Mitchell *et al.* (2014),[19] and Christidis and Boles (2008).[20]

each lineage evolved similar features in response to comparable selection pressures. The alternative scenario in which tinamous regained the ability to fly, while not impossible, is highly unlikely since there are no examples of birds (as far as we know) that have lost and regained flight. Indeed, flight has only evolved three times during 500 million years of vertebrate evolution: first by the pterosaurs (ancient reptiles that existed up until the end of the Cretaceous), then by birds, and finally by bats. In contrast, loss of flight is not uncommon in the avian world, having occurred in 18 extant families, especially the rails (family Rallidae), and many more times in extinct groups.

The final blow for the vicariance model came in 2014, with the publication of two studies that incorporated ancient DNA (aDNA). Intriguingly, both papers revealed that the ratites' geographical neighbours are not their closest evolutionary relatives. The first study, led by Alan Barker from the Royal Ontario Museum, included aDNA extracted from a small bone that belonged to the little bush moa.[18] After careful preparation in a specially designed clean room, the specimen yielded a small amount of purified aDNA suitable for amplification and analysis. The results were unexpected. Not only did the genetic sequences confirm that the tinamous are deeply nested within the ratites, but they also showed that their closest relatives are the New Zealand moas and not the South American rheas, as scientists had predicted. Ratites, therefore, are neither monophyletic nor polyphyletic but rather paraphyletic – a term used to describe a group that does not include all the descendants (in this case the tinamous) of their common ancestor (Figure 1.1).

The second study, led by Kieren Mitchell and Alan Cooper from the Australian Centre for Ancient DNA in Adelaide, analysed mitochondrial DNA from various ratite lineages, including samples from bones of two different species of elephant bird.[19] By aligning the genetic sequences, they were able to show that the nearest relatives of the herbivorous diurnal giants from Madagascar are not the African ostriches, but the New Zealand kiwis, a clade of secretive, shy, nocturnal omnivores. The resultant phylogeny implies that flightlessness must have evolved a minimum of six times and gigantism at least five times during the early phase of ratite evolution. Furthermore, the common ancestor of the kiwi and elephant bird existed millions of years after New Zealand and Madagascar had separated from Gondwana. Ratites, therefore, were not transported by continental drift following the break-up of Gondwana, but flew across the oceans to reach the southern continents and islands.

It is now believed that the flying ancestor of palaeognaths most likely evolved from an extinct group of birds known as the lithornithids, possibly within Western Gondwana. According to Kieren Mitchell, 'they would have been quite small, unassuming birds, probably the size of a chicken or quail.' Interestingly, lithornithids were once widespread and probably an extremely mobile group of species, since their fossils have been recovered from Europe and North America: places that lack ratites today.[21]

Finally, what is it about the ratites that made the species so prone to flightlessness? Professor Cooper believes that the group's anatomical modification relates to the ecological vacuum that followed the mass extinction event that wiped out the dinosaurs. Since the surviving mammals were small and unspecialised, and remained so for the next 10 million years, there was a unique opportunity for the evolution of huge, flightless herbivores among the continental birds. The abundance of early grasses and plants throughout the southern hemisphere and the absence of large herbivorous mammals offered rich pickings for any adaptable birds. The ratites rapidly filled the vacant niches and, in doing so, lost the ability to fly. It was a winning strategy that only failed with the devastating arrival of *Homo sapiens*. Later, the mounting competition from increasingly large herbivores prevented any further flightlessness, except on islands that remained devoid of mammals – the classic example being the Dodo on Mauritius. Cooper also has an explanation for why the kiwis and tinamous remained small, in contrast to their closest relatives. He believes that the first palaeognaths to arrive would have monopolised the available niches, forcing any subsequent arrivals to remain small and to adopt alternative lifestyles. It is for this reason that the tinamous retained their wings and the kiwis became insectivorous and nocturnal.

In his 1870 presidential address to the British Association for the Advancement of Science, Thomas Huxley stated that 'the great tragedy of science is the slaying of a beautiful hypothesis by an ugly fact.'[22] Little could he have known just how apt this statement would prove to be in the field of palaeognath evolution. For the Tinamou's story has destroyed the concept of Gondwanan vicariance, a theory cherished by many biogeographers for over 25 years.

But just how old are the palaeognaths? To try and answer that question, we need to take a trip to Antarctica.

CHAPTER 2

The Vegavis's Story

THE CRADLE OF MODERN BIRDS

After several years of planning, the 300-foot American icebreaker *Nathaniel B Palmer* cast off from its moorings and headed slowly southwards towards the Drake Passage, the roughest stretch of water in the world, and the frozen continent beyond. The previous few days had been hectic: survival courses to complete, safety drills to learn, cold-weather gear to select and research equipment to stow safely on board. For the senior scientists, Matt Lamanna and Julia Clarke, the routine was familiar, but for many of the team members it was an exciting prelude to their first Antarctic field trip. Experience aside, each scientist has the same overriding concerns. What would the weather be like 1,500 kilometres to the south, off the Antarctic Peninsula? Too much sea-ice, and the Zodiacs would not be able to reach the shore; too windy or foggy, and the helicopters would be grounded; too much snow, and the fossil-bearing rocks would remain hidden. For sure, the latest forecasts and satellite images appeared encouraging, but all were well aware of the vagaries of the Antarctic climate, cognisant that field work in their targeted area had not been possible for the last few years. It is February 2016, and the 12 palaeontologists and geologists are heading for several remote islands that lie to the east of the Antarctic Peninsula; the 50-kilometre-long James Ross Island, and the all-important Vega Island (Figure 2.1, Plate 4) – named after the vessel that made the first complete voyage of the Northeast Passage.

Sponsored by America's National Science Foundation, the team's field work is focused on areas that have already yielded a range of highly informative fossils. Indeed, the islands' palaeontological importance has been known ever since their discovery in 1903 by the Swedish South Polar Expedition 1901–1904.[1] Led by the explorer and geographer Otto Nordenskjöld, the Swedes had intended to spend only 12 months in the far south, but from the start, things went disastrously wrong. The expedition's wooden ship was crushed in the ice, and the scientists and crew were forced to spend two years

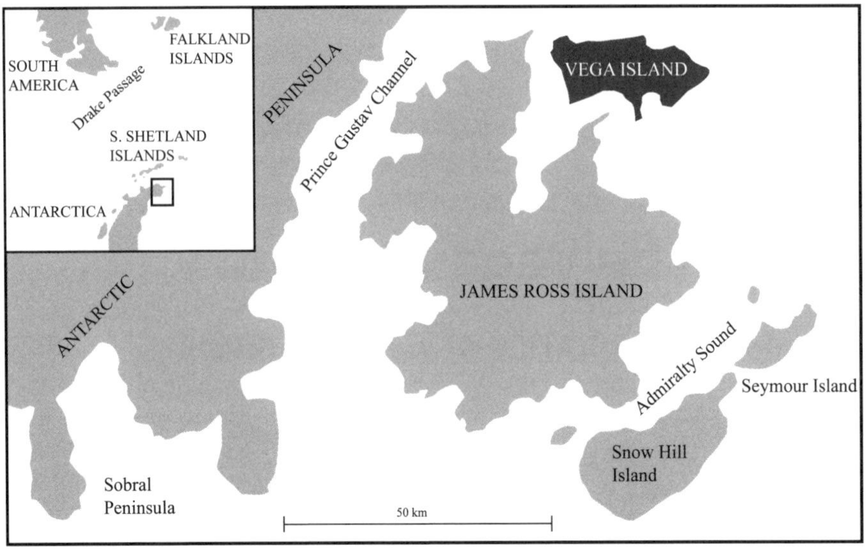

Figure 2.1 Location of Vega Island, site of the earliest modern bird fossil (*Vegavis iaai*), dating from 66.5 million years ago.

on a nearby island. Despite living in small stone huts and surviving on a diet of seal and penguin, the expedition's forced overwinterings were not in vain. They made several excursions across the sea-ice to nearby shores where they recovered fossils of not just penguins, but plants and trees. The latter included figs, laurel, sequoia, evergreens and the Antarctic beech *Nothofagus*, a species that still grows in Patagonia and Australasia today. The Swedes' hard-earned specimens turned out to be of considerable scientific value, for they indicated that Antarctica had once possessed a much warmer climate and that the southern continent had been joined in the past to South America and Australia. Later, palaeontologists discovered the islands' rich marine deposits, dating from the Cretaceous. These beds have yielded not only Antarctica's first dinosaur fossils but a wealth of marine fossils, including sharks, teleosts and shell-crushing fish, as well as plesiosaurs and marine lizards.

However, the area's most significant geological feature is the almost perfect exposure of the time at the end of the Cretaceous when one of the largest mass extinctions in history occurred, one that resulted in the extinction of the dinosaurs and up to three-quarters of all life forms.[2] Known as the K–Pg boundary, this geological marker can be discerned today as a dark line that stretches across the ice-free outcrops of both Vega and James Ross Island. Beneath this boundary lie older sedimentary rocks that are rich in fossils, while immediately above are younger deposits that are virtually fossil-free.[3] Indeed, the islands' exposure of the K–Pg boundary (or Cretaceous–early Palaeogene sedimentary sequence) is one of the most important on Earth, and its fossils have helped construct the picture of life at the end of the 'Age of Dinosaurs'.

While it may be an important K-Pg boundary site, it is far from ideal. Field work is hampered by the extensive glaciation, and the risk of hypothermia and frostbite, even when wearing several layers of protective clothing, balaclavas and hand warmers, is real. Furthermore, the islands were once part of the sea floor, and it was only after a local uplift that they rose above sea level. As a result, any non-marine fossils are likely to have lived over 100 kilometres away on the Antarctic Peninsula and been transported by rivers and streams before settling on the continental shelf. It is no surprise, therefore, that most fossils of terrestrial creatures are broken up and disarticulated, making their interpretation difficult.

Given the climatic and geological challenges of Antarctica, why are Lamanna and Clarke so keen to return to the islands' remote K–Pg boundary sites, and what is the relevance of their fossil hunting to the story of modern birds? The answer relates to a small 'bundle of bones' recovered from Vega Island during the austral summer of 1992.

That season, a party of Argentinian scientists, working at the island's southwest corner, discovered the partial skeleton of a bird among the rocks near the shoreline. Initial analysis suggested that the fossilised bones represented an extinct lineage of duck-like bird (Presbyornithidae), a species that was close to, but not part of, the radiation of modern birds. The specimen was duly described, catalogued (MLP 93-1-3-1) and deposited in the Museo de La Plata, without the scientific community realising that what had been unearthed would turn out to be one of the most significant avian fossils ever found (Plate 5).

Fortunately, 13 years later, the specimen was re-examined by a team of scientists led by Julia Clarke from North Carolina State University.[4] It transpired that the initial interpretation had been hampered by the specimen's rudimentary preparation and the degradation of its exposed and delicate bones. Instead, Clarke's team re-evaluated the fossil using the latest computed tomography and digital reassembly techniques, to obtain better views of the bone structure without inducing any further damage. The x-ray images, together with previously unknown latex casts taken shortly after its discovery, revealed numerous additional bones, with novel anatomical features, lying deep within the sample. The adult individual, named *Vegavis iaai* (the bird from Vega Island) was not a member of the Presbyornithidae as originally believed, but a new Magpie Goose-like species (it is even thought to have honked like a goose) that dated from at least 66.5 million years ago (Plate 6).[5] In other words, *Vegavis* lived more than half a million years before

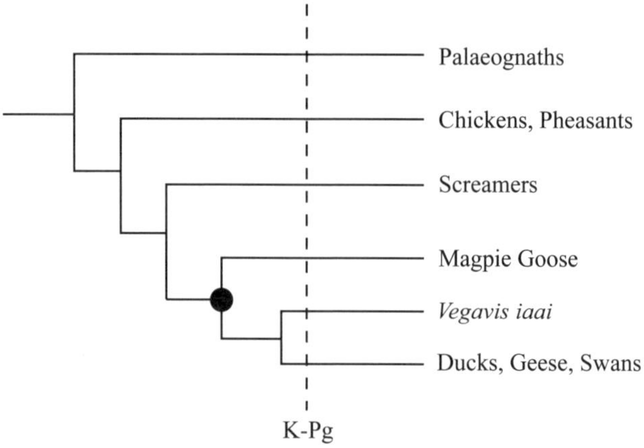

Figure 2.2 Phylogenetic tree showing the position of *Vegavis iaai*. The calibrated divergence date of the Magpie Goose and the lineage that gave rise to both *Vegavis* and the Anatidae (ducks, geese and swans) is estimated at a minimum of 66.5 million years ago (black circle). The dotted line represents the Cretaceous–Palaeogene (K–Pg) boundary. Adapted from Ksepka & Clarke (2015).[8]

the K–Pg boundary.[6] Clarke's unexpected conclusion generated considerable excitement, as *Vegavis* was the first Cretaceous fossil to be placed within the extant bird radiation. As she reported at the time, 'until now the fossil record has been ambiguous, but now we have a fossil which indicates that at least part of the diversification of living birds had begun before the extinction of non-avian dinosaurs.'[7]

Careful morphological comparison of *Vegavis* with other avian families enabled Clarke and her colleague Daniel Ksepka to construct a phylogenetic tree of early modern birds.[8] The analyses placed *Vegavis* between the older lineages of screamers (family Anhimidae) and the Magpie Goose (family Anseranatidae) and a younger lineage that gave rise to the world's ducks, geese and swans (family Anatidae). Clarke and Ksepka's phylogenetic tree implies that the common ancestor of all modern birds must have existed before the K–Pg boundary. Furthermore, the position of the Palaeognathae and Galliformes (including chickens, pheasants, grouse, and megapodes) suggests that these two groups evolved even earlier, maybe tens of millions of years earlier (Figure 2.2).

A Cretaceous origin of modern birds is further supported by the results of a recent international study, which utilised whole-genome analyses.[9] The constructed phylogenetic tree, the most comprehensive published to date, reveals that the Palaeognathae diverged from all other modern birds (Neognathae), around 100 million years ago. Then, sometime during the following 20 million years, the Palaeognathae split into several lineages that gave rise to the tinamous and ratites, while the Neognathae diverged to produce the Galloanserae (a group that includes the Galliformes and waterfowl) and the Neoaves. Of course, this does not imply that today's ostriches and chickens lived alongside the non-avian dinosaurs, but rather that the evolutionary lineages that gave rise to them did. Nevertheless, the implications are clear. The ancestors of many extant bird families inhabited the southern continent during the late Cretaceous and, unlike the dinosaurs, survived the massive K–Pg extinction event.

The Argentinian find (MLP 93-1-3-1), however, was only a partial fossil, and additional specimens are needed to confirm both the species' age and its phylogenetic position. Encouragingly, in 2006, Lamanna and Clarke recovered further *Vegavis* fossils during their first Antarctic expedition, and over 20 isolated bird bones and one partial skeleton on a second trip, five years later. Although formal analyses are awaited, the exciting point is that the only undisputed Cretaceous fossils of modern birds have all been recovered from Antarctica. This fact raises an intriguing question. Is it possible that every bird alive today, from the tiniest hummingbird to the largest ratite, has an ancestry that extends back to Antarctica, to a time when dinosaurs roamed the planet? In support of this idea, most early avian lineages have obvious Gondwanan distributions. Palaeognaths, for example, are restricted to the southern hemisphere. Furthermore, the three species of screamer are endemic to South America, while the unique Magpie Goose is confined to the coastal areas of northern Australia and southern New Guinea. Indeed, it is the fascinating possibility that Gondwana may have been the 'cradle of modern birds' that motivates scientists like Julia Clarke to visit Antarctica and seek further fossil evidence. One can only hope that the results of the latest expedition to Vega and the surrounding islands will further our understanding of this crucial phase in the evolution of modern birds.

The K–Pg boundary

I have repeatedly referred to the K–Pg boundary, the pivotal event in the *Vegavis*'s story, without discussing its cause. While many readers will be familiar with the serendipitous nature of its elucidation, it is nevertheless a story that is worth recounting. For the K–Pg boundary sets the scene for one of the liveliest debates in avian evolution. Why did the non-avian dinosaurs die out, and the modern birds survive?

The clue to solving the nature of the K–Pg boundary was iridium, a lustrous, silvery-white, transition metal related to platinum. Although hard and brittle, its corrosion-resistant properties and ability to withstand heat make it valuable as an alloying agent, and in the manufacture of spark plug tips, crucibles and aircraft engine components. Iridium is also a rare element, and its high density and tendency to bind with iron resulted in the majority of it sinking to the Earth's core when the planet was young and still molten. Indeed, it is the element's rarity that enabled the K–Pg event to be deciphered.

In the early 1970s, the American geologist Walter Alvarez was undertaking field work near the Italian hill town of Gubbio when he became curious about a thin layer of reddish clay that divided the Cretaceous and Palaeogene layers of limestone. Beneath the clay, he found that the lower whiter rock contained many different species of fossilised foraminifera, single-celled organisms with external shells. In contrast, he could not detect any foraminifera in the clay itself and only a single species in the darker limestone immediately above.

Walter wanted to know how long it had taken for the clay to form, so that he could estimate the length of time that the foraminifera took to recover. After discussing the problem with his Nobel-prize-winning father, the physicist Luis Alvarez, samples were sent to the nuclear chemists Frank Asaro and Helen Michel at the University of California, Berkeley, for rare-metal analysis. Luis knew that meteorites are rich in iridium, and reasoned that since their dust is deposited on the Earth's crust at a relatively constant rate, the quantity of iridium present could help determine the time taken for the layer of clay to form.

The results were a complete shock. The clay contained 300 times the iridium concentration of the limestone above and below, a result confirmed by analysis of similar samples from around the world. There was only one startling explanation. The high levels must have resulted from a large extraterrestrial object hitting the Earth, since iridium is common in asteroids and meteorites and rare in the Earth's crust. Given the global nature of the abnormality, the Alvarezes believed they had enough evidence on which to convict the prime suspect, a 10-kilometre wide asteroid (a guess based on the measured iridium in the clay and the average iridium content of meteorites). The results were published in the prestigious journal *Science* in 1980,[10] but the idea of an asteroid impact was ridiculed by many, not least because no tell-tale crater had been found (Walter predicted the crater to be 200 kilometres wide, since craters are usually about 20 times the size of the impact object). Detractors favoured alternative explanations for the K–Pg mass extinctions, including climate change and volcanic activity. Cold snaps in the general hothouse environment of the Cretaceous, aggravated by continental drift and the opening up of new ocean and air currents, would have stressed the biosphere and brought many species to the brink of extinction. But it was the idea of increased volcanism that had the greatest appeal. Indeed, the

Deccan Traps in India, the largest continental flood basalt province on Earth, covering over half a million square kilometres, was widely argued as the 'true' culprit for the K–Pg mass extinctions.

However, unbeknown to both the Alvarezes and their critics, evidence for a candidate impact crater was already in the public domain. In 1978, two geophysicists working for the oil company Pemex had detected anomalies in the Earth's magnetic and gravitational fields off the coast of Mexico. Further analyses suggested the presence of a large circular feature, 180 kilometres wide and 48 kilometres deep, centred near the small fishing village of Chicxulub (pronounced 'cheek-she-loob') on the Yucatán Peninsula. Glen Penfield, one of the scientists involved, was so convinced that his data reflected a cataclysmic event in geologic history that he obtained Pemex's permission to present the findings at the 1981 conference of the Society of Exploration Geophysicists. As it turned out, the symposium was poorly attended, since most of the big names in the science of impact craters were at an alternative meeting, and his findings were largely ignored. Understandably, Penfield became disillusioned and returned to his day job, searching for oil. Then, by chance, an asteroid expert, Alan Hildebrand, was told of the geophysicist's findings by a journalist from the *Houston Chronicle*. In April 1990, Hildebrand contacted Penfield, and the pair persuaded Pemex to release drill samples taken in the 1950s from the area of interest. The findings were conclusive. The samples contained shock quartz and small glass spheres, or tektites, features that only form in the heat of an asteroid impact or a high-yield nuclear detonation. After 10 years of searching, the 'smoking gun' for the Alvarez hypothesis had been found. The rest, as they say, is history.

Well, not quite!

In 2015, Professor Paul Renne, Director of the Geochronology Center in Berkeley, California, and his team provided a link between the asteroid impact and the Deccan Traps.[11] By using isotopic dating of lava collected from before and after the extinctions, they were able to work out rates of lava production over time. The results suggested that a sudden change in volcanic activity occurred simultaneously with the asteroid impact, and that the high levels of lava production continued for a further 100,000 years. The researchers postulated that the asteroid impact induced a massive seismic shock (equivalent to a magnitude 11 earthquake) that shook the core–mantle boundary plume, fundamentally altering the Deccan Traps' plumbing system. The result was the formation of much larger chambers that spewed out greater volumes of magma. According to Professor Renne, 'the time it took for marine fauna and many aspects of ocean chemistry to recover back to pre-extinction values is about half a million years, which happens to be the amount of time it took for the volcano to die down.'[12] In fact, the Deccan Traps had been active, albeit at a lower rate, for over 100,000 years before the bolide impact. Recent data from Seymour Island in Antarctica, for example, reveals that the Deccan Traps' volcanism had already had a deleterious effect on the Earth's climate and caused many ocean-floor species to become extinct well before the asteroid impact.[13]

Dinosaurs, therefore, appear to have been wiped out by a double-whammy – the Chicxulub impact coupled with the effects of the Deccan Traps' eruptions. As is so often the case with polarised scientific debate, the truth seems to lie somewhere between the extremes, with a coalition of salient features from both camps providing the best explanation.

What perturbed the extraterrestrial object from its presumably stable, but weakly bound, orbit in the outer regions of our solar system remains unclear. Nevertheless, recent observations from the Hubble Space Telescope and the Rosetta spacecraft offer clues. Images beamed back to Earth have revealed a strangely shaped asteroid, orbiting between Mars and Jupiter, that was most likely formed by a collision that occurred within the last few years. According to one of the team members, the combined data reveal 'an asteroid smash caught in the act for the first time.' This interpretation has fuelled speculation that a similar collision, over 100 million years ago, could have caused an asteroid from the same source, known as the Flora group, to change trajectory and head for a deadly rendezvous with Earth. A more speculative scenario has been suggested by Lisa Randall of Harvard University, one of the world's leading theoretical physicists. In her book *Dark Matter and the Dinosaurs*, Randall proposes that the gravitational pull from a disc of dense dark matter within the Milky Way dislodged a comet from the outer reaches of our solar system (from an area known as the Oort Cloud) as the Sun passed by. This chance interaction nudged the comet into a new orbit, one that ultimately changed our planet's history 66 million years ago.[14]

The obvious follow-up question is this: How did the combined effects of the Chicxulub impact and the Deccan Traps lead to the extinction of three-quarters of all life forms? While the exact details may never be known, there are some predictable consequences of a 10-kilometre asteroid hitting the Earth. A bolide the size of Manhattan, travelling faster than the speed of a bullet, would have delivered the equivalent energy of more than a billion atomic bombings of Hiroshima and Nagasaki.[15] This, in turn, would have induced the largest mega-tsunamis in the Earth's history and caused the devastation of coastal plains worldwide. The release of dust and particles would have covered the entire planet and blocked the sunlight for many years. As a result, photosynthesis would have been inhibited in both oceanic and terrestrial ecosystems, effectively shutting down large swathes of the food chain for months or even years. Volcanic ash ejected into the atmosphere would have added to the dust from the impact and increased the planet's albedo, the proportion of solar energy reflected back into space, to induce global cooling. As much as 100 billion tonnes of sulphur and 10 trillion tonnes of carbon are estimated to have been vaporised by the impact, causing the production of vast quantities of carbon dioxide and sulphur dioxide. These and other gases would have reacted with water vapour in the atmosphere and been precipitated as acid rain. The acidity of the oceans must have increased substantially, which in turn would have inhibited the growth of coral reefs and disrupted the marine ecosystem from the bottom up, features confirmed by the study of the fossils from Seymour Island off the tip of the Antarctic Peninsula.[16]

The most extreme of scenarios, however, has been promulgated by Douglas Robertson, from the University of Colorado, who envisages an incandescent Hadean landscape in which re-entry of ejecta produced a widespread infrared pulse sufficient to ignite global firestorms. He also calculates that such fires would have started almost immediately and burned for days or even weeks on end, leading to the complete destruction of terrestrial ecosystems.[17]

Whatever the effects of the Chicxulub meteor, the extinction event that ended the Cretaceous was not the most severe to have hit the Earth. That honour is held by the global annihilation that terminated the Permian, about a quarter of a billion years ago when more than 90 per cent of all species were lost. The end-Cretaceous event was nevertheless disastrous for life on Earth and resulted in the loss of three-quarters of known species. Terrestrial extinctions included the non-avian dinosaurs, pterosaurs, as well as many lizards, insects, and half the world's plants. Archaic birds were also wiped out, such as the Enantiornithes, which possessed teeth and clawed wings, and the aquatic, flightless Hesperornithines. Marine organisms fared little better, with the loss of ammonites, giant marine reptiles and plesiosaurs, while fish, sharks and plankton were all devastated.

Surprisingly, some species pulled through relatively unscathed: modern birds, crocodiles, turtles, snakes and mammals. For many of the surviving groups, the global destruction provided unique evolutionary opportunities, and several underwent remarkable adaptive radiations, sudden and dramatic divergences that produced a multitude of new species to fill the vacant ecological niches. One of the most dramatic radiations was that of placental mammals. Aided by their capacity for long-term tropical hibernation, the clade rapidly evolved to produce many novel forms, including horses, bats and whales and, luckily for us, humans.[18] Indeed, were it not for the cataclysmic impact 66 million years ago, you would not be here to read this book.

Modern birds' survival

And yet, of all the K–Pg events highlighted, it is the survival of modern birds that is the strangest and the most difficult to explain. One idea, however, can be immediately discounted; the possibility that today's 10,000 species might represent the fortunate descendants of some 'favoured' guild, a group of species that exploited the same resources. Such an explanation would require that the particular guild not only lacked all the archaic avian lineages but also contained every neornithine group that survived. An improbable event, I am sure you will agree.

So what could account for the survival of modern birds? Certainly, they had a greater mobility compared to most other species and could have flown to regions with more favourable resources and climate. Their feathers would have provided protection from the cold, especially during the months of darkness immediately after the impact, while the ability to swim, dive or build nests in burrows and tree holes would have helped. Neornithines, unlike theropod dinosaurs and archaic birds, evolved the ability to incubate their eggs by direct body contact. Such behaviour may have arisen because of their relatively small egg size, a fact that increased the likelihood of embryo loss from temperature fluctuations. The building of nests from feathers, grass and other organic material would have provided some insulation from the cold, making incubation by body contact all that more efficient.[19] In contrast, the larger dinosaur eggs, even those partially buried in the ground, would have perished.

Recently, a team led by Gregory Erickson of Florida State University reported that dinosaur eggs took much longer to hatch than modern birds' eggs.[20] By counting daily growth markers in embryonic teeth, Erickson discovered that dinosaur eggs took between three and six months to hatch. The study was only possible because, just like tree rings growing a new layer each year, teeth grow a new layer each day, which manifests as microscopic lines in the dentine. Ostrich chicks, in comparison, crack open their shells after only 42 days, while for smaller species the incubation time can be as little as 11 days. The longer it takes to reproduce, the longer it takes to replace a threatened population – a fact that would have worked against dinosaurs immediately after the impact. According to Erickson, the non-avian dinosaurs were clearly dealt a 'bad hand' in the game of survival.

An additional factor may have been the difference in the ratio of egg size to adult body mass between the dinosaurs and modern birds. Dinosaurs had comparatively small eggs, because of the physical constraints imposed by the need to maintain oxygen diffusion. According to Gary Kaiser, a biologist at the Royal British Columbia Museum, dinosaur hatchlings were small (a fact deduced from the fine structure of their fossilised bones) and it took them a long time to reach adulthood.[21] Indeed, the largest dinosaur chicks hatched from eggs weighing as little as 5 kilograms, and yet they had to bulk up to a staggering 90,000 kilograms. Such a slow and extended reproductive period would have been a major limitation during the end-Cretaceous, and would have been a significant factor underlying their demise.

Neornithines, however, did not become small overnight but were the result of tens of millions of years of downsizing. While the majority of dinosaurs continued to evolve on a massive scale, a subgroup of maniraptorans ('hand snatchers') became small again. These feathered meat-eating dinosaurs rapidly diverged while adapting to novel ecological niches. Although one lineage continued to produce large species, the second spawned progressively smaller forms. It seems that after 170 million years of divergence, only those feathered maniraptorans that had reduced to about 1 kilogram in weight – the neornithines – were able to survive.[22] Such small species would have had the benefit of reproducing more quickly, having a greater number of offspring and possessing a wider genetic diversity, features that aid survival under severe conditions. It is also possible that post-impact phenomena were concentrated in the northern hemisphere, which therefore favoured modern birds for the simple reason that they were mainly confined to the southern hemisphere.

Such explanations, however, cannot be the complete story, since many archaic birds and flying pterosaurs had similar physiological attributes and geographical distributions and yet were still annihilated.

Could it be that the superior brain power of modern birds was the key to their survival? It turns out that the widespread notion of birds as small-brained and dim-witted creatures could not be further from the truth. In fact, many birds appear to be rather smart at solving problems. New Caledonian Crows, for example, use modified tools to extract insects and other invertebrates, as well using sticks to inspect foreign objects that appear to pose a threat. Parrots have an enormous capacity for learning vocabulary. The *Guinness Book of World Records* lists the most proficient talker as an African Grey Parrot named Prudle. Captured in Uganda along the shores of Lake Victoria in 1958, Prudle had a vocabulary of nearly 1,000 words by the time he retired from public life in 1977. Pigeons are prodigious discriminators of complex visual stimuli and have been trained to differentiate images of benign and malignant tumours.[23] Even the humble farmyard chicken appears to think before acting, and can undertake simple arithmetical tasks.

According to Angela Milner and Stig Walsh, colleagues at the Natural

History Museum in London, it was the ability to solve problems that enabled modern birds to survive the mass extinctions. In a 2009 study, the two scientists used computerised tomographic (CT) scans to determine the brain size and structure of 55-million-year-old fossil brain cases from two species related to modern seabirds.[24] The results suggested that by the end of the Cretaceous, modern birds had already evolved their larger and more complex brains, especially in those regions that control sight, flight, and the higher-level functions (including the ability to learn and remember information). As Walsh states:

> In the aftermath of the extinction event, life must have been especially challenging. Birds that were not able to adapt to rapidly changing environments and food availability did not survive, whereas the flexible behaviours of the large-brained individuals would have allowed them to think their way around the problem.[25]

Finally, the development of a heightened sense of smell, or olfaction, may have added to the survival advantage of modern birds. Darla Zelenitsky and colleagues, while working at the University of Calgary, used CT scans to compare the size of the olfactory bulb (as a proxy for the ability to smell) in non-avian dinosaurs, fossil birds and living birds. Contrary to expectations, they found that the structure increased in size, relative to the length of the brain, during the emergence of modern birds. The results imply that an enhanced olfactory sense must have played a significant role in their evolution, since it is unlikely that olfactory bulb size would have continued to increase without conferring some selective advantage. Compared to their rivals, a more discriminatory sense of smell would have led to an increased efficiency at foraging, orientation and social interactions.[26]

Whatever the reasons for their survival, the *Vegavis* fossil tells us that modern birds made it through the K–Pg extinction event and gave rise to the great variety of species seen today.

CHAPTER 3

The Waterfowl's Story

REFUGIA, HIGH LIVING AND SEX

While devastating extinction events are bad news for the majority of species, they can provide unexpected opportunities for others. The ecological release caused by the loss of non-avian dinosaurs and archaic birds, for example, undoubtedly facilitated the adaptive radiation of modern birds. It was a rapid process, at least in geological terms, with a minimum of 36 extant lineages arising within the first 15 million years of the Palaeogene.[1] But before we hear their stories, let us discuss the emergence of ducks, geese and swans: a lineage known to have survived the Chicxulub impact. For the waterfowl's story provides us with insights into the role of refugia in speciation, the evolution of proteins at the functional level, and the unlikely mechanics of avian sex.

As discussed in the preceding chapter, molecular phylogenies and the *Vegavis* fossil suggest that the common ancestor of waterfowl lived during the late Cretaceous, probably alongside the early Palaeognathae and Galliformes on Gondwana. Although the family's taxonomy is complicated and disputed (ducks frequently interbreed, producing fertile offspring), the oldest lineage is that of the whistling ducks, a fact that may account for their predominance in the tropics and subtropics of the southern hemisphere.[2] All eight extant species are large, sociable ducks, with distinctive whistling calls, which congregate on lagoons and swamps to rest and preen. They differ from most other waterfowl in having an erect posture, relatively elongated necks and legs, and a distinctive habit of perching on trees.

A second offshoot gave rise to the Masked Duck and the stiff-tailed ducks (subfamily Oxyurinae), so called because of their firm tail feathers which are held erect when at rest. The Oxyurinae, which includes the Pan-American Ruddy Duck and the South American Lake Duck, are rarely seen far from water, as their feet are set far back, making walking on land rather awkward.

Following these early offshoots, the ancestral Anatidae underwent

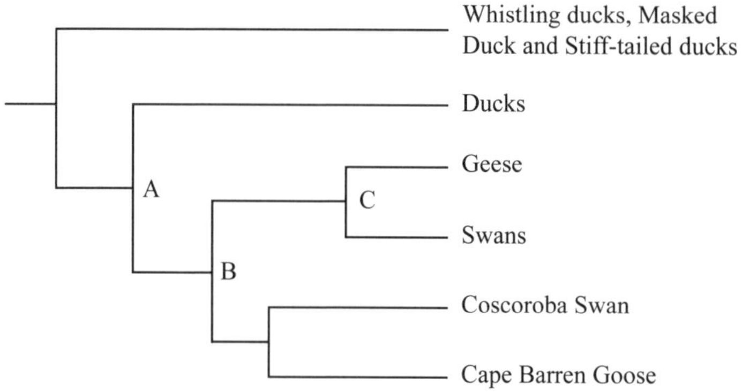

Figure 3.1 Phylogeny of ducks, geese and swans, with divergence dates: A = 28 Mya in the southern hemisphere; B = 23.5 Mya in the southern hemisphere; C = 20 Mya in the northern hemisphere. Adapted from Gonzalez *et al.* (2009).[3]

a bifurcation around 28 million years ago, somewhere in the southern hemisphere.[3] One lineage gave rise to the rest of the world's ducks (Anatinae) while the second produced the swans and geese (Anserinae) (Figure 3.1).

The basal radiation of the Anatinae took place between 23 and 5 million years ago during the Miocene, with the group's high dispersal ability resulting in a global distribution: hence their complex biogeography. Nevertheless, some conclusions can be drawn. The dabbling ducks, which include teal, wigeon, pintail and shoveler, are known to have evolved in the southern hemisphere before undergoing several dispersals cross the equator. The pintail and mallard clades, for example, originated in Africa around 1 million years ago and radiated northwards to give rise to American and Eurasian lineages.[4] The North American clade rapidly diverged during the Pleistocene to produce the American Black Duck, the Mexican Duck and the Mottled Duck. Although these three species are morphologically distinguishable, they are genetically similar because of ongoing genetic sharing due to a lack of complete reproductive isolation.[5] Indeed, since the speciation of most dabbling ducks occurred comparatively recently, many taxa remain compatible genetically, a fact that accounts for the high incidence of natural hybridisation. A further biogeographical complication is that several South American species, including the Yellow-billed Teal and the Chiloe Wigeon, evolved from ancestors that returned to the southern hemisphere, after leaving their kin behind in North America. It should be noted that the diverse array of domesticated ducks (e.g., Aylesbury, Peking, and Silver Bantam) derive mainly from the Mallard, a hugely plastic species that was first domesticated in China more than 3,500 years ago and then later in Europe.

A further dispersal of dabbling ducks warrants a mention: an early radiation that reached the islands of Hawaii around 3.6 million years ago. Sadly, their descendants, the *moa-nalo* ('lost fowl'), were wiped out by Polynesian settlers before the arrival of Captain James Cook in 1778. The ancestral *moa-nalo* were able to fly and spread from Kaua'i to the newer volcanic islands of Oahu and Maui as they emerged from the sea. However, the lack of mammalian predators on all the islands enabled these flying filter-feeders to become ponderous, flightless birds with tiny wings, heavy bodies and massive bills. They developed a browsing lifestyle and evolved the ability to ferment plant fibres in their hindguts, a trait that enabled the taxa to become the largest herbivores on the islands. Such phenotypic changes must have occurred at least 400,000 years ago, since the flightless species, weighing approximately 7 kilograms, were unable to reach the youngest island, Hawaii, which first emerged from the sea at this time.[6] As we will discuss below, the presence of *moa-nalo* on most but not all the Hawaiian islands influenced the evolution of the archipelago's geese, which arrived later.

The *moa-nalo*, however, were not the ancestors of the islands' two extant dabbling ducks, the Laysan Duck and the Hawaiian Duck. Instead, both species evolved from the mallard lineage, although by very different mechanisms. The Laysan Duck diverged from an east Asian ancestor of the mallard clade during the Pleistocene, while the Hawaiian Duck emerged much later, as the result of hybridisation between the Laysan Duck and the Mallard.[7] This conclusion is supported by fossil evidence: Laysan-like duck fossils date from the mid-Pleistocene, intermediate Laysan–Hawaiian forms derive from the early Holocene, and the fossilised bones of the Hawaiian Duck are the most recent. It is likely that the original mallard population arrived in Hawaii by chance during migration, since 'migratory dropouts', including dabbling ducks, continue to be recorded on the islands.

Sheldgeese and shelducks, which comprise the tribe Tadornini, also evolved within the southern hemisphere. Sheldgeese are South American grazing birds that occupy an ecological niche equivalent to that of the 'true' geese of the northern hemisphere. In contrast, shelducks are omnivorous, wading birds that feed on invertebrates, shellfish, algae and other plant material. According to fossil and phylogenetic evidence, shelducks evolved in Australia – a view supported by studies of the Raja Shelduck. For this distinctive species, which uses its beak to filter water in search of insects and seeds among the mangrove coastlines of Australasia, is the most basal of the Tadornini clade, being sister to all other shelducks.[8]

The seaducks, including mergansers, scoters and goldeneyes, had their origins in the northern hemisphere, where they remained, except for two independent transequatorial dispersals. The latter produced the Brazilian

Merganser, a critically endangered species, with fewer than 250 birds left in the wild, and the recently extinct New Zealand Merganser.[9]

The steamer ducks are restricted to Patagonia and the Falkland Islands and speciated during the Pleistocene as a result of glacial and sea-level fluctuations. Steamer ducks are robust, pugnacious birds, with very short wings, that escape predation by churning the water with legs and wing like paddle-steamers, hence their vernacular name (Plate 7). All four species (three flightless and one flying) evolved from a flying ancestor that lived in Patagonia around 1.4 million years ago.[10] This period coincided with the Great Patagonian Glaciation (GPG) that peaked approximately 1.1 million years ago. At its height, glaciers covered all, or nearly all, of the mainland, although an ice-free land bridge linked Patagonia to the Falkland plateau due to low sea levels. The increased glaciation forced the ancestral steamer ducks to seek more temperate conditions within two refugia, one located on the Falkland plateau and a second to the north of the glaciated area. Once the ice began to retreat, the rise in sea level effectively cut off the Falkland Islands birds from those in southern Argentina: a vicariant event that led to the speciation of the Falkland Steamer Duck. It was only much later, approximately 15,000 years ago, that the more extreme sea-level rises fragmented the mainland population and enabled the mainland's three species to evolve. The fact that two of these – the Chubut and Fuegian Steamer Ducks – are flightless, and the third is a weak flyer, suggests that the evolution of flightlessness is an ongoing process. Possible selective pressures favouring the non-flying state include an improved diving efficiency, the avoidance of energetically costly moulting, and enhanced thermoregulation.

The most recent common ancestor of the swan and goose lineages (Anserinae) inhabited the northern hemisphere around 20 million years ago. This finding raises two interesting questions. First, how is it that two atypical species, the Coscoroba Swan and the Cape Barren Goose (Plate 8), are found in the southern hemisphere? The former is endemic to South America, breeding from central Argentina to Tierra del Fuego and the Falkland Islands, while the latter is restricted to Australia's southern coastal areas and offshore islands. Intriguingly, both species have long been a taxonomic conundrum. The Coscoroba Swan, while swan-like, possesses a honking voice, a goose-like head and a duck-like bill. Also, unlike other swans, the male does not perform 'triumph ceremonies' when returning to his mate after seeing off rivals. The Cape Barren Goose is also a strange-looking species, being more shelduck-like than goose-like, with a relatively small head and a short black bill mostly concealed by a bright lime-green cere. They also have primitive-sounding calls that, in the case of the female, resemble the grunts of a pig, hence their colloquial name 'pig goose'. In hindsight, the species' southern distributions and unique phenotypic and behavioural features were phylogenetic clues. For recent molecular studies have shown not only that the Coscoroba Swan and Cape Barren Goose are sister species, but that they probably diverged in the southern hemisphere, around 23.5 million years ago, well before the swan–goose split had occurred in the northern hemisphere (see Figure 3.1).

The second question relates to the 'true' swans of the genus *Cygnus*. Given that the clade evolved in the northern hemisphere, when and where did the two southern species, the Black Swan of Australia and the Black-necked Swan of South America, evolve? To date, detailed phylogenetic studies incorporating both species have not been undertaken, but one might predict that both species dispersed southwards from their northern origins.

As previously indicated, the 'true' goose clade originated in the northern hemisphere: a conclusion supported by the discovery of the earliest goose fossils in western Europe (dating from the Miocene) and the knowledge that the basal species, including the Bar-headed Goose, the Brent (or Brant) Goose and the Red-breasted Goose, have a mainly Eurasian distribution. Approximately 9.5 million years ago, the ancestral population split to form two distinct lineages: the *Branta*, or black geese, with members in North America and Eurasia, and the *Anser*, which include the grey geese of Eurasia and the North American white geese.[11]

It was not until the late Pliocene and early Pleistocene, between 4 and 2 million years ago, that the major speciation of geese took place. The Earth at that time was experiencing a period of active cooling due to the closure of the Panama seaway and the uplift of the Tibetan plateau. Both events contributed to the formation of permanent ice sheets in the northern hemisphere, and the emergence of circumpolar tundra and extensive temperate grasslands. In turn, these novel ecological niches enabled the adaptive radiation of new groups of plants and animals, including birds. The tundra, for example, provided new breeding grounds for geese, while the temperate grasslands acted as wintering grounds where mate choice took place. During the Pleistocene ice ages, the ancestral *Anser* and *Branta* populations became separated into different breeding groups. As the polar ice sheets moved southwards, most geese were pushed to lower latitudes, only moving north again during the warmer interglacial periods.[12] Some areas remained ice-free, even during the times of maximal glaciation. These inhabitable 'refugia' included Beringia (a land bridge linking North America and Asia), Banks Island in Arctic Canada, Peary Land in northern Greenland, and several sites in the Pacific Northwest. There may have even been refugia within the ice sheets themselves that were capable of sustaining distinct plant populations.[13] While some birds survived in refugia during the summer months and even bred, others were forced southwards. It was the splitting of the ancestral populations by the repeated advances and retreats of the ice sheets, as well as the presence of refugia, that led to the genetic isolation and subsequent speciation of the various goose clades.

The Bar-headed Goose lies at the base of the clade's phylogenetic tree, having evolved around 4.5 million years ago. This medium-sized species, from the genus *Anser*, lives in central Asia and is one of the world's highest-flying birds (see below). Nearly 2 million years later, the Taiga Bean Goose and Pink-footed Goose emerged in Eurasia: the former having evolved a long thin bill for probing the soft substrates of bogs and marshlands, while the latter developed a short bill for grazing and stripping seeds.

Basal splits within *Branta* resulted in the circumpolar Brent Goose and the Red-breasted Goose: the latter being a small species that breeds in Arctic Siberia, mainly on the Taymyr Peninsula. In North America, another *Branta* population diverged to give rise to the Canada Goose in the west and the smaller Cackling Goose to the east. Later, around 2.5 million years ago, a population of Cackling Geese split off to produce the Barnacle Goose, a species which breeds in some of the remotest places on Earth (Greenland, Svalbard and Arctic Russia). Their genetic isolation enabled the development of a distinctive plumage pattern, although the body proportions remained similar to that of the Cackling Goose, which suggests that both lineages have experienced a stabilising selection on body form due to similar ecological conditions.[11]

Three million years ago, a population of Canada Geese colonised the Hawaiian islands. As we have discussed, this string of volcanic islands, islets and atolls was already occupied by the *moa-nalos*, a group of large flightless waterfowl that had become herbivorous. As a result of competition, the newly arrived geese remained small and evolved an independence of wetland habitats to give rise to several species, including the extant Hawaiian Goose, or Nēnē (pronounced 'nay-nay'). Later, around half a million years ago, the island of Hawaii emerged from beneath the sea and was soon colonised by geese from the older islands to the west. Since the *moa-nalos* were already flightless and unable to reach Hawaii, the newly arrived geese had no competition and evolved to become flightless giants, reaching up to four times the size of their ancestors. Sadly, only the Nēnē survives today, as Hawaii's other species of goose, as well as the *moa-nalos*, became extinct soon after the arrival of humans.[14]

So, the next time you encounter a gaggle of Canada Geese in the local park and bemoan their noisy and confrontational behaviour, can I suggest that you give a moment's thought to their family history? It is a pedigree that includes two of the world's most iconic species – the threatened Hawaiian Goose and the smart Barnacle Goose.

Adaptations for the high life

Bar-headed Geese are an unmistakable species, with two striking black and white bar patterns on their heads, from which they get their name. However, this medium-sized goose is best known for making the highest-altitude migration on Earth. Twice a year, the majority of the world's population travel from their breeding grounds in Mongolia and China to their wintering areas in India, a journey that involves traversing the Himalayas. Although they do not cross the highest peaks, Lucy Hawkes from Exeter University has shown that they can reach elevations of up to 7,290 metres (Plate 9).[15] Furthermore, GPS tracking indicates that the birds do not use the predictable tail winds that blow up the mountains during the day, but, surprisingly, fly at night when the winds have died down.[16] As a result, their nocturnal efforts involve the greatest rates of climbing ever recorded for a bird and, to cope, they have evolved a suite of adaptations not seen in any other species of waterfowl.[17]

The evolutionary modifications are both anatomical and physiological, and enable a sustained 10- to 20-fold increase in oxygen consumption rate when the geese are flying at altitude. Their wing span, for example, is larger than that of other waterfowl, a feature that produces a lower loading and a greater lift, resulting in a considerably reduced energy requirement during flight. Modifications to their cardiovascular and respiratory systems have evolved to allow higher metabolic rates, even in the presence of low oxygen levels. Their lungs are much larger, and the species can hyperventilate at up to seven times the basal rate when oxygen levels fall. For many organisms, including humans, hyperventilation causes a drop in blood carbon dioxide levels, a condition known as hypocapnia. Such gaseous alterations may lead to impairment in the transportation of oxygen to the brain, resulting in a loss of consciousness. The Bar-headed Goose, however, seems to be relatively immune to this condition and can undertake lofty flights with reduced risks of impaired brain function.

The birds also have an altered blood physiology that enables an increased amount of oxygen to be delivered to the vital organs. Their haemoglobin molecule, for example, has a higher affinity for oxygen, largely as a result of a single amino acid substitution that alters the protein's structural conformation. Furthermore, the low temperatures encountered at high altitude stimulate an increase in the protein's already high oxygen-binding ability. The species' heart receives a greater blood supply than in other geese, due to a higher capillary density, especially in the left ventricle, and a more homogeneous spacing of capillaries. The heart cells, or myocardiocytes, metabolise oxygen more efficiently since their mitochondria (cellular organelles that produce energy in the form of adenosine triphosphate, ATP) are closer to the cell membrane, and thus the capillaries. Also, an enzyme involved in the mitochondrial synthesis of ATP, cytochrome c oxidase, is structurally altered to reduce the chances of oxidative stress that would otherwise occur during high-altitude flying.

The ability of Bar-headed Geese to survive high altitudes and low oxygen levels is the result of many different adaptations accumulated over millions of years. But this finding presents biologists with a problem. Since each evolutionary modification is unlikely, on its own, to have enabled the species to cross the Tibetan plateau, how has selection pressure led to so many adaptive traits? The idea that all the anatomical and physiological adaptations occurred at the same time is so improbable that it can be ignored. Slightly less fanciful, but still inadequate, is the view that the bird's migration path changed as the result of being able to take higher and shorter flight paths as the adaptive changes accumulated.

Instead, the most likely scenario is that the goose's ancestors evolved the first physiological adaptation when the Tibetan plateau was lower and not so wide.[18] Even if the Himalayan mountains bordering the plateau's southern edge were as high as they are today, as some geologists believe, the early Bar-headed Geese might still have been able to make the crossing, although not quite so efficiently. Indeed, migrating herons and egrets have been detected flying at comparable altitudes over the Negev Desert, albeit for shorter distances and with the assistance of strong tail winds.[19] As the ancestral geese were gradually forced to fly higher and longer, due to the continuing upthrust of central Asia, the increasing selection pressure (including the resulting changes in habitat) would have induced further adaptive changes and led to eventual speciation.[20]

It is not certain which trait evolved first, although one might predict that it would be the most functionally important, probably the high-affinity haemoglobin. This idea is supported by the results of Kevin McCracken's studies on eight different species of Andean duck that have evolved at high altitude for between 10,000 and 1 million years.[21] Although each species has a different set of adaptations, they all possess a high-affinity haemoglobin. The simplest explanation for this finding is that haemoglobin is more susceptible to selection pressure than other proteins – a conclusion supported by the molecule's relatively simple gene structure, and the fact that different genetic mutations, or amino acid substitutions, can produce the same beneficial effects.

Contingency and protein evolution

Let us explore avian high-affinity haemoglobins in a little more detail, as they have provided scientists with unique insights into how proteins may evolve. The major haemoglobin of birds, haemoglobin A, is made up of four polypeptide chains – two alpha and two beta – that fold together to produce a three-dimensional structure capable of binding oxygen. The Bar-headed Goose's alpha chain differs from that of the Greylag Goose, a closely related species which lives at sea level, by only four amino acids. Two of these substitutions, located on the protein's exterior, have no functional effect and are known as neutral mutations. The other two are located at the alpha–beta chain interaction site, of which one, a proline-to-alanine switch at position 119 (α119), is mainly responsible for the protein's high-affinity state. These observations support the neutral theory of protein evolution, which holds that the majority of mutations in a protein sequence have no functional effect, while one or two changes at critical sites can dramatically alter the activity. High-affinity haemoglobins can also result from other amino acid substitutions that affect the alpha–beta binding site. For example, the Andean Goose, which spends its whole life at altitude, has a high-affinity haemoglobin due to a mutation in the corresponding position of the beta chain (β55). Both types of mutation, neutral and functional, persist for different reasons: the neutral changes are the result of genetic drift, or random selection, while the functional mutations are maintained by natural selection. But can neutral changes have long-term evolutionary consequences?

High-affinity haemoglobins are examples of convergent evolution in which the same environmental condition produces similar phenotypic adaptations across different species. An extreme example is presented by the Andean hummingbirds, in which two identical amino acid substitutions evolved in all the high-altitude species; an event that occurred on at least 17 separate occasions, in different clades and at various times.[22]

Recent studies, however, suggest that the molecular evolution of such phenotypes is not as predictable as it might seem and may be dependent on the historical accumulation of many apparently 'neutral' changes in the protein. In other words, mutations that produce an adaptive change in one species may not have the same effect on others because of differences in their genetic backgrounds. To test this hypothesis, a team headed by Jay Storz from the University of Nebraska used genetic engineering to construct, or 'resurrect', haemoglobins from both ancestral hummingbirds and the common ancestors of all modern birds, species that lived more than 100 million years ago.[23] They then inserted the amino acid that gives Andean hummingbirds their high-affinity haemoglobin (β83 serine) and measured the effect on oxygen binding. As expected, when the critical amino acid was created in the low-affinity haemoglobin of ancestral hummingbirds, a high-affinity state resulted. However, when the same substitution was inserted into the haemoglobin representing the common ancestor of all extant birds, no functional change was detected. It seems that the high-affinity substitution is only effective when placed amid the many other amino acid changes, 18 in all, that differentiate the haemoglobin molecules of ancestral hummingbirds from those of the modern birds' common ancestor. According to Storz, these results imply a potentially important role of contingency in adaptive protein evolution. As different species adapt to the same selection pressure, the set of possible amino acid substitutions that have beneficial effects may be contingent on the set of antecedent changes that have independently accumulated in the history of each lineage. In other words, the possible options for adaptive change in one species may not be available for other species. Of course, in the case of the hummingbirds' high-affinity haemoglobin, the same 18 'neutral' amino acid substitutions were already present in all clades.

While natural selection can predictably produce similar beneficial phenotypes in different species (convergent evolution), at the molecular level these changes tend to be idiosyncratic and are much less predictable. Maybe Stephen Jay Gould was right after all when he stressed that 'any replay of the tape [of life] would lead evolution down a pathway radically different from the road actually taken.'[24]

Loss of the avian penis

One of the most baffling events in vertebrate evolution is the reduction and loss of the avian penis. Despite all birds requiring internal fertilisation, only 3 per cent of species have retained the ancestral copulatory organ capable of female insertion, or intromission.[25] Such 'endowed' taxa lie at the base of the avian phylogenetic tree and include the Palaeognathae as well as some members of the Galloanserae, such as screamers, guans, curassows, waterfowl and the Magpie Goose (Figure 3.2). Structurally, the avian phallus superficially resembles that of their closest living relatives, the Crocodilia (alligators, crocodiles and gharials), and it is also possible that the theropod dinosaurs had something similar, although no fossilised specimens have yet been identified. All other birds, over 10,000 species, have completely lost the penis and possess instead a cloaca (derived from the Latin *cloaca* for 'public sewer or drain'): a multipurpose orifice used for defecating, urinating and reproduction. Mating, for the majority of birds, takes only a few seconds and involves

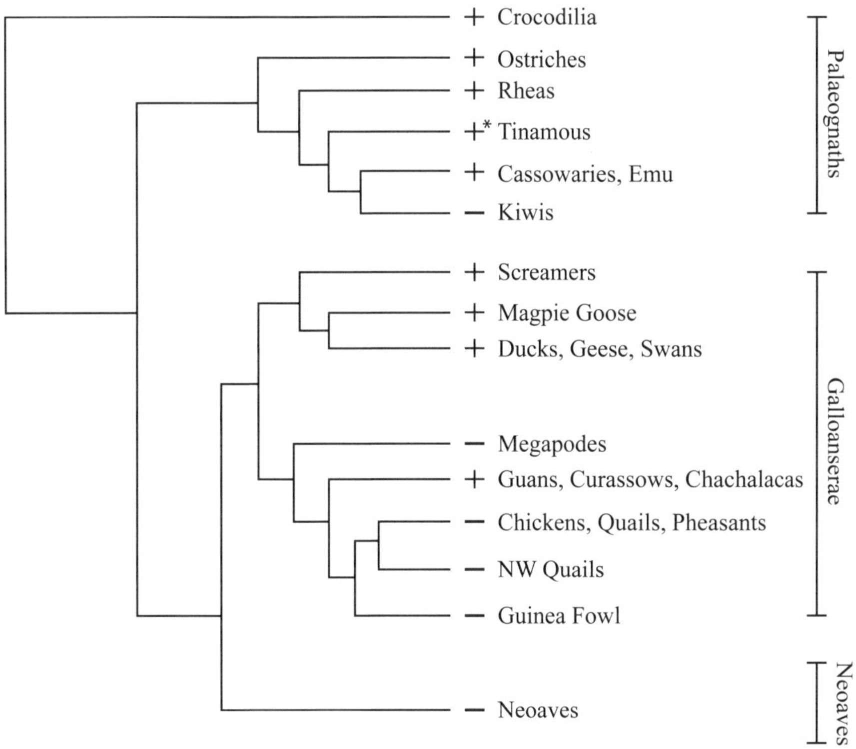

Figure 3.2 Phylogenetic distribution of penis loss in birds. + indicates present, – indicates absent or reduced in the majority of the clade. * Most tinamous possess a penis, the exception being the genus *Crypturellus*. Adapted with permission from Herrera *et al.* (2013).[26]

the brushing together of cloacae while the sperm is transferred from male to female – an act often referred to as a 'cloacal kiss' by ornithologists.

The lack of a penis in the majority of birds poses intriguing evolutionary questions: what, for example, were the selective pressures that led to the loss of an organ that seems so crucial for internal fertilisation, and how exactly did so many species lose their penis? Before attempting to answer these questions, let us first explore the evolutionary pathways that led to some of nature's most bizarre reproductive organs.

Avian penises are not homologous to the mammalian penis, but are formed from an eversion of tissue from the cloacal wall. Erection mechanisms are also different in birds. Scientists from Yale University have shown that, rather than the phallus filling with blood as in most other vertebrates (including humans), all avian erections result from the release of lymph fluid.[27] This clear-to-white bodily fluid is produced from a series of unique structures located alongside the cloaca and is released into the penile lumen during copulation. Compared to the blood circulation, the lymphatics are a low-pressure system and not ideal for maintaining erections. As a result, ostriches and rheas have evolved extra muscles and fibrous tissue to help maintain a rigid phallus and facilitate insemination, while ducks force lymph fluid at high pressure to achieve 'explosive' but short-lived erections (see below). The major transition, from blood to lymph, must have occurred in the earliest common ancestor of modern birds, since all the reptiles so far studied have a vascular system. In other words, the avian system evolved more recently than the vascular arrangement found in other vertebrates, although the selective pressures that prompted the switch remain a mystery.

The size of the avian phallus varies considerably, with the largest occurring in waterfowl, especially ducks. Indeed, the male Lake Duck, a small South American stiff-tailed species, holds the record for the largest phallus relative to an animal's length in the entire vertebrate world. Initially thought to be a mere 20 centimetres in length, after studies of dissected specimens, its penis is now known to reach an impressive 42.5 centimetres when fully everted (Plate 10).[28] That is longer than the bird itself. Even the 3-metre-long penis of the mighty Blue Whale doesn't come close to the Lake Duck in comparison to body size. This remarkable discovery took place in April 2001 at a lakeside in Argentina's Córdoba province. A team of zoologists, led by Dr Kevin McCracken of the University of Alaska Fairbanks, was seeking research specimens when they shot a male Lake Duck that was obviously in an 'aroused' state. Their observations were reported in the prestigious journal *Nature*, and, not surprisingly, attracted considerable attention in both the popular and academic press. McCracken speculated that the duck's organ could have evolved through sexual selection. Groups of drakes, for example, might display their everted members to attract females, with those males sporting the most impressive organs having a competitive advantage. The unusual anatomy of the Lake Duck's penis also suggested another function: success in sperm competition. The corkscrew-shaped phallus is covered with spines, while the tip is soft and brush-like. Before ejaculation, drakes probably employ their penises like bottle-brushes to remove sperm stored in the oviduct by the female's last consort. The larger the bottle-brush, the more successful the male will be in inseminating the female and passing on his genes.[28]

Further evidence that waterfowls' impressive phalluses may relate to sperm competition was provided by Christopher Coker and colleagues, from the University of British Columbia. Waterfowl are noted for their high levels of promiscuity, resulting primarily from extra-pair copulations when sex is forced on females of pair-bonded couples by other males. Such extreme behaviour suggests that sperm competition is important in the group, although hard evidence to support this hypothesis has been lacking. By studying over 50 museum specimens, Coker showed that the penis size of waterfowl is proportional to their testicular mass. More convincingly, those waterfowl that indulge in coercive sex had larger phalluses and larger testes, and their penises have not only more ridges and bulbs, but they protrude more.[29]

The unusual phalluses of waterfowl intrigued Professor Tim Birkhead, an avian biologist at the University of Sheffield, and caused him to ponder on the implications for females. Together with Patricia Brennan and her co-workers from Yale University, he examined the vaginas and corresponding phalluses of 16 species of wildfowl and obtained unexpected results.[30] Unlike most birds, in which the vagina is a simple, tube-like organ, they found in certain waterfowl that the vagina is exceptionally complex, possessing a spiral section and up to eight dead-end side branches. The oviduct was always found to spiral in a clockwise direction, in contrast to the universal anticlockwise spiralling of the everted phallus. The more elaborate the male member, the longer and more complex the vagina. While such structures do not prevent a drake from ejaculating, they do limit how far the semen is deposited along the reproductive tract. Once a female is receptive, however, copious amounts of lubricating mucus are secreted, while the reproductive tract's muscles relax, to allow full penetration. According to Tim Birkhead, the sexual organs of ducks are 'assumed to have coevolved through an arms race – allowing females to retain some control over fertilisation in the face of forced inseminations.'[31] The fact that only 3 per cent of duck offspring are born from forced extra-pair matings suggests that females have evolved a winning strategy. It seems that ducks are providing biologists with unexpected opportunities to explore the evolutionary consequences of sexual conflict, situations where the reproductive interests of the sexes differ. Such studies may help explain the relative importance of female preference and resistance, as well as the advantages to males of forced copulations.

Brennan and her colleagues then investigated how duck insemination was achieved, and the role played by the unique anatomy of the female's vagina. The study required considerable ingenuity on the part of the research team since, unlike humans who get erections before sex, drakes only evert their penis directly into the female's vagina. Using farmed Muscovy Ducks, high-speed video recorders and transparent glass vaginas, the scientists demonstrated that as soon as the drake's cloaca touches the female's, the penis is everted explosively, within half a second, at an average velocity of over 1.5 metres per second! However, when confronted by a glass vagina spiralling in the opposite direction to the penis, insemination was prevented.[32] These dramatic results support the idea that waterfowls' genitalia may have coevolved as a consequence of the sexual conflict between males and females.

The loss of the male organ in the majority of birds is an evolutionary paradox. How can the reduction or absence of a structure that enables internal fertilisation exert a positive effect on reproductive fitness? Several explanations have been proffered, although none is entirely convincing. For those species that copulate on water, an intromittent organ could prevent damage or dilution of the sperm (most species of waterfowl mate while swimming). Sexual selection is also a strong possibility.[33] Female birds may have been more willing to mate with males lacking penises, since they would be better able to avoid unwanted advances and have a greater say in which male fathers their offspring. Indeed, the case for penis loss evolving to enhance the sexual autonomy of females is explored in detail by Richard Prum, Professor of Ornithology at Yale University, in his recent book *The Evolution of Beauty*.[34]

Hygiene could also be relevant. Birds face a higher disease risk than mammals because the cloaca is used for both defecation and sex. As a result, a 'cloacal kiss' may reduce the likelihood of sexually transmitted diseases (STDs), which can be a source of mortality and morbidity, especially in domestic fowl. Goose venereal disease, for example, can cause up to 10 per cent mortality, as well as resulting in weight loss, reduced fertility and poor egg production. Lessening the incidence of STDs may have been important in promiscuous species, such as some megapodes and tinamous. While it is true that reptiles also have a single orifice, they are cold-blooded, and their lower body temperature probably reduces any infection risk. It has also been proposed that the loss of the phallus represents a weight-saving adaptation, but this is questionable, as the weight loss would have been minimal. Indeed, some species of Neoaves, including a variety of parrots and passerines, have evolved non-homologous intromittent structures that are surprisingly large.[35]

If you had asked scientists several years ago how birds lost their penises, the universal answer would have been that the genetic machinery controlling their growth had become inactivated. It was a surprise, therefore, when a team from the University of Florida, led by Martin Coln and graduate student Anna Herrera, announced in 2013 that the opposite was true.[26]

The American research group compared the embryonic development of species that lacked penises (domestic chicken and Common Quail), with species that possess a fully intromittent copulatory organ (ducks and geese). Using electron microscopy, they noted that during the early stages of development, male embryos of both groups start evolving identically, with paired genital swellings fusing to form a single stump or genital tubercle. The same process occurs during the development of the mammalian penis.

Within a short time, however, the chicken tubercle appears to suddenly stop growing, while it continues to enlarge in ducks and geese to form their characteristically large copulatory organs.

This initial observation tallied with the accepted view. But what growth factors were involved? An early breakthrough was the discovery that the mechanism of genital development is evolutionarily conserved in birds and mammals. Could it be that the growth factors involved in mammalian penile growth might also be essential to birds, and that a deficiency in one of them might account for the failure of organ development? It soon became apparent, however, that all the likely candidate proteins, including the oddly named 'sonic hedgehog factor', were present in adequate amounts. Furthermore, the target cells in ducks and chickens responded equally well to all the growth factors. This was a surprising result and suggested an alternative mechanism. Were the cells at the end of the chicken's tubercle undergoing apoptosis, a controlled and coordinated form of cell death?

In embryonic development, apoptosis or 'programmed cell death' is known to play a vital role in tissue patterning, since growing structures need sculpturing and moulding before their final form is achieved. It is an energy-consuming process, encoded by genes that dismantle cells in an orderly fashion, without the release of toxic enzymes that occur when cells die from injury. A good example of the effects of apoptosis is the structural difference between the feet of ducks and chickens. The developmental pathway in both species is identical until the last step, when apoptosis kills off the cells that connect the digits in chickens. In duck's feet, however, the cells are maintained as webbing. Similarly, apoptosis determines the shape of human fingers and toes by carving out the interdigital webs during early fetal growth.

To explore the role of apoptosis further, Cohn and his colleagues analysed developing avian penises for bone morphogenic protein 4 (Bmp4), a protein known to trigger a cascade of signals causing apoptosis in mammals. To their surprise, the protein was present along the whole length of the chick's genital tubercle, compared to only minute amounts at the base of a duck's organ. Furthermore, by applying a Bmp4 inhibitor to the genital tubercle, the researchers were able to prevent cell death in chickens, allowing the birds to grow a penis. In contrast, the treatment of duck genital stumps with Bmp4 resulted in apoptosis and penile shrinkage, a state that mimicked the development of the penis in male chickens. In other words, the evolutionary loss of the penis in chickens occurs not by disruption of outgrowth signals, as expected, but by the activation of apoptosis.

While the above experiments have revealed the mechanism of penile loss in chickens and quails, they do not explain why it should have occurred.

Cohn and colleagues have speculated that the function of Bmp4 could be a secondary, or bystander effect, as the result of its expression elsewhere in the body. Bmp proteins are implicated in at least three morphological innovations in birds: the development of feathers, toothlessness, and beak shape. In other words, the modulation of Bmp expression in other key structures could have had a collateral effect on bird's genitalia, a biological process known as pleiotropism.

To my mind, it seems unlikely that one of the most significant events in avian evolution, the loss of the phallus in nearly all species, is the result of a collateral or secondary effect of a protein whose primary function lies elsewhere. Others, including Bob Montgomerie, an evolutionary biologist at Queen's University in Ontario, agree.[33] Instead, most academic money rests with sexual selection. As previously discussed, female chickens may have chosen males with smaller penises to give them greater control in partner selection and to escape forced copulations. Over time, such female preference would have selected for individuals with increasing amounts of apoptotic protein in their genital tubercles, and this selection, over many generations, would have resulted in the disappearance of the copulatory organ.

The evolution of the apoptotic mechanism of penile loss in chickens must have occurred after their lineage diverged from the ducks, since low Bmp levels characterise the genital tubercles of crocodiles and palaeognaths. Crucially, further independent penile losses took place during the early evolution of modern birds, including at least once in each of the tinamou (genus *Crypturellus*) and megapode lineages, and a complete loss in the common ancestor of Neoaves.[36] Whether all these evolutionary events can be explained by altered Bmp expression, however, remains to be determined.

Without a doubt, the most significant of penile losses was the one that occurred in the ancestral Neoaves: a population that lived during the late Cretaceous alongside the non-avian dinosaurs. For this clade gave rise to the vast majority of the species alive today and explains why nearly all extant birds lack a penis.

So let us move on and hear their story, beginning with the remarkable Hoatzin.

CHAPTER 4

The Hoatzin's Story

AN IMPROBABLE VOYAGE

Any doubt one may have about the dinosaurian origin of birds will be dispelled when confronted with the sight, sound and even smell of the extraordinary South American Hoatzin (pronounced 'hwat-sin'). With its punk hair-do, maroon button eyes, blue facial skin, long neck and pheasant-sized body, the Hoatzin's prehistoric mien is utterly convincing (Plate 11). Nestlings even possess two hooked claws on each wing, rather like those of the 150-million-year-old *Archaeopteryx*. These features are essential for the survival of the chicks, as they assist in hatching and in climbing branches to escape predators. Hoatzins' nests are typically constructed over rivers and lakes, and when threatened, the youngsters jump into the water for safety. After the danger has passed, the hatchlings swim ashore and use their claws to clamber back to the nest like small monkeys. Once adulthood is reached, however, and the birds can fly, the appendages are lost. These strange anatomical structures are not thought to reflect an ancient developmental pathway that has lain dormant for millions of years, nor is it likely that the Hoatzin is a sole survivor of a claw-possessing lineage. Indeed, claws are not unique to Hoatzins: some species of geese retain them on their wings, while young flamingos, coots and ducks grow them on each wing. Rather, the Hoatzin's wing structure is the result of convergent evolution, an adaptation to a rather precarious lifestyle among the swamps and mangroves of the Amazon and Orinoco deltas.

Hoatzins are noisy, in a reptilian way, continually wheezing, hissing and grunting as they lumber in their ungainly way through the trees. I well recall the hoots and yelps of a dysfunctional family of three that had taken up residence in the mangroves outside our Ecuadorian lodge on the Napo. Furthermore, the birds exude a characteristic manure-like smell, a fact that accounts for their alternative name, 'stink bird'. Such a bizarre appearance and behaviour, so redolent of an ancient past, was recognised by the American ornithologist, explorer and writer William Beebe (1877–1962) when he encountered the species in the rainforests of Guyana:

> The young hoatzin stood erect for an instant, and then both wings of the little bird were stretched straight back, not folded, bird-wise, but dangling loosely and reaching well beyond the body. For a considerable fraction of time he leaned forward. Then without effort, without apparent leap or jump he dived straight downward, as beautifully as a seal, direct as a plummet and very swiftly. There was a scarcely-noticeable splash, and as I gazed with real awe, I watched the widening ripples which undulated over the muddy water – the only trace of the whereabouts of the young bird.
>
> It seemed as if no one, whether ornithologist, evolutionist, poet or philosopher could fail to be profoundly impressed at the sight we had seen. Here I was in a very real, a very modern boat, with the honk of motor horns sounding from the river road a few yards away through the bushes, in the shade of this tropical vegetation in the year nineteen hundred and sixteen; and yet the curtain of the past had been lifted and I had been permitted a glimpse of what must have been common in the millions of years ago. It was a tremendous thing, a wonderful thing to have seen, and it seemed to dwarf all the strange sights which had come to me in all other parts of the earth's wilderness. I had read of these habits and had expected them, but like one's first sight of a volcano in eruption, no reading or description prepares one for the actual phenomenon.[1]

The Hoatzin's taxonomy has baffled scientists more than most species, and despite the recent advances in DNA analysis, its position remains unclear. The latest study, based on whole-genome sequencing, places the Hoatzin as a sister taxon to the Gruiformes (a diverse order including coots, cranes, and rails) and the Charadriiformes (an order containing mainly shorebirds).[2] However, because of its anatomical distinctiveness, it is placed in an order of its own, Opisthocomiformes (from the Greek *opisthes,* 'behind', and *kome,* 'hair', referring to their shaggy neck and head plumage). To gain a sense of just how unusual this is from a taxonomic point of view, the Opisthocomiformes is one of only 40 orders that encompass the world's 10,694 species.[3]

Foregut fermenters

The source of the Hoatzin's 'bovine' odour is its large fermentation chamber or crop located between the upper and lower oesophagus.[4] This distensible structure, which makes up 18 per cent of the bird's weight, is analogous to the rumen, or major stomach compartment, in cows. To accommodate the crop when full, the Hoatzin has evolved modified chest muscles and a displaced breastbone: features that restrict the species' ability to fly. Despite these anatomical modifications, adult birds still have to rest their laden chambers on branches, causing marked skin callosities, or pads, to develop on their chest walls (Figure 4.1).

The Hoatzin's crop contains a cocktail of fermenting microbes that include novel bacteria, archaea, fungi and ciliated protozoa.[5] This unusual mixture of organisms enables the Hoatzin to survive on a diet of otherwise unpalatable

Figure 4.1 The anterior gut of an adult Hoatzin showing (A) crop, (B) oesophagus, (C) gizzard. The anterior sternum is reduced to accommodate the large fermentation chambers, resulting in drastic reduction in area available for flight muscle attachments to (D) the sternal carina; (E) 'resting' pad at the base of the sternum, used while perching with a full crop. Modified with permission from Grajal (1995).[4]

vegetation – woody plant tissue and toxic leaves that contain harmful alkaloids and hormones. Once the leaves are fermented, the resultant product, a nutritious mixture of smelly volatile fatty acids and associated microbes, passes down the gut to be absorbed in the intestines. The unique relationship between Hoatzin and microbe is a symbiotic one: the Hoatzin provides food for the bacteria, which, in turn, provide nutrition for the host. As a result, the Hoatzin has been able to adopt a unique browsing lifestyle, relying on a cellulose diet that is resistant to digestion by all other species of bird.

Foregut fermentation is not unique to the Hoatzin, and has evolved independently in several mammalian lineages. The most widespread species are the ruminant herbivores: animals that regurgitate and chew their cud, such as cattle, sheep and deer. Fermentation is also undertaken by colobine monkeys, including the Asian langurs and the African colobus, which diverged from the Old World monkeys around 15 million years ago. It is now recognised that the Hoatzin and the mammalian fermenters utilise the same salmagundi of novel cellulolytic microbes to aid digestion, despite millions of years of independent evolution. Remarkably, the evolutionary convergence between the Hoatzin and the two groups of mammals is also present at the molecular level (discussed later in the chapter).

The early foregut fermenters faced a formidable challenge that threatened the viability of their novel lifestyle – the loss of crucial nutrients assimilated by the vast numbers of microorganisms. Nature, therefore, required a means of breaking up and digesting the microbes, once they have served their biological usefulness, and recycling their valuable constituents. The solution was the synthesis of a series of proteins called lysozymes. But how did these enzymes evolve?

Nature rarely starts from scratch, since the chances of producing a novel protein with a required function at random is vanishingly small. Just consider, for a moment, the following: proteins are made up of hundreds of amino acids, joined to form chains. Since there are 20 possible amino acids for each position in the protein, the total number of theoretical proteins with a length of only 100 amino acids is astronomical (more than 10^{130} – that is a 1 followed by 130 zeros, a number many times greater than all the atoms in the universe). If a quadrillion different Hoatzins and colobus monkeys had experimented with a different protein every second since they evolved, their combined efforts would have evaluated only a minuscule fraction of the 10^{130} possible proteins.[6] However, evolution innovates and takes short-cuts to solve adaptive challenges. Life, it seems, can be conservative and progressive at the same time by duplicating essential and well-tested genes. This allows the original function to be retained while random mutations in the extra copies have the potential to lead to an expanded repertoire of activities. According to the Japanese biologist Susumu Ohno, duplication creates redundancy, and redundancy provides fuel for innovation.[7] Amazingly, both the Hoatzin and the mammalian fermenters have adopted the same evolutionary short-cut and duplicated similar genes to solve their recycling problems.

The story of the genes' identification began in 1922 when Alexander Fleming, the discoverer of penicillin, was suffering from a heavy cold. Being a perennial scientist, the Scottish biologist added a few drops of his nasal secretions to a bacterial culture. After a few days, he was surprised to find that something in his nasal discharge was killing the bacteria.[8] The substance turned out to be lysozyme C, a ubiquitous bacteriolytic enzyme since identified in virtually all animals and expressed in macrophages (a type of white cell), tears, saliva and mammalian milk. It is also found in egg white, where it functions to help keep eggs bacteria-free while the embryos develop.

There are now known to be two major classes of lysozyme C: a conventional type and a calcium-binding type. The conventional type is widespread and is the form discovered by Fleming. In contrast, the calcium-binding lysozyme is relatively rare, being known only from the milk of a few mammals (e.g., horse, cat and dog), and from pigeon eggs and the Hoatzin. Both genes appear to have arisen from an ancient gene duplication that preceded the divergence of birds and mammals some 300 million years ago. In mammals, the conventional lysozyme gene was further duplicated and independently recruited by ruminant herbivores and colobine monkeys to work in their stomachs. Surprisingly, the Hoatzin also duplicated and modified its lysozyme genes to solve the same digestive challenge, but this time from the calcium-binding group and not from the conventional group like the mammals.[9] Janet Kornegay, from the University of California, believes that the duplication of the Hoatzin's lysozyme genes occurred during the last 12 million years, well after its divergence from other birds.[10]

At this point, the story becomes even more intriguing. Quite independently, the new lysozymes, whether derived from a conventional lysozyme gene (ruminant herbivores and monkeys) or a calcium-binding lysozyme gene (Hoatzin), underwent identical structural modifications to ensure optimal function in the gastric environment. In all three lineages, amino acids containing acid-sensitive bonds were replaced, while other amino acids were substituted to confer resistance to their intestinal enzymes. It should be stressed that lysozymes are inactive when secreted into the fermenting chamber, to prevent the killing of microbes upon which the animals depend, and only become functional once they pass further down the gastrointestinal tract. The probability that these identical substitutions occurred by chance in lineages separated by 300 million years is negligible. Instead, the identically altered lysozymes must have arisen by convergent evolution, with the specific gene mutations being selected for by natural selection. As we have discussed, nature frequently solves adaptive challenges by refashioning genes that had originally evolved for different functions.

Let us press on, for the Hoatzin's ancestors have an even stranger secret to reveal, one that has only come to light during the last five years.

Transatlantic rafting

Hoatzins were widely believed to have originated in South America, given their geographical restriction to the Amazon and Orinoco river basins. However, in 2011, Gerald Mayr, a senior palaeontologist and Curator of the Senckenberg Research Institute in Frankfurt, re-examined nine fossilised bones that had been collected several years previously. The 17.5-million-year-old specimens, which belonged to the shoulder complex and wing of several individuals, were recovered from alluvial deposits near the mouth of the Orange River in southern Namibia. Mayr realised that the fossils had been misclassified: they did not come from a member of an extinct seriema-like family as first reported but belonged, instead, to a species of hoatzin, a novel representative of the Opisthocomiformes.[11] This reassessment was big news in the palaeontological world, for if the Hoatzin had African ancestors, it could no longer be regarded exclusively as a South American species. Luckily, it did not take long for confirmatory evidence to emerge. In 2014, Mayr reported further Hoatzin-like fossils from Africa and Europe. A leg bone from Lake Victoria in Kenya was identified as a close relative of the Hoatzin that lived 15 million years ago,[12] while fossils collected in France belonged to a much older ancestor, *Protoazin parisiensis* ('proto-Hoatzin from Paris'), dating from the late Eocene, around 34 million years ago.[13] According to Mayr, the Hoatzin now belongs to a growing list of South American relictual species which, like hummingbirds, potoos and seriemas, was once far more widespread than they are today. But how can the Hoatzin's transatlantic presence be explained, and what led to their disappearance from the Old World?

The notion that South American and Old World hoatzins split during the Cretaceous as the result of continental drift can be dismissed. The separation of South America and Africa was complete around 100 million years ago, well before the existence of not just the Hoatzins but probably all modern birds (except palaeognaths). In other words, dispersal not vicariance must account for their transatlantic distribution. However, it is highly unlikely that ancestral Hoatzins could have made their way to South America, using either a northern route, via Greenland, or a southern route, via Antarctica, as they were tropical birds, with a limited tolerance for the cold and a requirement for a continuous supply of edible plants. Also, no fossils have been found to support a polar route. Equally improbable would have been a transatlantic flight, even if the ocean's width was less during the Miocene, at around 2,000 kilometres, or if island stepping stones had existed in the South Atlantic.[14] The argument that ancestral Hoatzins might have been better flyers is not supported, as 22- to 24-million-year-old fossils from Brazil indicate that they had already evolved a large crop and some degree of folivory, features that would have impaired their flying ability.[11]

The only logical conclusion, no matter how strange it might seem, is that a population of early Hoatzins drifted across the Atlantic on floating vegetation. Of course, such rafts would not have been small structures but by necessity enormous floating islands – vast assemblages of greenery, tree trunks, sections of river bank and sediment. Every flood season, trees are swept into rivers and carried out to sea, and vegetative islands are known to have been transported by currents from Africa to the coast of Brazil. If Miocene rafts possessed trees, then they could have acted as sails. According to Alain Houle, a Canadian primatologist, floating islands could have crossed the Atlantic in two weeks, given the wind and currents that existed 30 million years ago.[15] Indeed, support for this hypothesis came unexpectedly in the summer of 1995. A party of at least 15 Green Iguanas survived Hurricane Marilyn by clinging to a collection of uprooted trees after being swept from Guadeloupe's shores. The reptiles drifted across the Caribbean seas for over three weeks before successfully colonising Anguilla to become potential founder members of a new species. Local fishermen reported that the arboreal mat was so big that it took two days to pile up on shore.[16] Such flotsam is typically washed into oceans from the mouths of large rivers, so the Hoatzin's riparian lifestyle and folivorous diet would have favoured their dispersal by vegetative rafts.[11]

The Hoatzin's dispersal route was almost certainly from Africa to America, since non-flying vertebrates are not known to have voyaged eastwards against the prevailing Cenozoic winds and currents. In contrast, many different vertebrate groups have crossed the Atlantic from east to west, including species of gecko, worm lizard, blind snake and thread snake.[17] Surprisingly, the last three groups are burrowers, spending most of their lives underground – a lifestyle that implies that the rafts must have contained sufficiently large clumps of earth to resist becoming waterlogged.[18] The most remarkable voyages, however, were those of mammals: rodents and monkeys. Recent molecular studies have shown that all of South America's caviomorph rodents (guinea pigs, capybaras, chinchillas, and agoutis) are derived from a single ancestral stock that floated across from Africa 40 million years ago. Similarly, a single colonisation by ancestral Old World monkeys, possibly concomitantly with the rodents, gave rise to all the American species alive today – marmosets, tamarins, night monkeys, squirrel monkeys, howlers, capuchins and uakaris (Figure 4.2). Like most readers, my initial reaction to these ideas was one of incredulity, but the evidence for rafting is irrefragable. It turns out that an image of rodents scuttling beneath the swaying trees,

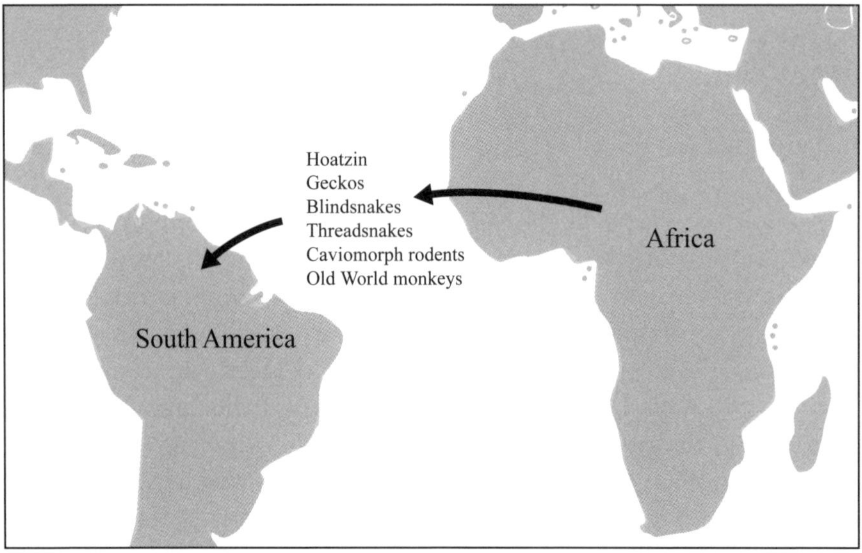

Figure 4.2 The Hoatzin, together with monkeys, caviomorph rodents and various species of snakes, reached South America by rafting across the Atlantic Ocean on floating vegetation.

dodging the breaking waves while a party of monkeys scans the horizon for salvation, may not be as far-fetched as one might have originally thought.[19]

Let us return to the Hoatzin's story and see why their ancestors died out in Europe and Africa. Fossil evidence indicates that the northern Hoatzins became extinct much earlier than those in Africa, where the latest fossils date from the Miocene (around 15 million years ago). In part, this may be related to the increasingly temperate climate and the development of open landscapes, as well as the closure of the Turgai Strait. This vast swathe of salt water, which stretched from the present-day Caspian Sea northwards to the Arctic, isolated European lands from continental Asia for millions of years. Once the water barrier disappeared, a migration route opened up, giving mammals from Asia and America access to Europe. A marked turnover of vertebrates resulted, with many extinctions – an event termed the 'Grande Coupure' (the 'Great Break') by the eminent Swiss palaeontologist Hans Stehlin.[20] Arboreal carnivores, such as cats, civets, and genets, entered Europe for the first time, and it is likely that the Hoatzins struggled to compete. Not only would Hoatzin chicks have been vulnerable in their exposed nests, but adult birds would have made easy prey, given their limited flying capabilities. Africa, in contrast, did not receive a similar influx of mammals for at least another 10 million years.

Whether judged by their symbiotic relationships, chance dispersals, or just comic embonpoint, I think you will agree that Hoatzins are undoubtedly one of the world's most bizarre species of bird.

CHAPTER 5

The Penguin's Story

PHENOTYPE AND ENVIRONMENT

Quintessentially Antarctic and instantly recognisable by all, penguins (family Spheniscidae) have captured our imagination like no other avian species. Indeed, penguins have been a perennial source of inspiration for advertisers, with their iconic 'tuxedo-clad' image adorning many a global product, from chocolate biscuits to paperback books, all unashamedly aimed at our anthropomorphic proclivities. Who would have predicted that *March of the Penguins*, a film documenting the annual trek of the Emperor Penguin to their breeding grounds deep in the Antarctic, would have been a box office hit, grossing over £100 million within the first few years? Despite their obvious appeal, the evolutionary basis for the penguins' endearing phenotype – black plumage above, white below, an upright waddling gait and wings reduced to flippers – has only recently received scientific scrutiny, and the findings have been instructive.

Early history

Before discussing how penguins evolved to survive the extreme polar environment, let us start at the beginning of their story, at a warmer time immediately before the K–Pg boundary. Scrutiny of the 'bird tree' reveals a major branch at 79.6 million years ago, which, if followed, gives rise eventually to a large seabird clade: one that includes the divers (Gaviiformes), penguins (Sphenisciformes), albatrosses, shearwaters and petrels (Procellariiformes), storks (Ciconiiformes) and pelicans (Pelecaniformes). At some stage in the late Cretaceous, between the splits that led to the diver and storm petrel lineages, the ancestors of extant penguins took to the skies somewhere in the southern hemisphere. These birds, estimated to have been around 1 kilogram in weight and with the ability to fly and swim underwater as well as in the air, evolved through a long series of now extinct species to produce all 18 penguin species we recognise today.

The earliest penguin fossil, *Waimanu manneringi*, dating from 61.6 million years ago,[1] was recovered from a rocky outcrop in the Waipara Valley, a wine-growing area on New Zealand's South Island. Although the bones were located just above the K–Pg boundary, *Waimanu* was already surprisingly penguin-like, having developed a large body size, upright stance, short stubby feet and relatively small wings (Plate 12). These features, coupled with a dense bone structure and limited flexion at the elbow joint, rule out the possibility of aerial flight and suggest instead that *Waimanu* swam, loon-like on the surface using its feet, and dived, using its modified wings for locomotion. *Waimanu,* derived from the Māori words *wai* (water) and *manu* (bird), was only the first of a series of fossils that have been found in New Zealand, implying that the common ancestor of all penguins evolved on those islands. According to Daniel Ksepka from the Bruce Museum, Connecticut, 'the location was great in terms of both food and safety. Most of New Zealand was underwater at that time, leaving isolated, rocky landmasses that kept the penguins safe from potential predators and provided them with a plentiful food supply.'[2]

Dispersal played a key role in penguin speciation and distribution, a process aided by the establishment of strong currents in the Southern Ocean and the emergence of new islands and continental coastlines. The break-up of Gondwana was almost complete by the time the first penguins appeared in New Zealand, and the separation of Antarctica, with the establishment of the Tasman and Drake Passages, seems to have been a critical factor facilitating their radiation. These geological events created not only extensive coastlines that lacked terrestrial predators, but also a means of dispersal, with the formation of the Antarctic Circumpolar Current. Later, the formation of the polar ice cap led to a marked cooling of the surrounding seas, allowing the region to become rich in nutrients that gave support to huge populations of fish and crustaceans. The early penguins evolved rapidly to cope with these deteriorating climatic conditions and became widely distributed throughout the southern hemisphere. The largest living descendant, the Emperor Penguin, for example, has approximately 40 isolated breeding colonies, spread out over all sectors of Antarctica's coastline. Several of these sites have never been visited, since their locations have only recently been identified by the British Antarctic Survey, using satellite images to detect their tell-tale guano stains from space.[3]

The fossil record also suggests that penguins dispersed from New Zealand to colonise South America and Australia by the late Eocene. While many of these forays ended in extinction, repeated dispersals ensured the wide austral distribution observed today. Barriers, formed by the warmer tropical seas and the southerly drifting currents, prevented colonisation north of the equator, with one exception: the Galápagos Penguin. This species is thought to have separated from its Peruvian relatives approximately 4 million years ago, after being caught up in the cool Humboldt Current and conveyed to the remote Pacific islands where they thrive today. Africa hosts only one species, the African Penguin (formerly known as the Jackass Penguin), but this was not always the case. Indeed, Africa, the last continent to be colonised by penguins, received multiple waves of immigration from South America, all of which became extinct by the end of the Pliocene, owing probably to falling sea levels and loss of offshore nesting sites. Like all penguins, their arrival was assisted by ocean currents, in this case by the system known as the South Atlantic Gyre (Figure 5.1). Daniel Ksepka, in collaboration with fellow palaeontologist Daniel Thomas, recently proposed that this current system, which creates a huge anticlockwise flow, may have served as a 'penguin conveyor belt' from South America to South Africa.[4] This hypothesis would also explain why there are no penguins on Madagascar, since the island is surrounded by unfavourable currents that push southwards and away from its coastline.

The plausibility of these dispersal scenarios is strengthened by the knowledge that penguins are naturally great roamers. Several species travel hundreds or even thousands of kilometres throughout the year, moving from feeding grounds to breeding colonies. In 2011, a juvenile Emperor Penguin got lost and turned up well over 3,000 kilometres from home on the Kapiti

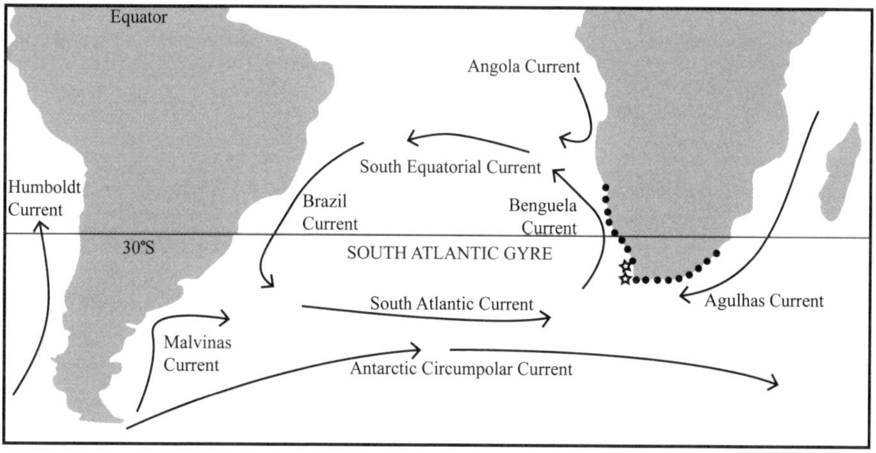

Figure 5.1 The South Atlantic Gyre. Black dots indicate present-day breeding sites of the African Black-footed ('Jackass') Penguin; stars indicate the location of fossil finds from two extinct penguin species. Since neither extinct species was a close relative of the extant Black-footed Penguin, it suggests that Africa's penguins arrived independently by the 'penguin conveyor belt'. The Agulhas Current would have prevented penguins colonising Madagascar. Modified from Ksepka & Thomas (2012).[4]

coast of New Zealand's North Island – to become an immediate international celebrity. After eating sand, which it probably mistook for snow, it was nursed back to health and eventually returned to the Southern Ocean. While such occurrences are usually detrimental for an individual bird, penguins' inherent wanderlust has clearly been a significant factor underpinning their speciation.

Polar adaptation

The perceived 'cuteness' of penguins results from their unique anatomical and behavioural traits, features that have evolved over millions of years to enable their survival in the most extreme conditions on Earth. For example, to function in seas as cold as –2.0 °C, penguins have acquired a gamut of insulating features: a thick layer of subdermal fat, a rotund body shape, specialised short, tightly overlapping feathers of low thermal conductivity, and an under-layer of fine woolly down.

Surprisingly, the Emperor Penguin does not have the densest outer, or contour, feathers in the avian world; that prize goes to the White-throated Dipper, a passerine that dives in fast-flowing streams, which has six times as many feathers per square centimetre. Instead, the Emperor Penguin possesses a markedly increased density of downy feathers (plumules), which not only provide insulation but assist the bird's rapid underwater ascent and exit onto the sea-ice.[5] The small pockets of air trapped within the downy layer are released during the ascent and, by collecting at the boundary layer, reduce drag to allow the penguins to reach high speeds. Indeed, the released bubbles are so tiny that it appears as though a trail of smoke is coming from the feathers. The density of contour feathers is greatest on the penguin's front, which provides a cushion when tobogganing, as well as extra insulation when resting on the ice. All surviving penguin species possess dark-coloured dorsal feathers that aid heat absorption from the sun, and white ventral feathers to help them hide from predators, especially while swimming near the surface where it is hard for Killer Whales and Leopard Seals to discern their white underbelly. Interestingly, a fossil dating from 36 million years ago, *Inkayacu paracasensis*, reveals that penguins evolved their specialised feathers deep in the past, although the colour then was not the characteristic black and white of extant species, but a reddish brown or grey. The evidence for this remarkable finding came from *Inkayacu*'s well-preserved melanosomes, tiny pill-shaped cellular organelles that contain melanin, the pigment responsible for colour and photoprotection in animal tissues.[6] The Avian Genome Consortium have recently begun to explore the molecular basis for such adaptive changes.

It appears that penguins have an increased number of genes that code for beta-keratin, a key component of feathers, compared to other bird species. Such gene amplification, coupled with a positive selection for other dermal-related genes, may well have enabled penguins' unique skin and feathers to have evolved.[7]

The vertebrate tongue possesses five different types of taste receptor that detect sweet, sour, salty, bitter and savoury, or umami, flavours. While all birds (except hummingbirds) lack sweet receptors, penguins have also lost the ability to taste bitter and umami. A team of scientists from the United States and China believes that the two relevant receptors became non-functional sometime after the penguins' common ancestor separated from the tubenose seabirds around 60 million years ago.[8] Functional sweet receptors appear to have been lost much earlier. The researchers also found a lack of the signalling protein Trpm5 in penguins' taste perception. This interesting molecule is required for the transmission of sweet, bitter and umami taste signals to the brain, as well as for the optimal secretion of insulin. However, Trpm5 cannot function at the low temperatures experienced by the penguins' taste buds, approximately 0 °C. As a result, natural selection may have sacrificed its gustatory function while retaining its more important metabolic role in the body (39 °C in Emperor Penguins). Of course, such a loss of taste may not matter, since penguins swallow fish whole and are not exposed to the bitter toxins found in the plant world.

Ancestral penguins, inhabiting the balmy Eocene, had hyper-elongated, spear-like, beaks (Plate 12), while extant species have evolved relatively small bills (Plate 13).[9] These evolutionary changes appear to coincide with the cooling of the high latitudes, a finding predicted by Allen's rule, which states that appendages of warm-blooded creatures become smaller in colder climates to minimise heat loss.[10] There's plenty of evidence for this. The Polar Bear and Arctic Hare, when compared to their more temperate cousins, have stockier bodies and shorter ears, leaving the least possible amount of core temperature exposed. Recently, studies have confirmed that Allen's rule also applies to the size of birds' beaks and supports the conjecture from fossil and palaeoclimatic data that thermoregulation may have played a role in the development of the size and shape of penguins' bills.[11] However, it is also possible that dietary changes could have had a contributory effect.

Heat dissipation from flippers and feet is prevented by an important anatomical structure, the humeral arterial plexus or *rete mirabile* (Latin for 'wonderful net'). This system of arteries and veins acts as a countercurrent heat exchanger and significantly reduces the risk of hypothermia. Outgoing arterial blood warms cooler incoming venous blood so that heat is returned to the body core, rather than continuing to the periphery to become lost. Arthur P. Fraas, a mechanical and aeronautical engineer, described the penguin's blood vessel structure as 'one of the world's most effective regenerative heat exchangers', since in winter it enables penguin's feet to be maintained a few degrees above freezing, minimising heat loss and preventing frostbite.[12] Indeed, its design is so ingenious that human engineers have copied it.

Fossil evidence reveals that the vascular structure was not present in the earliest 'proto' penguins, but evolved much later, around 49 million years ago, during the 'Greenhouse Earth' interval, when temperatures were relatively mild.[13] This apparent paradox is resolved if the plexus had initially developed as palaeontologists now suspect, to prevent hypothermia while feeding in localised cold-water upwellings that would have supported rich food supplies. At the same time, it may also have helped offset the energetic costs incurred during the long foraging excursions required to reach such sites.[14] The absence of an arterial plexus in *Waimanu* and related early penguins suggests that the basal species were restricted to inshore feeding and did not stray far from coastal waters. Irrespective of the original selection pressure, the presence of heat-retention structures undoubtedly facilitated the successful invasion of Antarctica's icy waters by penguins, many millions of years later. Indeed, recent fossil finds reveal that some early species reached striking dimensions, with *Palaeeudyptes klekowskii*, known as the 'colossus penguin', standing approximately 2 metres in height.[15]

A further polar adaptation is that the muscles operating penguins' flippers are not located in their limbs, as one might expect, but are found deep in the warmer parts of their bodies. The flippers are manipulated instead by tendons that pass right along their length, attaching to the ends like a sort of remote-control wire. It does not matter, therefore, if the flippers become cold, since they can still be operated by muscles that remain fully functional at body temperature.

Penguins have not only evolved to survive the world's coldest environment, but they have also adapted to withstand immense pressures encountered during foraging dives. These can be up to 500 metres in depth and last up to 23 minutes. At such depths, the pressure is 40 times that at the surface and would lead to structural damage, or barotrauma, in most terrestrial creatures. Humans can only work at these depths by using specialised equipment and undergoing prolonged periods of decompression when returning to the surface. To cope, penguins have evolved solid bones, rather than the marked skeletal pneumaticity, or air-filled bones, which characterise flying species. This adaptation not only reduces their buoyancy to assist underwater 'flying', but also eliminates the risk of mechanical trauma from repeated exposure to high pressures. How penguins avoid nitrogen narcosis, or the 'bends', despite the accumulation of nitrogen in their bloodstream, remains unclear.

When penguins return from extended fishing excursions under the sea-ice, their lung and blood oxygen content may fall to levels not seen in any other living animal. Indeed, the Emperor Penguin can use up all its lung oxygen without it resulting in immediate tissue damage. This remarkable feat reflects evolutionary changes to their haemoglobin molecules, modifications that produce a much higher affinity, or binding capacity, for oxygen. Similar changes in haemoglobin are found in geese and hummingbirds that live at high altitude (see *The Waterfowl's Story*). However, since the oxygen-carrying proteins of penguins and geese evolved at different times and in response to different environmental challenges, it is maybe not surprising that the same functional effect results from alternative amino acid substitutions – evidence of convergent evolution at the molecular or genetic level.

The Emperor Penguin is the only vertebrate to breed during the brutal Antarctic winter when temperatures of –60 °C and winds up to 100 kilometres per hour are commonplace. The payoff for adopting such an extreme breeding approach is the lack of predation, but the challenges are immense. Male penguins, who take sole responsibility for incubating the single egg, are unable to feed for up to 110 days, since the colonies are located miles from the open sea, or open areas within the pack-ice, known as polynias, where they feed. Energy conservation is vital, and to achieve this they have evolved a unique and complex form of social behaviour, best summarised as 'huddling with shuffling'.

Unlike other penguin species, Emperors are not territorial and gather together to produce large huddles, or 'turtles', during extreme conditions, a feature which halves their overall energy requirements while maintaining a constant core temperature of around 37.5 °C (Plate 14). All birds must get an equal chance of protection so that no one individual becomes exposed on the periphery for too long. However, how is this achieved? If the huddle is too loose then individuals will freeze, too tight and no movement is possible, and those on the outside will succumb. This apparent impasse is resolved using a remarkable adaptive strategy: Emperor Penguins shuffle collectively in a highly coordinated manner every 30–60 seconds. Whenever a small group take a few steps, their neighbours follow suit and the resultant wave passes through the entire huddle. Eventually, this behaviour leads to large-scale re-organisation so that all birds get a fair share of shelter in the centre. Such changes are too slow to be seen by the naked eye, and it is only recently that scientists have revealed these previously hidden and seemingly choreographed manoeuvres with the use of high-resolution time-lapse photography. It remains a mystery how the periodic waves are initiated and whether there is a hierarchy among the individuals. In the end, although each penguin is out to look after itself, the whole colony benefits from the collective behaviour, as the heat dispersal ends up being shared by all.[16]

Loss of flight

To march, or rather waddle and toboggan on their bellies, for days on end to reach their feeding grounds is an inordinately demanding undertaking and raises the question of why penguins ever lost the power of flight. The biological advantage, for there must be one, has only recently been explored, and the answer has come, not from the study of penguins, but from their northern equivalents, the flying auks, or alcids.

The alcids, including puffins, guillemots and Razorbill, evolved in the northern hemisphere much later than the penguins, around 35 million years ago during the late Eocene. The diversity of present-day species is the result of two main evolutionary events: a major diversification during the Miocene, associated with an increase in cold-water upwellings (i.e., foraging areas), and a differential cull around 2.5 million years ago due to a marked decline in marine vertebrates and invertebrates.[17] All living alcids use a rare form of locomotion, wing-propelled diving, that probably evolved soon after their lineage diverged. How this innovative behaviour came about, however, remains unclear. It is possible that their ancestors moved from coastal regions to feed in deeper and deeper waters and, in doing so, became adapted to exploit benthic resources. Such a transition could then have encouraged free-swimming in the water column, a form of locomotion exhibited by most alcid species today. But how can the evaluation of auks explain the evolution of flightlessness in penguins?

The study in question, undertaken by an international research group headed by Kyle Elliott from the University of Manitoba, set out to prove that flightlessness in penguins evolved to produce efficiency gains in other locomotive skills – for example, diving.[18] A colony of Brünnich's Guillemots (Thick-billed Murres) in Nunavut, Canada, was selected for study, since the birds' physiology and energy consumption was thought to match those of the last flying ancestors of penguins. The research team injected the guillemots with radio-labelled water, to allow energy expenditure to be monitored, while, at the same time, time-depth devices were attached to track their movements. By correlating energy requirements with activity, it was shown that guillemots dive more efficiently than any other flying species, and are only bettered by penguins. In contrast, their flight costs are the highest recorded for any vertebrate, a level that would be impossible to maintain if the bird's weight increased further. Indeed, the Great Auk, an extinct relative of the guillemots, supports this conclusion, since the species was both heavier and flightless.

Elliott's findings imply that as penguins became more efficient at diving, their energy requirements for flying became more and more demanding. At some point in their evolution, it simply became impossible to sustain the high energy requirements of flight, and they became flightless. Clearly, the greater efficiency at swimming and the resultant proficiency at catching prey and avoiding predation must have outweighed any limitations imposed by flightlessness. Indeed, competition for marine resources was fierce after the K–Pg boundary event with the rapid radiation of mammals, especially cetaceans and pinnipeds. The biomechanical trade-off, diving versus flight, resulted from a reduction in wingspan, the development of enlarged wing bones, and an increased body mass – features that were already evident in the earliest penguin fossils. Furthermore, the reason why penguins became bigger-bodied so soon after becoming flightless is that their larger frames boosted swimming efficiency and allowed for longer and deeper dives.

'The worst journey in the world'

The egg of the Emperor Penguin provides a poignant footnote to the history of evolutionary theory, for it was believed that its contents would provide proof for the once popular, but now discredited, 'theory of recapitulation', also known as the biogenetic law. Best encapsulated by Ernst Haeckel's phrase, 'ontogeny recapitulates phylogeny', the hypothesis holds that a developing embryo will go through various stages that mirror its entire evolutionary history. Since the Emperor Penguin was once thought to be the most primitive bird, it was hoped that the study of its embryo would reveal an evolutionary link between birds and dinosaurs. The dilemma for the early scientists was that Emperor Penguins only reside in the Antarctic and lay their eggs in the depth of winter. In 1911, three members of Captain Scott's expedition set off from their base camp, on a 210-kilometre round trip to the rookeries at Cape Crozier, suffering throughout from sleep deprivation, perpetual darkness and unimaginably low temperatures. Their ultimately futile but truly heroic winter journey, recounted in Apsley Cherry-Garrard's two-volume work *The Worst Journey in the World*,[19] resulted in three eggs being brought back to England. Examination of their embryos was delayed by the outbreak of the First World War, and by the time the results were published in 1934, science had moved on, and the theory of recapitulation had been rejected. While the three eggs may not have produced the expected scientific breakthrough, they remain carefully stored in London's Natural History Museum as fragile reminders of our ongoing quest to unlock the secrets of avian evolution.

CHAPTER 6

The Storm Petrel's Story

SYMPATRY VERSUS ALLOPATRY

Type 'storm petrel' into the OneZoom Tree of Life search box and you will find two closely spaced offshoots, one at 58.5 and the other at 53.8 million years ago, which, if followed, give rise to all the extant storm petrels.[1] The 27 species are consequently non-monophyletic; in other words, they are not a natural group, but rather consist of two clades, one in each hemisphere (Figure 6.1). The 18 northern storm petrels (family Hydrobatidae) tend to have pointed wings and relatively short legs, while the nine austral storm petrels (family Oceanitidae) have rounded wings and longer legs. They are the smallest of all pelagic birds and obtain their food by hovering and fluttering on the surface, taking mostly crustaceans, although they will feed on fish, molluscs, and even droplets of oil. The name 'petrel' is believed to be a diminutive form of 'St Peter's bird', an allusion to St Peter's walking upon the Sea of Galilee, though it may reflect an attempted onomatopoeic rendering of their 'tiptoeing' activities.

Little is known of storm petrels' behaviour or distribution at sea, where they can be difficult to locate and even harder to identify. Indeed, there is no better illustration than the remarkable story of the New Zealand Storm Petrel, a species which had not been seen for over 150 years before its

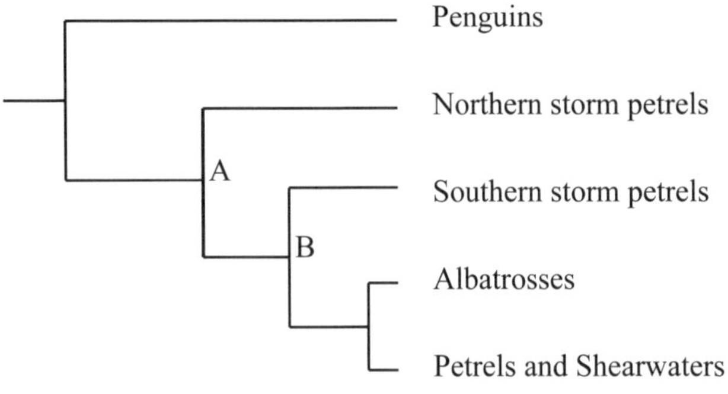

Figure 6.1 A phylogeny of storm petrels. A = 58.5 Mya; B = 53.8 Mya.

spectacular rediscovery in 2003 by two enthusiastic birders.[2] Their report, the first of a freshet of writing on this Lazarus species, was prosaically entitled, 'The New Zealand Storm Petrel is not extinct', a statement that belied the initial scepticism of the country's ornithological hierarchy. Since then, an intensive research programme, involving radio-tagged birds, led, in 2013, to the discovery of their breeding sites on Little Barrier Island, a mere 50 kilometres from Auckland City.[3]

Monteiro's Storm Petrel

Fascinating as these vignettes are, the importance of storm petrels for our story of birds relates to one northern family member in particular: the Band-rumped, or Madeiran Storm Petrel. This species has recently been shown to have undergone sympatric speciation: the situation in which two species have evolved from one, in the absence of any known geographic, or physical barrier.[4] Sympatry differs from allopatry, the predominant mode of speciation in birds in which geographical isolation leads to a gradual and random accumulation of genetic changes that result from different selection pressures – climatic, predatory and resource availability. Eventually, such gene mutations are sufficient to prevent the populations from interbreeding, even in the event of the barrier being removed. In contrast, sympatric speciation does not require large-scale geographical distances to limit genetic exchange, although it is uncommon and has been difficult to prove. The tale of the Madeiran Storm Petrel, therefore, is one that warrants being told, not least because of its scientific scholarship, serendipity, and unexpected tragedy.

Luis Rocha Monteiro, a Portuguese doctorate student at the University of the Azores (Plate 15), was studying mercury levels in the islands' seabirds when he noticed that the seasonally segregated populations of breeding Madeiran Storm Petrel looked slightly different. Birds that bred in the spring, or hot season, appeared to be a little smaller, with a more pronounced forked tail and a thinner bill, than the autumn (cool-season) breeders. Other differences emerged. The spring breeders had lower mercury levels and were thought to forage in the local seas all the year round, while the autumn breeders had high metal levels, indicating an alternative and, at the time, unknown feeding area.[5] Was it possible that the two groups were ecologically isolated and were, in fact, different species? Luis Monteiro thought so. However, several pieces of the jigsaw were missing, including the timing of the autumn breeders' moult and the whereabouts of their winter feeding grounds. If the moult of these birds could be shown to be out of phase with the spring breeders, then the likelihood of there being two species would be strengthened. Unfortunately, the missing data seemed unobtainable: the moult of the autumn breeders had never been witnessed, since the birds had already dispersed far out into the Atlantic Ocean and could not be examined. This seemingly insurmountable problem was about to be resolved. Enter Barry Sweatt.

Sweatt, an experienced fisherman working out of Orange Beach, Alabama, was accustomed to saving exhausted birds that landed on his vessel, while far out in the Gulf of Mexico. One spring day in 1998, a moribund female Madeiran Storm Petrel landed on his boat and survived long enough to be taken to the Wildlife Sanctuary of Northwest Florida for attempted rehabilitation. Unfortunately, the bird died before it could be released, but staff noticed it had been ringed, not with the expected US Fish and Wildlife band, but with one that had been used by ringers in the Azores. It turned out that the bird had been mist-netted six years previously on Praia islet, a rocky outcrop lying off Graciosa Island in the Azorean archipelago. Monteiro was understandably elated, particularly since the Gulf bird turned out to be a cool-season individual with evidence of an active primary moult. The existence of two species, therefore, seemed increasingly likely. Unfortunately, while these observations were being collated, Monteiro was killed in a plane crash in the Azores, and did not live to witness the widespread acceptance of his idea; nor was he aware that the spring breeders would be given his name in recognition of his pioneering field work.[6] While the Band-rumped or Madeiran Storm Petrel breeds on islands in the warmer parts of the Atlantic and Pacific Oceans, Monteiro's Storm Petrel is confined to the Azores, where they are currently known to nest on just two small neighbouring islets. Despite conservation efforts, the Monteiro's Storm Petrel remains vulnerable, with an estimated population of only 300 pairs (Plate 16).

A picture has now emerged of how Monteiro's and Band-rumped Storm Petrels might have undergone sympatric speciation.[7] Mitochondrial DNA analysis indicates that the two species diverged around 100,000 years ago and that there has been little, if any, subsequent genetic exchange. Intriguingly, sympatric seasonal populations of storm petrels are also found on other islands, including the Galápagos, Cape Verde and Desertas, but these have diverged more recently and are not yet genetically differentiated. In other words, they have not had time to evolve into full species. Any explanation of storm petrel speciation, therefore, needs to take into account the ubiquitous nature of the divergence of their breeding seasons. A clue to the puzzle was that the dates for the separation of each sympatric pair correlate well with interglacial, or postglacial, periods. It is now recognised that cold-water upwellings, with their rich nutrient content and associated marine productivity, would have increased during periods of extensive glaciation, leading to increased foraging opportunities and expansion in both bird numbers and range. Conversely, warmer upwellings, with their reduced marine productivity, characterised the interglacials, and the increased competition for their limited food supply would have favoured a divergence in the bird's breeding times.

For speciation to be complete, non-random (assortative) mating is required, to ensure that cool- and hot-season breeders do not exchange genetic material. One might imagine that population separation by breeding time alone, or allochrony, would in itself be sufficient. But picture what would have happened if a few of the early diverging populations had mated and produced intermediates or 'hybrids'. Such offspring, together with any that arose from subsequent back-crossing, would have led ultimately to widespread genetic mixing and caused the incipient diversity to collapse back to a single homogeneous population. So what prevents this from happening? In the case of storm petrels, seasonal 'hybrids' are thought to have intermediate breeding times and, as a result, have a relatively low fitness because of their failure to compete for nest sites. In some areas, Band-rumped and Monteiro's Storm Petrels not only use the same nesting sites but may even share the same burrow: one species in, one species out! It is this unique combination of features that has underpinned storm petrel speciation, and makes allochrony a 'magic trait' – a term coined by Sergey Gavrilets to describe any feature that, singularly, can influence pre-mating isolation and the fitness of intermediates.[8] Later, the evolution of subtle differences in vocalisation would have helped to maintain speciation, as both groups of storm petrels rely heavily on acoustic signals to locate potential mates. Playback experiments have shown that hot-season birds, when they return to their breeding sites at night, fail to recognise the calls of cool-season birds, thus reinforcing their genetic isolation.

Sympatric speciation

Sympatric speciation has been at the centre of one of the most protracted controversies within evolutionary biology, not least because any assessment of its true contribution requires the complex synthesis of ecology, genetics and behaviour. Proponents and sceptics alike have filled many a journal page since Darwin first suggested that species are not static, but tend to diverge when faced with empty niches, even in the absence of geographical separation. Early geneticists weighed in and argued that new species arise instantaneously, from within their geographic range, via spontaneous mutations. These views prevailed until the influential biologist Ernst Mayr refuted the concept in the 1940s and promulgated the idea that initial separation was a prerequisite for speciation. As a result, allopatry became the dominant dogma for the next four decades. However, Mayr, cognisant of the vagaries of scientific progress, added a prescient rider, 'the issue will be raised again at regular intervals. Sympatric speciation is like the Lernaean Hydra, which grew two new heads whenever one of its heads was cut off.'[9] He was right to do so. During the 1990s, a plethora of theoretical models indicated that speciation in the same location is possible under certain conditions, while experimental studies provided examples for which a sympatric origin appeared to be the most parsimonious explanation.

The argument for sympatric speciation shifted from plausibility to one of frequency. Jerry Coyne and Trevor Price, colleagues at the University of Chicago, sought an answer for birds by searching for pairs of endemic sister species on remote and relatively small oceanic islands. They compiled species lists for 46 such locations and found only a handful of islands that contained two endemics belonging to the same genus. However, because one species appeared to be more closely related to a mainland relative, or congener, than the other, it was concluded that such cases were better explained by separate long-distance invasions from the continent, rather than by sympatric speciation.[10] This study highlights an issue that faces all investigators: many species may have shifted their ranges significantly in the past, making proof of sympatry exceedingly difficult. A recent collaborative study, headed by Albert Phillimore, a doctorate student from Imperial College London, approached the problem by applying simulations and modelling techniques, before incorporating the best fit to study the distributions of nearly 300 paired sister species. The team concluded that sympatry occurs but is uncommon in birds and contributes to only 5 per cent of speciation events.[11]

This last study, however, was based on statistical modelling and, as such, has little value when determining speciation mechanisms for individual organisms. In reality, any claim of sympatric speciation must demonstrate, not only that both species overlap geographically, but also that they have a sister relationship, exhibit reproductive isolation and lack a previous allopatric phase in their evolutionary history. Few examples convincingly satisfy all four criteria, and the exclusion of past allopatry is the most difficult to assure. To illustrate this point, consider the following three case histories that have been proffered in the past as exemplars of sympatric speciation: the Junin Grebe, crater-lake cichlids (a suborder of fish) and the apple maggot fly.

The flightless Junin Grebe is restricted to a high-altitude lake of the same name, in the *puna* zone of western-central Peru, where approximately 200 individuals remain. In contrast, its flying relative, the Silvery Grebe, is widespread, frequenting lakes throughout the Andes, from southern Chile to northern Peru – including Lake Junin. It has been widely held that the Junin Grebe evolved, *in situ*, from Lake Junin's Silvery Grebe population by sympatric speciation. However, the idea that a flying form remained on the lake throughout the flightless grebe's evolution has now been challenged. The evolutionary biologist Trevor Price thinks it more likely that geographical separation occurred during the speciation phase, and that the Silvery Grebe recolonised the lake only after the development of reproductive isolation and ecological differentiation.[12] Unfortunately, it has not been possible to determine which of these scenarios is correct.

The remarkable richness of cichlid (pronounced 'sick-lid') species in the east African lakes, which between them may contain more than 1,500 endemic species, has long been regarded as a paradigm of sympatric speciation. This view is based on detailed genetic analyses which suggest that each primordial lake was seeded by one, or at most a few, ancestral fish from the surrounding river systems.[13] Scientists, however, have recently provided evidence of significant water-level changes over the last 100,000 years, a fact that may have led to the fragmentation of the early fish populations. Initial drought conditions would have produced many separate, smaller lakes, with opportunities for allopatric speciation, while the subsequent dramatic rise in water levels enabled the shallow-water species to become isolated around new rocky islands and undergo further diversification.[14] Elucidating the mechanisms of cichlid speciation requires a detailed knowledge of past climatic and geological conditions, a fact germane to the understanding of most species evolution.

The story of the American apple maggot fly (*Rhagoletis pomonella*) is even more labyrinthine than that of either the Junin Grebe or the cichlids.[15] When farmers began planting apple orchards over two centuries ago, a population of flies switched from hawthorn to apple trees to lay their eggs, despite apple trees having different fruiting times. Since the flies mate only on trees where they feed, the switch is pushing the two gene pools apart – an example of evolving sympatric speciation, one might think. However, recent genetic research has revealed an unexpected twist in the apple maggot's story. It appears that the ancestral hawthorn-infesting population became temporally divided between Mexico and North America several million years ago, which allowed genetic variation to evolve in isolation. These changes included chromosomal rearrangements that pre-adapted certain flies for apples long before the fruits were available as hosts. Does this fact now render the appellation of sympatric speciation inappropriate, given that the preference for apples by a subpopulation of flies evolved in allopatry?

A final thought before we park the contentious topic of sympatry. Given the increasing awareness of the complexities of speciation, it has been questioned whether the current obsession with procrustean categorisation, sympatric or allopatric (at its simplest), is helpful. Since each species has travelled a different evolutionary path, it may be more informative to stress the contribution of each factor involved – historical biogeography, contemporary distribution, natural selection and genetic influence – rather than attempt the assignment of category labels to particular case studies. Process, not classification, is the field's emerging mantra.

We have dallied too long. Let us turn to *The Albatross's Story*, for it is an interesting one that raises the perennial problem of what constitutes a species.

CHAPTER 7

The Albatross's Story

THE SPECIES PROBLEM

The adoption of Samuel Taylor Coleridge's artistic representation of the albatross as a metaphor for a burden to be carried, a portent of doom, is unfortunate. To many, it is their unrivalled mastery of the wind, their improbable wanderings across the Earth's most forbidding oceans – a life of apparent freedom, liberated from the shackles of earthly constraints – that would make a more fitting basis for allegory. For albatrosses (family Diomedeidae) are nature's supreme flyers, the ultimate gliding machines, capable of staying aloft for months on end, only coming ashore to breed on some of the world's remotest islands. Indeed, albatrosses can travel up to 15,000 kilometres in a single journey and circumnavigate the globe in 46 days. Their story, one that stretches back to the early Eocene, approximately 50 million years ago, is far from complete, given the paucity of informative fossils and the ambiguities of molecular detective work.

Albatross phylogeny

The evolutionary stem of the albatross branch, with its terminal 21 living species, is depicted on the 'bird tree' as nestling between an earlier branch giving rise to the southern storm petrels and a later branch of 'true' petrels and fulmars. While this schema appears reassuringly authoritative, it fails to convey the current controversies that relate, not just to albatross phylogeny, but also to the number of accepted species alive today. A recent study, for example, reported albatrosses to be more closely related to the northern storm petrels, or Hydrobatidae, rather than to the family of austral storm petrels, or Oceanitidae.[1] Others believe that albatrosses and 'true' petrels (family Procellariidae) are sister groups and that both are somewhat distant from storm petrels. A third view places albatrosses within the same clade as northern storm petrels, 'true' petrels and diving petrels, but with the southern storm petrels lying outside. Confused? Despite this complex mélange of phy-

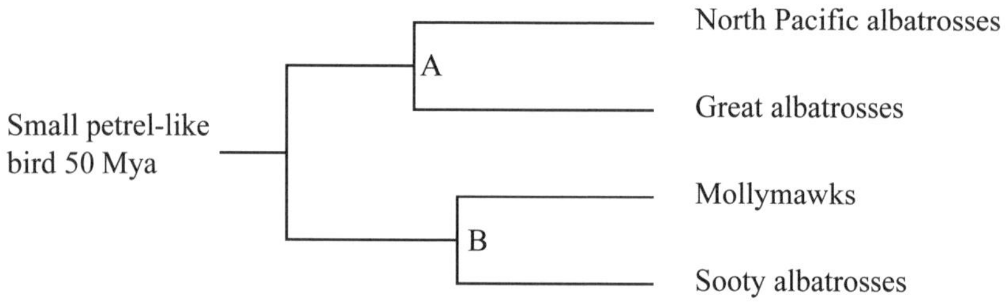

Figure 7.1 A phylogeny of albatrosses. A = at least 15 Mya; B = at least 10 Mya.

logenetic possibilities, most likely a result of the different methodological approaches, one conclusion appears inescapable. The four genera – North Pacific albatrosses (*Phoebastria*), great albatrosses (*Diomedea*), mollymawks (*Thalassarche*) and sooty albatrosses (*Phoebetria*) – have all evolved from smaller petrel-like birds that nested in burrows around 50 million years ago (Figure 7.1).

Birders, however, are unlikely to lose sleep over such arcane phylogenetic deliberations, and would much prefer that the current confusion over the number of extant species be resolved. Indeed, depending on which authority is consulted, there are between 14 and 24 different species of albatross inhabiting the world's oceans, an inconsistency that relates to the troublesome question of what constitutes a species.

The 'species problem', as it is prosaically termed, has taxed biologists for decades and has resulted in the use of at least two dozen different definitions. All are variations on one of two themes: the 'biological species concept' and the 'phylogenetic species concept'. In the past, ornithologists have embraced the former, first proposed by Ernst Mayr in 1942, in which a species is defined as a population that is reproductively isolated and unable to mate with other groups.[2] Distinctive geographical forms of the 'same' bird are usually lumped as one species, since it is assumed that they would breed if they were given a chance. Detractors, not surprisingly, argue that such populations rarely provide us with evidence of their willingness to reproduce. In contrast, the phylogenetic species concept defines a species as the smallest group of organisms that share an ancestor and that can be distinguished from other such groups, by whatever means, phenotype or genotype.[3] Identifiable geographic forms of the same 'kind' of bird, therefore, are treated as distinct species, since they have evolved separately and have unique histories (whether or not the taxa would interbreed if they met is, conveniently, immaterial). Put another way, a species defined according to the biological species concept only exists when the lineage separation is complete, while those fulfilling the phylogenetic species concept are 'made' as soon as their evolutionary lineage has started to separate. Of course, a major limitation of the phylogenetic species concept is that, by dispensing with the need for reproductive isolation, the threshold for the separation of species becomes entirely subjective. If the most extreme view of the phylogenetic species concept were adopted, it could be argued that a single nucleotide substitution represents a uniquely derived trait. At this level, nearly every individual could be classified as a separate species.[4]

This dichotomy of opinion, 'lumping' versus 'splitting' in birding parlance, accounts for the lack of agreement concerning the number of extant albatross species. The Wandering Albatross complex best illustrates this confusion. Lumpers regard the complex's five subspecies, each with its particular nesting area (the Sub-Antarctic Islands, Tristan de Cunha, and the islands off the south of New Zealand) as just that, subspecies, not species. Splitters, in contrast, recognise five different taxa: the Wandering Albatross, the Antipodean Albatross, the Tristan Albatross, the Amsterdam Albatross and the Gibson's Albatross. These splits, as well as those proposed for the other albatross genera, are argued on the basis of subtle genetic differences and varying degrees of neoteny, the retention of dark juvenile, or intermediate, plumage colouration into adulthood. Both lines of evidence, however, are critically dismissed by the Australian ornithologist John Penhallurick, an acclaimed authority on albatrosses and one of the world's top birders. He believes that the genetic differences are too small to warrant species status and that the plumage changes, resulting from different levels of brown colour deposition, are best explained by environmentally induced epigenetic changes that alter the expression of genes, but not their DNA sequence.[5]

Allegiance to the phylogenetic species concept is not always dependent on scientific rigour, and its widespread acceptance may, in part, be related to concern about the survival of albatrosses, especially from long-line fishing. Governmental conservation policies are often determined by species, not subspecies, and since the phylogenetic species concept enables more taxa to be regarded as endangered, its adoption permits a greater number of species to qualify for legal protection, a view considered by Penhallurick as taxonomically inconsistent and blatantly biased.

A universally agreed solution to the 'species problem' is probably unattainable. All species evolve, no matter how slowly, and many genera, like those of the albatross family, form part of a spectrum, a continuum of change, that does not lend itself to punctuated classification. While current efforts are undoubtedly proving revelatory, there is unlikely to be 'a royal road to unequivocal truth in taxonomy', to paraphrase the science writer Colin Tudge.[6] Indeed, despite years of research and a wealth of morphological, physiological and genetic data, the number of extant species and their precise phylogeny remain stubbornly unresolved. Given this scenario, what can palaeontologists deduce about albatrosses from a small collection of fragmented fossilised bones? The answer is more than one might expect.

Fossil evidence

Early fossil discoveries have helped clarify the divergence dates of the four albatross genera. Excavations of the middle Miocene deposits at Sharktooth Hill in California, during the early 1960s, produced fossil bones from two species of albatross that, although extinct, clearly belonged to the genera *Phoebastria* and *Diomedea*. These specimens, as well as similar finds in the southern hemisphere, suggest that the split between the great and the North Pacific albatrosses had already occurred by 15 million years ago, and that the separation of sooty albatrosses from mollymawks was evident 10 million years ago.[7] While these finds have been informative, the fossil remains from another site, the Lee Creek Mine, have had a more significant scientific impact.

For the last 50 years, the tiny coastal town of Aurora, in North Carolina, has lived in the shadow of Lee Creek Mine, the world's largest open-cast phosphate mine. Huge bucket-wheel excavators dispose of vast areas of topsoil, while draglines, fitted with garage-sized scoops, strip out the overburden to expose the extensive phosphate deposits that are craved by the agricultural industry. The enormous scale of the undertaking, together with its supporting railroad, can best be appreciated by scanning Google Earth, where the mine blights the coastal strip to the south of the mighty Pamlico River. While the removal of the pit's overlying fossil-rich Pliocene and Miocene layers has added to the countryside's disfigurement, there is no doubt that it has provided rich pickings for both professional and amateur palaeontologists. Indeed, the whole enterprise has been a metaphorical gold mine, yielding untold thousands of nuggets – sharks' teeth, fish and whale bones, turtles, corals and shells. Crucially for our story, a wealth of avifaunal fossils has also been recovered: a haul that includes grebes, auks, albatrosses, shearwaters, puffins, pelicans, gulls and gannets, and indicates that the community was overwhelmingly dominated by pelagic, piscivorous species. Collectively, these fossils reveal that the region contained one of the most spectacular assemblages of marine vertebrates the world has ever seen. But, of all the specimens, it is the fragmented albatross bones, retrieved from the lower Pliocene, around 4–5 million years ago, which are the most exciting and also the most puzzling.

Storrs Olson and Pamela Rasmussen, from the Smithsonian Natural History Museum, concluded that the albatross bones represented five different species.[8] Two are now extinct and are known only from the North Atlantic, while the remaining three closely resemble species that are confined today to the Pacific Ocean: the Short-tailed Albatross, the Laysan Albatross and the Black-footed Albatross. The significance of these findings is not that several species living 5 million years ago would have been recognisable by today's birders, but that at least five albatross species inhabited the North Atlantic during the Pliocene. This observation was unexpected and has significant biogeographical ramifications since, except for a few annual vagrants, members of the Diomedeidae are not found in the North Atlantic today. Additional fossils have since been described from Florida, Bermuda and southern England, suggesting that albatrosses were once common throughout the Atlantic. So how can their disappearance from the North Atlantic be explained?

The onset of widespread glaciation is one possibility. Around 2.5 million years ago, at the commencement of the Pleistocene, extensive ice sheets, up to several kilometres thick, developed in the northern hemisphere, ice that advanced during cooler periods and retreated during the warmer interglacials. It is likely that these enlarging ice fields led to a southerly retreat of all seabirds, and that albatrosses, unlike other species, were unable to return to the north once conditions improved, owing to the lack of forceful winds in the equatorial regions. In support of this hypothesis, the doldrums, or more accurately the intertropical convergence zone, is thought to account for the long-staying singletons that have become 'stuck' in the North Atlantic in recent times. The most famous of these was a Black-browed Albatross, affectionately nicknamed 'Albert' by the birding fraternity, that had the dubious distinction of being the most observed and photographed bird in the British Isles. He is thought to have been carried over the equator by a storm in the 1960s and, being unable to return, was condemned to wander the northern Atlantic for the next 40 years. Adding extra poignancy to his plight, he was spotted forlornly wooing gannets during each breeding season, on remote islands off the Scottish coast, a fact beloved by both tabloid and broadsheet editors. Albert, however, was not the first albatross to get trapped in the north. In 1860, a female Black-browed Albatross joined a gannet colony on Myggenaes Holm in the Faroe Islands and returned each summer for the next 34 years – and might have continued to do so had she not been shot in the spring of 1894. The combined effects of a deteriorating northern climate and unfavourable equatorial winds are likely to have been contributory factors, but they cannot be the complete story, since albatrosses are still found in the North Pacific. The population, therefore, must have been under stress from other factors.

An unexpected breakthrough came in 2003, with the discovery of a fossilised breeding colony of albatrosses among the limestone deposits along Bermuda's southern coast.[9] The remains, including bones from both juveniles and adults, as well as a collection of eggs, belonged to the Short-tailed Albatross, a critically endangered species found today on only a few islands off Japan. Sadly, the species was all but exterminated by feather collectors a century ago. Olson's research team speculated that approximately 400,000 years ago, during the middle Pleistocene, a massive storm and its associated tsunami entombed the colony under several feet of sand. This terminal event was made possible by a dramatic rise in sea level at the time, due to an increase in world temperatures and the melting of the Greenland and the western Antarctic ice sheets.

Although such palaeoclimatic deductions were initially met with scepticism, subsequent research has confirmed both their timing and extent. In 2014, a team from the University of Wisconsin, led by Alberto Reyes, reported that Greenland's southern ice sheet completely melted for 10,000 years during the mid-Pleistocene, a finding that fully supports Olson's thesis.[10] Based on the knowledge that glaciers erode underlying rock and dump sediments into the sea, Reyes and his colleagues demonstrated an absence of glacial silt in 400,000-year-old marine samples collected off Greenland's southern coast. Overall, the collapse of ice sheets from Greenland and Antarctica is estimated to have resulted in the sea levels being up to 20 metres higher than they are today. Albatrosses choose their nesting sites with care, as they require islands where the wind conditions allow gliding take-offs and landings and which are also free of mammals and other predators. Olson believes such islands were scarce in the North Atlantic, whereas in the Pacific there were many suitable sites that could have survived the rising sea levels. The Bermudan fossils, however, also held another clue – concerning not what was found, but what was missing.

The great diversity of seabirds in the North Atlantic during the early Pliocene contrasts markedly with the impoverished present-day state. Over the past 5 million years, many species have either retreated from the North Atlantic or have become extinct. While the causes are unknown, the Bermuda finds reveal that the Short-tailed Albatross inhabited the area for 4 million years after the early Pliocene, since it was breeding up until 400,000 years ago. Today, Black-footed Albatrosses are known to nest in proximity to Short-tailed Albatrosses, so the lack of any additional species in the Bermuda deposits suggests that other albatross species were already extinct in the North Atlantic by the middle Pleistocene. Their disappearance may relate to the increasing competition from gulls and skuas that appeared in northern waters during the Pleistocene. It is also possible that the mainly pelagic species, such as the Black-footed Albatross and Laysan Albatross, were more susceptible to changes in North Atlantic currents that occurred at the end of the Pliocene, compared to the inshore-feeding Short-tailed Albatross.

Fluctuations in the Pleistocene air currents, caused by the closure of the Panama seaway and rising sea levels, could have affected the birds' distribution, speciation, and even survival, since their search for food is entirely dependent on the prevailing winds. Indeed, recent observational studies lend support for this view. In 2012, Henri Weimerskirch and colleagues analysed decades of data on the feeding habits of the Wandering Albatross and concluded that the birds are exploiting the increasingly strong westerly winds which are slowly moving to higher latitudes as a result of global warming.[11] Consequently, albatrosses are flying faster and covering greater distances than they did 30 years ago and now spend less time foraging for food. Their breeding success has improved, and the birds have significantly increased in weight, by as much as 10 per cent. While this seems a positive effect, the authors warn that it is an ongoing process and that the benefits are likely to be short-lived. Climate projections indicate that the westerlies will continue to move to even greater latitudes and that the presently favourable winds will deteriorate dramatically within the next 60 years.

In 2012, two palaeontologists, Gerald Mayr and Thierry Smith, analysed a selection of fossils that had lain unstudied for over a century in the Royal Belgian Institute of Natural Sciences.[12] Of particular interest were several specimens that had been recovered near Antwerp, amid sediments from a shallow open sea that formed part of the North Sea Basin. The fossils – pectoral girdle and wing bones from two individuals – were readily identified as belonging to a novel species of albatross, one that lived in the early Oligocene, around 30 million years ago. Named *Tydea septentrionalis* (from Tydeus, the father of Diomedes in Greek mythology, after which the genus is named), it is the earliest well-established record of an albatross. With an estimated size of a Black-browed Albatross and a 2-metre wingspan, the fossils reveal that albatrosses had already evolved the ability to travel long distances by the early Oligocene. However, the bone structure also showed that the species was not as efficient a flyer as the modern albatross.

The establishment of the Antarctic Circumpolar Current at the beginning of the Oligocene created the world's largest ocean current, a mass of cold water that travels west to east, encircling the southern continent. To the north, where the current's cold waters meet the warmer sub-Antarctic waters, a zone of nutrient-rich upwellings occurs that support high levels of phytoplankton. The resultant food chain (krill, fish and copepods, especially squid) provides an important food source for many species, including albatrosses. To locate such productive areas, which vary both temporally and spatially over vast areas of seemingly featureless ocean, albatrosses have evolved to become the most efficient flyers on Earth.

Dynamic soaring

To reduce the energy demands of flight, modern albatrosses, unlike *T. septentrionalis*, have evolved a shoulder-locking mechanism, consisting of a sheet of tendon, or fascial 'strut', which fixes their shoulder joint when fully extended. A special muscle releases the lock by sliding the head of the humerus back into the correct position for folding the wings. This adaptation enables their wings to be maintained in the horizontal position and allows fine control when confronting the buffeting of ocean winds during manoeuvres such as banking and turning. Furthermore, their wing muscles have evolved a high content of slow fibres: cells that are specialised for sustained contraction with a high resistance to fatigue.[13] Indeed, albatrosses have become such specialised 'gliding machines' that their heart rates during flight remain close to their basal, or resting rate. Such novel adaptations, however, come at a cost: the inability for sustained powered flight. It is for this reason that albatrosses are restricted to high latitudes, where there are predictable wind speeds and waves, and this is why they have such difficulty crossing the doldrums to reach the North Atlantic.

The Wandering Albatross, with the largest wingspan of any living bird, at approximately 3 metres, is the elite of avian gliders. But how can these birds soar for thousands of kilometres across open water without the help of thermal uplifts? By combining computer modelling and GPS tracking, scientists have revealed their secret: a technique known as dynamic soaring that depends on the existence of different wind speed layers above the ocean waves (Figure 7.2). The lowest layer, nearest the water surface, incurs friction and slows down, which then becomes an obstacle for the air layer immediately above, which also slows down. The result is a 10- to 20-metre region, known as a boundary layer, across which the wind speed increases smoothly the higher you go above the waves.[14] This explains why albatrosses repeatedly dive into the valleys of ocean waves and then wheel back up into the air. As they pull up into the wind and transfer from the slow-moving air in the lee of a wave, they experience uplift due to the increasing air speed over their wings. By turning in the opposite direction, with the wind behind, they can further increase their speed and dive back into the shelter of the wave. By repeating this pattern of 'wheeling', albatrosses can continue flying almost indefinitely

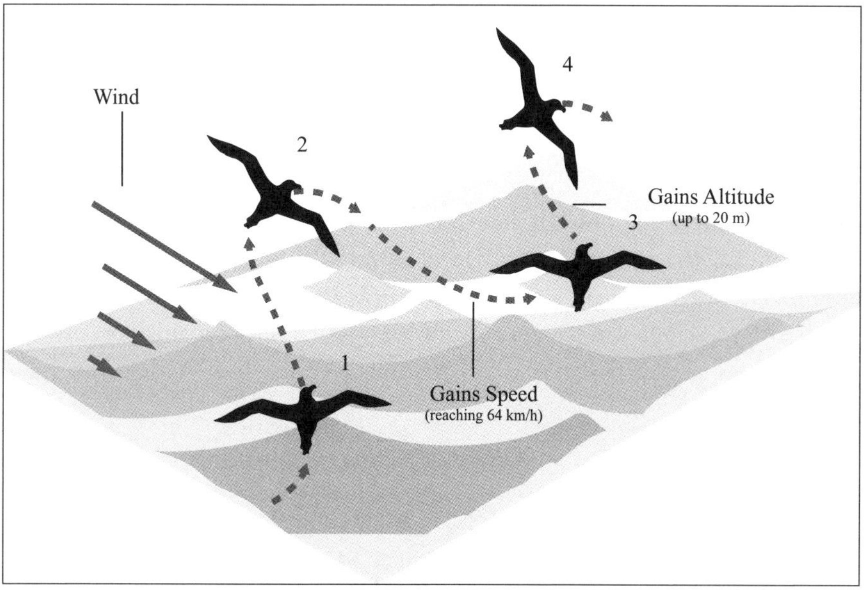

Figure 7.2 Dynamic soaring. Albatrosses repeatedly turn into the wind, gaining height, before gliding back down to the sea to gain speed. Wind speed is slowed by friction near the surface and begins to speed up about 5 metres above it. This gradient lifts the bird, as airflow is slower over the bottom than the top of the wings. 1. Climb to windward; 2. change flight direction to leeward; 3. turn back to windward; 4. turn back to leeward.

without having to put in any effort other than steering. In effect, the birds have evolved a means of harnessing energy from the wind gradients that exist above the ocean waves – an adaptation made even more efficient by their thickened and streamlined leading wing edges, which reduce frictional drag.

Albatrosses spend most of their lives at sea and only come ashore to breed once they have reached sexual maturity – which, for some species, may take up to 10 years. Consequently, most species have lifespans ranging from four to six decades. Females typically lay one egg a year, or one every other year, and remain faithful to their partner and nest site throughout their adult life. As a result, albatrosses put more effort into raising fewer chicks than other bird species, an adaptation to a stable environment and limited availability of resources. Furthermore, both parents undertake shifts, lasting up to a week, incubating and tending the chick, while their mate forages over thousands of square kilometres in search of prey. Such a lifestyle presents considerable physiological challenges, since the chick and its tending parent have to survive without fresh food for days on end, while the partner out foraging needs to bring back sufficient food without running out of energy itself.

To solve these problems, albatrosses have evolved a digestive tract that functions differently from that of most other birds. The adults can separate off the fatty components of their diet (wax esters and triglycerides) and store them, in liquid form, in a special stomach chamber called a proventriculus. The remainder of their diet then continues down the digestive tract to free up space for more food and further fat storage before the bird returns home. In effect, the rich content of the proventriculus acts as a long-term energy store, with a calorific value only slightly less than that of diesel fuel. The survival advantage for the albatross is clear: the liquid oil provides more energy than undigested prey and can sustain both chick and adult between food runs. Furthermore, whenever a parent needs a boost, it allows a small amount of the liquid fat to pass from the proventriculus to the rest of its digestive tract.

Olfactory cues

The question of how albatrosses locate their food has dominated Gabrielle Nevitt's research career, and her findings have been illuminating. Given that albatrosses have prominent nostrils and unusually large olfactory bulbs, Nevitt set out to prove that smell was an important factor. With the use of sophisticated GPS tracking techniques to check the bird's position every 10 seconds, and stomach temperature transmitters to record every meal, her research team monitored the behaviour of a cohort of nesting Wandering Albatrosses.[15] Nevitt predicted, given that odours from prey will disperse laterally and downwind, that the most efficient foraging strategy would be for the birds to fly crosswind to detect a scent, and then to adopt a zigzag flight, casting from side to side, to locate its source. The experimental results exactly matched the scientists' predictions. In nearly half of the prey-capture events, the birds flew crosswind, before zigzagging towards their prey. Albatrosses, however, do not rely entirely on olfaction, but incorporate visual inputs, especially during the final stages of foraging when the prey is captured. Their retinae are also sensitive to motion on the horizon, a facility that enables the location of food by detecting the activity of other seabirds.

The olfactory cue used by albatrosses is not the same as that used by most other tube-nosed species (Procellariiformes), an order that includes prions, storm petrels, diving petrels, fulmars and shearwaters. Most of these species locate their prey, consisting mainly of small crustaceans known as krill, by detecting the chemical dimethyl sulphide (DMS) they release when feeding on phytoplankton. Since DMS in sea water is readily released into the air, it serves as an olfactory marker for the presence of krill collecting near the surface. Albatrosses, however, detect a different set of odours, fishy-smelling ones associated with macerated krill or fish, including pyrazines and trimethylamine, in conjunction with visual cues. This finding raises an interesting evolutionary question. Since all tube-nosed birds evolved from a common ancestor that used DMS detection, why did albatrosses change their foraging method? Nevitt believes that the answer lies with the early environment experienced by the different species' chicks.[16]

Most 'tube-nosed' seabirds, like their common ancestors, rear chicks in dark underground burrows to lessen the risk of avian predation. Since they do not emerge until fully fledged, odours, especially parental scent, dominate the youngster's experience. Once learned, the familial scents are used in later life in the context of kin recognition and mate choice. In contrast, albatrosses evolved a surface-nesting strategy, possibly because their chicks were larger and less vulnerable to predation and heat loss. This significant behavioural innovation, which must have occurred early in their evolutionary history, presented new opportunities for natural selection and resulted in profound alterations in their foraging ecology. Their use of sensory modalities changed, and their search for prey became more dependent on vision, while olfactory cues were relaxed and switched from DMS to pyrazines.

CHAPTER 8

The Godwit's Story

QUANTUM COMPASSES

The first winter snow flurries only add to the birds' restlessness as they anxiously wait for their departure cues. But the southerly winds are not yet strong enough, and they continue probing and prodding the intertidal mudflats with their long bills, piling on valuable grams of body weight from a diet rich in worms and crustaceans. Many species have already left, abandoning the Alaskan estuaries for the warmer climes of the coastal Americas. But not the Bar-tailed Godwits (Plate 17). For they have much further to go and need to continue adding to their fat reserves for the arduous flight ahead. The last few weeks have been frenetic, with body weights doubling and internal organs atrophying to make room for the extra fat, while their proteins are diverted to strengthen the muscles of the heart and wings.[1] Many individuals are first-year birds with no experience of the hazards to come, but they too, like their parents, feed around the clock. Instinctively, they commence their pre-flight checks, pacing up and down the shore, fluffing their wing feathers, bathing in rock pools, and calling to their neighbours as if to say, 'Do we go yet?'

Then, all of a sudden, it's time. A cyclone in the Gulf of Alaska has created the ideal flying conditions along the Yukon–Kuskokwim Delta, and the flock takes to the air. After circling their breeding grounds for one last look, they head south, riding the strengthening winds on a bearing of 185 degrees south. At speeds of up to 60 kilometres an hour, each bird takes its turn leading the group out across the Bering Sea and over the remote Aleutian Islands, before taking on the vastness that is the Pacific Ocean. Eighty thousand other godwits are making the same perilous journey, although they are all out of sight, spread out over thousands of kilometres of nothingness. Despite efficiently shaped wings and sleek flight feathers, the birds are constantly buffeted by storms and cyclones that carry them ever southwards. After 40 hours, some 600 kilometres to the north of Hawaii, their inbuilt navigation systems indicate that they have strayed too far east and that a course correction is needed to make landfall in New Zealand. But the equatorial winds are lessening, and they have to use more and more energy to maintain their demanding schedule, and, as a result, all are rapidly losing weight. Fatigue is setting in, and each bird sleeps for multiple short spells or power naps. Each eye is closed in turn, while the corresponding half of the brain is switched off so that they can still remain alert and not become disorientated and fall into the sea. Occasionally, they slip into a state of drowsiness, characterised by a partial closing of both eyes that still allows for some visual processing.[2] But they cannot stop to rest or feed, and the weakest succumb. By day five, the birds pick up the southeast trades and austral westerlies and experience once again the all-important tailwinds that will assist them on the final leg of their journey. Further adjustments are made to their course, using the southern stars, which many have never seen before, and the Earth's magnetic field.

After nearly nine gruelling days and 11,500 kilometres of non-stop flight, the godwits spot the northern shores of New Zealand in the far distance. Exhausted, they make for the shallow estuarine waters of North Island, where they drop into the bordering marshlands and immediately set about rehydrating their bodies before falling asleep.

Ten years ago, this account would have been met with ridicule, as scientists assumed that the race of Bar-tailed Godwits (*baueri*) that arrived in New Zealand each September followed the Pacific's coastal rim. But in early 2007, satellite-tracking transmitters, implanted into 16 birds from New Zealand, gave results that challenged conventional wisdom.[3] Of the six individuals successfully tracked to their breeding grounds in eastern Siberia and Alaska, one, prosaically dubbed E7, confounded the experts (Figure 8.1). She left early one morning in March and, after flying 10,000 kilometres non-stop, spent the next five weeks refuelling at Yalu Jiang, near the border between Korea and China. Once her strength was restored, she set off again and, though unseen by human eyes, was regularly monitored from space by an orbiting satellite. Flying eastwards over the Korean peninsula and Japan, E7 eventually reached the Yukon Delta, where she nested – a journey of over 17,000 kilometres, accomplished over 14 days of flying. Again she eluded humans, including one scientist from New Zealand who could only get to within 30 kilometres. Thankfully, E7's battery worked much longer than expected and scientists were able to monitor her return journey, a flight that set a world record for a migrating bird – a staggering non-stop journey from Alaska to New Zealand.

The Bar-tailed Godwit's remarkable story demands answers to three related questions. Firstly, what is the evolutionary history of godwits, and

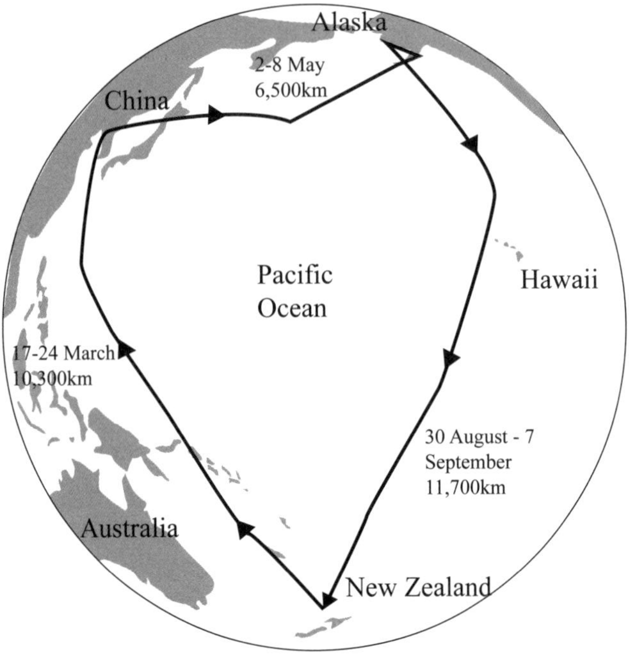

Figure 8.1 The trans-Pacific migratory route of the Bar-tailed Godwit was only revealed when 16 birds, including the record-breaking 'E7', were tagged with small satellite transmitters. Source: PRBO Conservation Science/USGS Alaska Science Center.[3]

how did they reach the remote Arctic shores of Alaska? Secondly, how did their record-breaking long-distance migrations develop? And lastly, and the most challenging of all, how did godwits evolve the navigational skills to cross thousands of square kilometres of featureless ocean? The search for the answers to these fundamental questions will take us from the established fields of phylogenetics and palaeontology to the counterintuitive and weird world of quantum mechanics.

Shorebird evolution

The shorebirds (Charadriiformes) are one of the largest and most complex of avian radiations, not just regarding the number of species, over 550, but also in terms of their morphology, ecology, behaviour and geographical distribution. Their last common ancestor lived around 90 million years ago during the Cretaceous, a population that diverged before the K–Pg boundary to give rise to three major clades, or suborders, consisting of 14 lineages. The ancestors of each of these lineages inhabited the Gondwanan supercontinent and, unlike the dinosaurs, survived the Chicxulub impact (see *The Vegavis's Story*). They included the buttonquails, seedsnipes (plus the Plains-wanderer), terns and gulls, alcids and skuas, coursers, pratincoles, sandpipers and curlews (including godwits), jacanas, plovers, oystercatchers, sheathbills and the Magellanic Plover, thick-knees, and the Egyptian Plover.[4] The survival of so many ancient lineages from the late Cretaceous may account for why some 'oddball' monotypic families (e.g. Crab Plover, Plains-wanderer, and Magellanic Plover) have been so difficult to place phylogenetically. In reality, they are all ancient relicts of a much earlier radiation.

The southern origin of shorebirds is backed up by the Gondwanan signature of many basal lineages: buttonquails (southern hemisphere), seedsnipes (South America), the Plains-wanderer (Australia), sheathbills (sub-Antarctic islands and Antarctica) and the Magellanic Plover (Argentina and Chile). Furthermore, phylogenetic studies place the buttonquails and the Plains-wanderer as the most basal species, sister lineages to all other shorebirds,[5] while recent palaeontological discoveries also favour a southern origin. In 2016, an early Miocene fossil, *Hakawai melvillei,* was found in New Zealand which indicated that the common ancestor of seedsnipes and the Plains-wanderer must have inhabited East Gondwana before the supercontinent's fragmentation.[6] Regarded as an evolutionary 'missing link', *Hakawai melvillei* was not a typical shorebird but possessed both Plains-wanderer and seedsnipe-like features. It inhabited subtropical wetlands, and its aquatic lifestyle implies that seedsnipes and the Plains-wanderer developed their grassland adaptations independently, as the result of convergent evolution.

Most shorebird speciation events, however, occurred after the K–Pg boundary and were triggered by the warming climate of the Eocene when ecosystems were highly productive. They include the plovers, whose non-migratory ancestors emerged in the temperate and tropical latitudes of South America.[7] In contrast, the explosive radiation of sandpipers, godwits and curlews occurred in the northern hemisphere and was linked to climate cooling during the Neogene and the emergence of grasslands, steppes and tundra. Later, during the glacial maxima, when ice sheets covered much of the northern landmasses, remnants of tundra were limited to refugia, while the warmer interglacials produced a northward expansion of forests and a rise in sea levels. Each of these geographical changes led to the isolation of shorebird populations and their allopatric speciation. Indeed, the genetics of several Arctic species, including the Red Knot, Dunlin and Ruddy Turnstone, hide a history of vicariant fragmentation that strongly supports a link between Pleistocene glaciations and speciation. For example, the five recognised subpopulations of Dunlin correlate well with documented fluctuations in climate and ice coverage. The oldest clade, a population breeding in central Canada, split from its ancestors around 220,000 years ago, during a warm interglacial period when tundra was restricted and fragmented. A second split occurred approximately 100,000 years later, the result of a subsequent interglacial, while the remaining three lineages evolved during a time of extensive glaciation.[8]

Godwits and curlews share an ancestry with dowitchers, suggesting that their northern ancestor was probably a medium-sized and straight-billed species. So why did the Bar-tailed Godwit evolve a slightly upturned bill? Experimental studies on curlews, all of which possess strongly decurved bills, show that they are capable of a greater rotation at maximum penetration than straight bills, and are better suited for grasping prey within a confined space.[9] Indeed, the increased manoeuvrability of curved bills applies to both curlews and godwits, as the same advantage applies irrespective of whether the bill curves downwards or upwards. In contrast, the Black-tailed Godwit maintained the ancestral state and, being longer-legged and larger, evolved the ability to feed in deeper waters, frequently submerging its belly and its entire head.

Evolution of long-distance migration

While there are many unanswered questions relating to shorebird migration, we can be sure of two facts: the migratory route of Bar-tailed Godwits started from Alaska rather than New Zealand, given the species' northern origins, and their journeys did not arise by accident.

Godwits have had to evolve a raft of anatomical, physiological and behavioural adaptations over millions of years to accomplish such extreme journeys. For example, they can increase the bulk of their flight muscles, build up large fat stores quickly, and auto-digest their liver, kidney and bowel tissue to reduce weight and provide extra space for fuel. Furthermore, they can metabolise non-essential proteins as an energy source, use specialised neurosensory systems for orientation and navigation, and cope without water for the flight's duration. Once landfall is made, preparations are undertaken immediately for the return flight; a process that involves a transient recovery of their internal organs and a full moult to enable the growth of a new set of flight feathers. The genes associated with these migratory adaptations remain unknown, although preliminary research has identified a potential gene linked to the migratory behaviour of songbirds.[10] The record-breaking journeys of the Bar-tailed Godwit, therefore, do not depend on a single 'magic' adaptation, but on the integration of a wide range of modifications. As a result, the bird's migration route must have evolved in stages.

The migration of the godwits probably owes its origin to the minor seasonal movements of their ancestors, as wintering grounds were gradually shifted southwards. This view is supported by the finding that songbird migration was triggered by global cooling and the development of increasingly distinct seasons at high latitudes.[11] However, such migrations were not new evolutionary behaviours, for seasonal movements, in some form or other, trace back much deeper and may even have been a feature of the giant sauropod dinosaurs of the late Jurassic.[12] The sequence of glacial and interglacial events that occurred during the late Pliocene, and the northward recolonisations by populations after the glacials, have been considered a major factor initiating the long-distance migration of shorebirds. However, it is also possible that their migrations were already fully established, and that the glaciations merely resulted in modifications to their routes.

According to the Swedish biologist Anders Hedenström, two hypothetical scenarios could explain the course of the godwits' record-breaking autumn migration.[13] The first assumes that there was already a migrating population breeding within central Siberia and wintering in south Asia. If this population then expanded slowly eastwards, its migration route could have increased in distance to involve wintering sites in the Philippines, the Indonesian islands, New Guinea, Australia, and eventually New Zealand (Figure 8.2A). The second scenario is that there was an initial Alaskan breeding population with a short-range migration to wintering sites in northeast Asia that, over time, extended to south Asia. As long-distance migrations became established, a continuous shift to wintering sites further east and south could have eventually created a direct route to New Zealand (Figure 8.2B). As we have discussed, godwits split their return migration into two stages, stopping in the Yellow Sea area to refuel, before flying on to their breeding grounds in Alaska. Such a looped journey argues in favour of the second scenario, and for the evolution of a direct migration route in the autumn.

Why godwits migrate, given the journey's extraordinary risks and high energy demands, is an easier question to answer. All birds require a year-round food supply as well as suitable nesting sites, and the short Arctic summer provides both. The tundra vegetation of moss and grass, interspersed with bogs and freshwater pools, is home to swarms of insects. These invertebrates provide a rich food source for the newly hatched chicks, a 'high-energy soup' not available elsewhere in the world. On the other hand, shorebirds cannot survive the rigours of the long Arctic winters and need to move south to warmer wintering grounds.

But to do so, they need to be able to navigate.

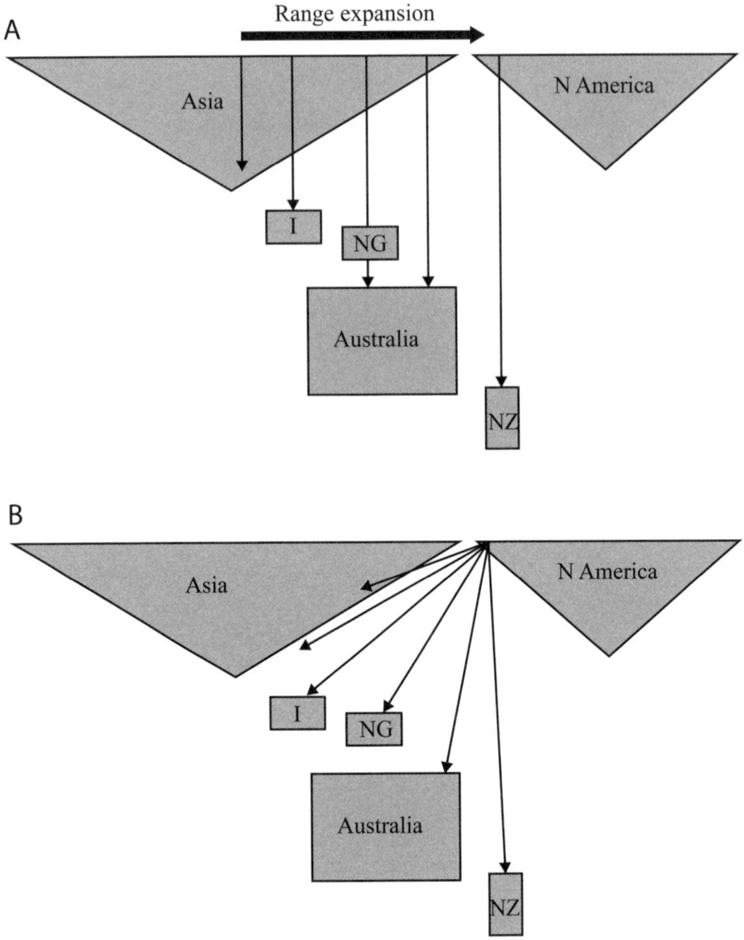

Figure 8.2 Two hypothetical scenarios to explain the evolution of the long-distance autumn migrations of Bar-tailed Godwits. (A) A breeding-range expansion with a maintained migration direction, but increased distances. (B) An increased migration distance coupled with a shifted migration direction. I = Indonesia; NG = New Guinea; NZ = New Zealand. Modified from Hedenström (2010).[13]

Quantum compasses

The extraordinary navigational ability of birds has challenged scientists ever since their long-distance migrations were documented. First recognised in the eighteenth century, migration is now thought to involve a combination of cues: the position of the sun, the moon and stars, as well as the incorporation of olfactory, visual and auditory signals. Nevertheless, each of these mechanisms has its limitations.

The difficulty of using a celestial body as a compass, for example, is that its position continuously changes throughout the day. When setting off from Alaska, godwits would need to keep the easterly sun to their left in the morning, head directly towards the sun at midday, and keep the westerly sun to the right during the evening. Young birds also learn a 'sky map' as they observe the rotation of the stars while growing, and recognise, in particular, the area of least apparent movement around the Pole Star. Consequently, conceptually difficult computations need to be undertaken to determine a stable compass direction in space. To do so, birds rely on their internal clock, or endogenous circadian rhythm, which varies according to the light–dark cycle of the environment. The required compass bearing is then calculated by determining the time of day, in relation to their circadian rhythm, and the celestial body's altitude and azimuth (lateral angle from due north). An obvious drawback is that celestial cues can be obscured by clouds and may be unavailable for days on end. Other aids such as sounds and smells are mainly of local value and do not help in transcontinental or transoceanic migrations. To overcome these deficiencies, birds have evolved an additional navigational tool: the geomagnetic compass.

In the 1960s, the German ornithologist Wolfgang Wiltschko demonstrated that songbirds were sensitive to the direction of an applied magnetic field and surmised that they could detect the Earth's magnetism. Robins, captured just before they set off on migration, were placed in a specially designed cage that enabled the researchers to see in which direction they preferred to hop or flutter (known as 'migratory restlessness'). Wiltschko demonstrated that the birds' favoured direction changed if the surrounding magnetic field was altered. In other words, the birds shifted their orientation in parallel with the altered magnetic field.[14] Several years later, while working with his wife, Roswitha, he discovered that the birds' compasses did not work as expected. To his surprise, they were unlike those used by hikers and sailors, known as polarity compasses, but were inclination compasses, which measure the angle of the magnetic field in relation to the Earth's surface.[15] The angle of inclination is vertical at the poles and horizontal at the equator, since the Earth's magnetic field emerges from the South Pole and radiates outwards, curving around in loops to enter at the North Pole. As a result, an inclination compass cannot discriminate between north and south but only indicates how far away one is from the equator. Such a mechanism means that birds can rely on the same migratory program irrespective of whether they are in the northern or southern hemisphere. However, long-distance migrants, such as the Bar-tailed Godwit, face the challenge that their magnetic compasses will give confusing bimodal information at the magnetic equator. At this point, it is likely that alternative cues are integrated in a hierarchical fashion until the angle of inclination moves from the horizontal.

At the time of the Wiltschkos' findings, no one had the faintest idea how such a geomagnetic compass might work. All sensory inputs, whether visual, olfactory, auditory or touch, are dependent on receptor-induced chemical reactions. Once activated, each specific receptor generates an electrical impulse that is transmitted along a connecting nerve to the brain for interpretation and perception. The amount of energy available in the Earth's magnetic field, however, is a billion times too weak to cause any known cellular reaction.[16] While the Wiltschkos' experiments conferred a degree of respectability to the discipline, scientists did not know which organ contained the receptors, let alone understand how a cell might react to a magnetic field. But that was all about to change.

In 1976, Klaus Schulten, a brilliant German-American computational biophysicist, proposed that the generation of radical-pairs during chemical reactions might be influenced by magnetism. Radical-pairs are two simultaneously created molecules, each of which possesses a single unpaired electron whose spin is correlated to that of the other – a state physicists call 'quantum entanglement'. Spin is a strange quantum property that is completely unlike the angular momentum we associate with ice skaters or gyroscopes and has no analogy in our everyday world. Schulten realised that during such an event, the quantum state of both electrons would make them susceptible to the effects of an external magnetic field. Later, he realised that the generation of radical-pairs by light could provide the basis for a functional magnetoreceptor, although he knew of no molecule that would fit the bill. Such a speculative theory was deemed so improbable that his paper was rejected by the prestigious journal *Science* with a note that read 'A less bold scientist would have designated this piece of work for the wastepaper basket.'[17] Convinced that his idea was a scientific breakthrough, Schulten quickly published his work in an obscure German journal and, undeterred, continued to develop and modify his ideas of how such biocompasses might work.[18]

Meanwhile, Wiltschko's team had demonstrated that the ability of birds to detect the Earth's magnetic field is dependent on ambient light conditions: homing pigeons transported in total darkness have difficulty finding their way home, while European Robins show good orientation in green and blue light, but become disorientated in red light.[19] In other words, the avian magnetoreceptor is blue-light-dependent.

Further research by Thorsten Ritz and Klaus Schulten concluded that the light-sensitive protein cryptochrome, found in the retinal cells of birds, can produce just the sort of radical-pair needed to detect a magnetic field. Their proposal, published in 2000, would become one of the classic papers in the emerging field of quantum biology.[20] At the centre of cryptochrome is a cofactor known as flavin adenine dinucleotide (FAD) that is constrained by several of the protein's amino acids. When a photon of green or blue light enters the eye and hits the cryptochrome–FAD complex, an active state is created in which each molecule has an unpaired electron, producing a radical-pair. Each electron will experience a slightly different magnetic field, since they have become separated by a minuscule distance. Depending on how the magnetic field affects the electrons' spin, a range of chemical reactions could be produced in the bird's eye. (For an analogy of how a weak magnetic field interacting with a radical-pair could produce a biochemical response, see Figure 8.3). In theory, the outcome of many similar reactions across the hemispherical retina could provide a migratory bird with information relating to the angle of the Earth's magnetic field and hence aid its navigation. The perceptual consequences, however, remain a mystery. Some scientists postulate that birds may perceive dots in their peripheral vision that move according to the direction they're flying, while others think birds might 'see' magnetic fields as a range of different colours or hues.

Support for Ritz and Schulten's theoretical radical-pair model has come from two distinct experimental approaches: magnetic field manipulations and gene transfer studies. Ritz and colleagues, for example, reported that migratory birds could be prevented from using their compasses when exposed to weak oscillating magnetic fields, but only if the frequency resonates with that of the spins of the proposed radical-pairs.[22] The second approach involved the transfer of cryptochrome genes from butterflies to fruit flies. The experiment was based on two biological discoveries: mutant strains of

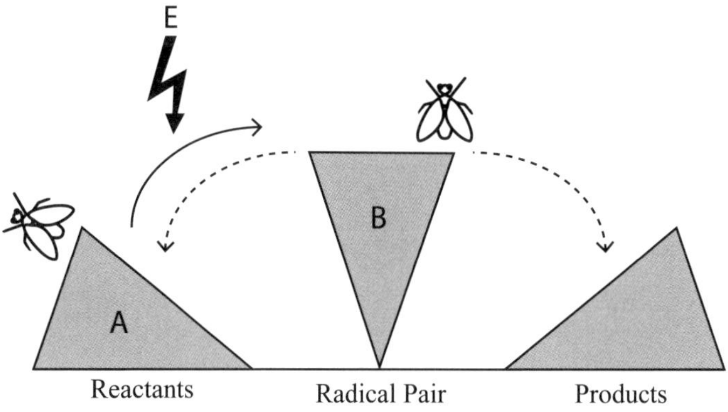

Figure 8.3 An analogy of the radical-pair mechanism. Insight into why the outcome of a radical-pair reaction can be dramatically influenced by the interaction with the Earth's extremely weak magnetic field. (A) A fly landing on the side of a heavy stone pyramid would have a negligible chance of tipping it over. (B) However, if the stone is first prepared in a highly non-equilibrium state (solid arrow), then the minuscule amount of energy imparted by the fly could profoundly alter the chance of the pyramid falling to the right rather than reverting to its original position (dashed arrows). E = energy input (a photon, in the case of cryptochrome). Adapted with permission from Hore (2011).[21]

the fruit fly *Drosophila melanogaster* deficient in cryptochrome lose their magnetoreceptive behaviour, while migrating Monarch Butterflies (*Danaus plexippus*) possess functional cryptochrome-based magnetoreceptors in their antennae. Amazingly, the transfection of intact cryptochrome genes from the butterflies to the mutant *Drosophila* fully restored the flies' magnetic sensitivity.[23]

While it is recognised that birds use a combination of celestial and magnetic compasses, the initial direction that an individual sets off appears to be innately represented in their central nervous systems. Indeed, the genetic programming can be so sophisticated as to include not only the appropriate direction but also an awareness of the length of time required for the different stages of their migrations.

Birds also possess a second means of detecting magnetism, one that consists of magnetite-based magnetoreceptors close to the skin in their upper beaks. Such cellular structures are not compasses, but function as sensors of magnetic intensity and thus provide one component of a multifaceted 'navigational map'. They are linked to the brain by the trigeminal nerve and allow the bird to evaluate the strength of the magnetic field, which is weaker at the equator and stronger at the poles. By integrating both types of information, together with many other cues, migrating birds can navigate their way, whether it is across the vast Pacific Ocean or the extensive Tibetan plateau.[24]

But how did a biocompass, based on the quantum effects of cryptochrome, evolve? While the precise details remain unknown, we can hazard a guess. As already discussed in *The Hoatzin's Story*, nature rarely innovates from scratch and, instead, it is often the accidental duplication of essential and well-tested genes that provides the scope for future experimentation. Duplicated genes allow the original form and function to be maintained, while random changes in the extra copies can lead to novel phenotypic outcomes. This is an important point. New activities must arise by random mutations; they cannot be developed to meet a need, since evolution has no foresight and is unable to predict the future. Of course, new proteins that affect an individual's phenotype, whether behavioural or structural, will be subject to the omnipresent pressures of natural selection.

The evolution of avian cryptochromes is no exception. The proteins first appeared over a billion years ago as photolyases, enzymes that function as bacterial DNA repair molecules (although it is possible they may have evolved even earlier as DNA replicating enzymes). All photolyases are activated by blue light and repair ultraviolet-induced DNA damage by removing pyrimidine bases. Over millions of years, and multiple gene duplications later, a series of proteins emerged with different DNA repair activities that required a FAD molecule to function. Still further gene duplications gave rise to cryptochrome proteins that lacked the capacity for DNA repair but which acquired, instead, the ability to coordinate an animal's circadian clock. This biochemical process, present in most animals and plants, evolved between 1,000 and 540 million years ago and made it possible for organisms to coordinate their cellular activities with diurnal rhythms. The protein's altered function was probably an adaptation to the reduced levels of harmful ultraviolet radiation and the increasing length of day that occurred at this time in the Earth's history.[25]

Eventually, the ancient photolyases gave rise to a series of proteins that, in association with FAD and blue light, were able to generate radical-pairs that were susceptible to the Earth's magnetic field. After aeons of structural alteration, an ancient DNA repair enzyme unwittingly provided evolution with the building blocks to produce a biological compass. The emergence of light-sensitive cryptochromes was a key evolutionary event, as magnetoreception has been demonstrated in a broad range of organisms including bacteria, arthropods and molluscs, as well as in members of all the major groups of vertebrates.[26] It seems that biocompasses and their radical-pairs may have facilitated the orientation and navigation of life around our planet for well over 500 million years.

A sad caveat to this remarkable story is that the New Zealand population of Bar-tailed Godwits is under severe threat and is predicted to halve within the next decade because of extensive habitat destruction along the Yellow Sea's coastline.[27] Both China and South Korea are constructing seawalls and reclaiming wetlands at an unprecedented rate to support their rapidly growing economies. The loss of these critical 'fuelling stations' is also having a marked effect on many other migratory shorebirds, most notably the Red Knot and the Great Knot.

CHAPTER 9

The Buzzard's Story

ACCIDENTAL SPECIATION

After the break-up of Gondwana, landbirds diverged into two major clades – Afroaves and Australaves – that then underwent marked radiations to give rise to nearly all the landbirds living today.[1] Since carnivorous species occupy the basal branches of both groups – diurnal birds of prey in the case of Afroaves (Figure 9.1), and seriemas for the Australaves (Figure 13.1) – it is likely that the ancestor of all the core landbirds was an apex predator, and that the raptorial trait was lost twice during their evolution. The birds of prey (family Accipitridae) comprise a global radiation of 256 birds that includes kites, hawks, buzzards, eagles and harriers. They all belong to a single clade or order, known as the Accipitriformes, whose earliest branches spawned the New World vultures, the Secretarybird, and the ospreys (Figure 9.1).

Falcons, however, are not part of the clade, as they evolved later and are sister species to the parrots and passerines (see *The Parrot's Story*). Indeed, the evolutionary distance between the Accipitridae and falcons should not

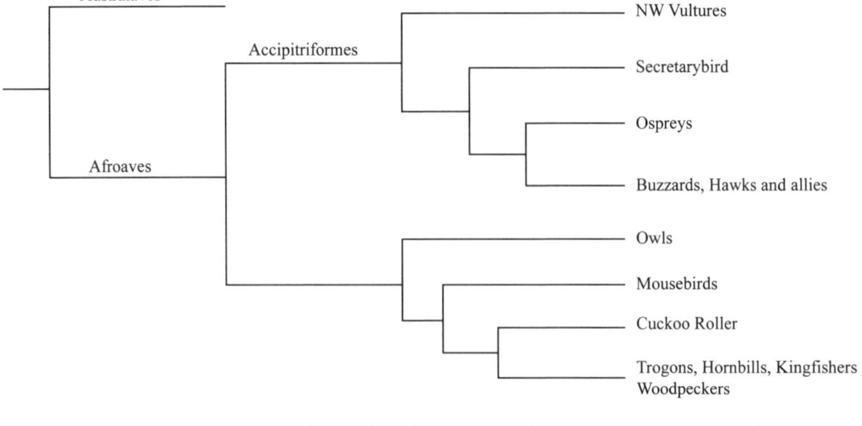

Figure 9.1 Phylogenetic relationships of the Afroaves. It is likely that the ancestor of all landbirds (Afroaves and Australaves) was an apex predator. Modified from Prum *et al.* (2015).[2]

be a surprise, since the two families are very different in morphology and behaviour. Falcons lack the distinctive brow ridges of the hawks and tend to have longer and pointed wings. Accipiters also tend to prefer forests, where they capture and subdue struggling prey using their taloned feet, while falcons favour open country and kill swiftly with their beaks.

One of the morphologically most primitive groups of Accipitriformes to evolve was that of the New World vultures (family Cathartidae). They first appeared in the early Palaeogene, although their crown group did not diversify until much later, during the middle Miocene, around 14 million years ago. At this time, there were many more species than today, as the presence of large herbivores and predatory mammals in open environments provided an abundant supply of carrion. During the late Pleistocene, a dramatic decline of the American megafauna occurred, which reduced the opportunities for scavenging and led to the extinction of many species.[3] As a result; there are only seven extant taxa: two condors and five vultures. The California and Andean Condors are the biggest flying birds, with wingspans of around 3 metres and weights of up to 12 kilograms. Perhaps the most remarkable vultures were the teratons (albeit usually placed in their own family, the Teratornithidae), which included *Argentavis magnificens* from the late Miocene of Argentina. This species weighed up to 70 kilograms and possessed an extraordinary wingspan of 5–6 metres. It is by far the largest known flying bird and exploited the thermals for soaring across the pampas in search of its large prey. Although the Cathartidae resemble the Old World vultures, they are not closely related, and any phenotypic or behavioural similarities are the result of convergent evolution.

The monotypic Secretarybird (family Sagittariidae), unlike most birds of prey, is a terrestrial species that hunts its prey of lizards, snakes and small mammals on foot.[4] They have long, pink, scaled legs and typically kick and stamp on their prey's head until it is killed or incapacitated, particularly if it is a large lizard or a venomous snake. Despite this behaviour, Secretarybirds can fly and soar very well, especially during their nuptial displays, when they may undertake acrobatic flights at a great height. The species' common name derives from the crest of long black-tipped feathers that gives it the appearance of an old-fashioned clerk with pens tucked behind the ear. An endemic of sub-Saharan Africa, the Secretarybird diverged around 40 million years ago, after the New World vultures and before the osprey and Accipitridae lineages. While confined to sub-Saharan Africa today, fossil evidence from France and the Middle East indicates that Secretarybirds had a much wider distribution in the past.

Ospreys (family Pandionidae) have recently been split into two species –

the Western and Eastern Osprey – although their global population consists of four clades with clear genetic differences. This knowledge has enabled biologists to work out the birds' evolutionary history and to determine their route of global spread. A specialised, fish-eating raptor, the osprey appeared in North America during the Miocene, although there are older fossils from the late Eocene and early Oligocene of Europe and Africa.[5] The New World population first dispersed, via the Pacific coast of Asia, to colonise the Indo-Australian regions. Then, from Pleistocene refugia located in Indonesia–Oceania, a rapid range expansion occurred, with populations settling in eastern Asia and the western Palaearctic.[6] Interestingly, each of the four clades exhibits a low genetic variability, which suggests that the colonisation of each new area involved only a few individuals with a limited number of genetic variants compared to the ancestral pool (known as a 'founder effect').

The phylogeny of the Accipitridae has been difficult to resolve, but it is likely that the basal branch gave rise to the kites. The latter group includes the genus *Elanus* (Black-winged Kite, Black-shouldered Kite, White-tailed Kite and Letter-winged Kite) that inhabit savanna-like habitat in temperate and arid areas. Like owls, they possess a velvety-comb structure to their upper wing feathers for dampening flight sound, vibrissae around the beak, and disproportionately large, frontally placed eyes. The four species disperse over long distances to feed on cyclic populations of small mammals, and produce many broods each year. These typical owl-like traits are unusual among the Accipitridae and are again the result of convergent evolution.[7]

The largest raptors belong to the harpy eagle clade (Crested Eagle, Harpy Eagle and Papuan Eagle), found in the Americas and New Guinea. The Philippine Eagle was once regarded as a sister species, but a study in 2005 found that this Asian raptor evolved much earlier and that its nearest relatives are the snake eagles, such as the Bateleur.[4] All harpy eagles are immense, hunting carnivores that feed on monkeys and other medium-sized mammals, and, as such, lie near the apex of their respective food chains. The largest of all is the Harpy Eagle, which weighs up to 10 kilograms, and, with a wingspan of over 2 metres, possesses the power and lift to carry a 7-kilogram monkey. As with many birds of prey, the females are larger than the males and tend to take larger prey. And yet, surprisingly, these impressive birds are not the largest eagles to have ever lived (see below).

'True' or aquiline eagles diverged after the harpy eagles, around 12–15 million years ago.[8] Originating in the Old World, the ancestral population gave rise to a single colonisation event of the Neotropics, via Asia and North America, and produced the four hawk-eagle species in the genus *Spizaetus*. The only other American dispersal was a recent one by the Golden Eagle, which entered the continent via Beringia. Over time, the New World Golden Eagles differentiated into the subspecies (*canadensis*), which is fully diagnosable from its Palaearctic relatives by molecular analysis. However, the main diversity of aquiline eagles was centred in Africa and Asia, from which two lineages entered Australasia at the end of the Pliocene to give rise to the Little Eagle and the much larger Wedge-tailed Eagle.

Island gigantism

In the spring of 1871, Frederick Fuller, a taxidermist working at Christchurch Museum, New Zealand, recovered some bones in a dried-up swamp that belonged to a large unknown raptorial bird. The fossils, including a femur, a few claw bones and a rib, were associated with a considerable quantity of moa remains. Fuller passed on his finds to the curator of Canterbury Museum, Julius von Haast, who published the first scientific description of the extinct species and named it *Harpagornis moorei*.[9] Subsequent discoveries from dozens of sites in the South Island have added to our knowledge and allowed scientists to reconstruct the raptor's anatomy and lifestyle.

Despite spanning 3 metres, the wings of Haast's Eagle, as it is now called, were relatively short for its massive weight (up to 18 kilograms) and better suited to flying among trees than soaring in search of carrion. Its tail was relatively large, and the extra surface area partly compensated for the reduced wing size. Although it possessed a sharp, vulture-like beak, its talons were as large and as lethal as modern tiger claws. The predicted wing aerodynamics suggest that it probably hunted from a forest perch, waiting for a moa to walk by, before swooping down like a hawk and hitting its prey at great speed from the side (Plate 2). Damage on moa bones indicates that its bill was able to reach the ratite's internal organs, while its talons could grasp their pelvic bones. Indeed, *Harpagornis moorei* has the distinction of being the only known raptor to have become the top predator in a complex ecosystem. Such a conclusion supports Māori mythology of the legendary *pouakai* or *hokioi*, a huge bird that attacked mountain people and could kill small children. And yet, by 500–600 years ago, the largest eagle the world had ever seen disappeared for good as Polynesian settlers destroyed its habitat and hunted its prey to extinction.

Since many New Zealand birds have sister species in Australia, it was assumed that Haast's Eagle had evolved from one of the continent's large raptors, most probably the Wedge-tailed Eagle. However, in 2005, Michael Bunce and Richard Holdaway analysed DNA from two 3,000-year-old *Harpagornis moorei* bones, housed in the Museum of New Zealand.[10] The results were a surprise. The genetic sequences bore little resemblance to the Wedge-tailed Eagle's but were remarkably similar to the DNA from two *Hieraaetus* eagles that weigh only a kilogram: the Little Eagle from Australasia and the Booted Eagle from Europe. The small number of base changes between the three species' DNA indicates that the Haast's Eagle diverged from the *Hieraaetus* clade as recently as 0.7–1.8 million years ago. These results imply that the ancestors of Haast's Eagle were small when they arrived in New Zealand and increased rapidly in size, by a factor of 10 to 20, during the next million years. Such gigantism was only possible because the ecological niche usually occupied by large meat-eating mammals was vacant in New Zealand. While moas occupied the place of grazing ungulates such as deer or cattle for millions of years, the Haast's Eagle became the apex predator that hunted the grazers. According to researchers Paul Scofield and Ken Ashwell, the increase in the eagle's body size was so rapid that the development of the brain and some sensory systems, including sight and smell, lagged behind – evidence, they believe, that the species was a killing machine and not a vulturine scavenger as originally thought.[11]

The rate and magnitude of the Haast's Eagle's increase in size are unique within the vertebrates, particularly as this occurred in a species that retained the ability to fly. While other large predatory birds have evolved on islands free of competitors, notably the extinct giant eagles and owls on Cuba, the dramatic phenotypic changes of Haast's Eagle remain unrivalled. Given its evolutionary origins, it seems sensible that the species' Latin name should be amended to *Hieraaetus moorei*, since, in the words of Colin Tudge, 'the world's mightiest eagle belongs among the little ones.'[12]

Accidental speciation

The genus *Buteo* was the final Accipitridae clade to evolve. With origins in South America, it gave rise to a large group of soaring hawks with long, broad wings and relatively short tails and legs. They spread rapidly throughout the Americas to reach the Arctic regions, where a population became isolated in a Beringian refugium during the Pleistocene. Later, under the influences of cyclical climatic changes during the ice ages, descendants of this population invaded the Old World. This eastern offshoot gave rise to two African branches (Madagascan and Red-necked Buzzards), a cluster of Asian species (Eastern, Himalayan and Upland Buzzards) and the most recent offshoot, a superspecies that includes the Common and Steppe Buzzards.[13]

The story of the genus *Buteo*, including their relationships and geographical dispersal routes, was deduced by Martin Riesing at the Museum of Natural History in Vienna, after comparing the base sequences of their mitochondrial DNA. But, unbeknown to Riesing and his colleagues, an unexpected evolutionary clue lay hidden among their raw data – one that was revealed eight years later by Professor Mark Pagel's team from Reading University.

The evolutionary relationships of the Accipitridae, as well as those of most of the bird families discussed in this book, have been inferred from phylograms: family trees drawn by comparing the DNA sequences from different species. When constructed, the root of the tree represents the common ancestor, while the branch tips depict that ancestor's descendants. As you proceed from the root to the tip you are moving forward in time, and the longer the branch, the greater the amount of genetic change and the longer the species' evolutionary history. The unit of time, therefore, is not a direct measurement but usually inferred from the number of nucleotide substitutions per site and presented either as the number of base changes divided by the sequence length or as the percentage change. Phylograms and their branch lengths are valuable tools, as they provide evolutionary biologists with information about diversifications of lineages, patterns and rates of trait evolution, and the timings of speciation. But, if Mark Pagel is correct, branch lengths also provide valuable information about not just when and where taxa arose, but also how they evolved.

One of the great mysteries of evolution is how subsets of species can suddenly become sexually incompatible and form new species. While mechanisms leading to reproductive isolation are known, such as vicariance (*The Manakin's Story*), allochrony (*The Storm Petrel's Story*) and hybridisation (*The Sparrow's Story*), they are unlikely to account for the majority of speciation events. Instead, it has long been accepted that natural selection is key: a mechanism that leads to the gradual accumulation of many small changes that will, at some critical point, result in a population that can no longer mate with its kin. Although Darwin made a convincing case for natural selection over 150 years ago, no one has been able to devise a means to prove it: that is, until Mark Pagel.

Ten years ago, Pagel realised that if speciation results from the addition of lots of small changes, then there should be statistical evidence hidden within the species' phylograms. His reasoning was as follows. If a large number of small factors summate to produce an outcome – for example, the combination of environmental and genetic factors that influence blood pressure, height, weight or blood sugar levels – then the population values for each of these measurements will describe a bell-shaped curve when plotted against frequency. In other words, such biological values are normally distributed. Similarly, if speciation is the outcome of many small evolutionary changes, as was believed, then Pagel predicted that the individual branch lengths in a given phylogram should also be normally distributed and describe a bell-shaped curve.

To see if this was correct, Pagel and two colleagues, Chris Venditti and Andrew Meade, plotted a histogram of branch lengths from a wide range of published phylogenetic trees. This task was no easy matter, as it involved obtaining and analysing the primary DNA sequence data that had been used to construct the original phylograms. Importantly, Pagel did not choose the trees randomly but selected only those that contained a narrow taxonomic range of species with similar life histories, morphology and ecology, to reduce any bias from different rates of speciation. Eventually, the available trees were whittled down to 101, including data sets for bumblebees, cats, turtles and roses – as well as birds. In addition to Riesing's data on buzzards, there were trees for frigatebirds, shags and cormorants, swiftlets, vangas, thrushes and seedeaters. Working with each phylogram separately, Pagel and his team noted the number of base changes between each successive speciation event and used this as a measure of its branch length. If speciation results from many small changes, as they expected, then the branch lengths should fit one of two forms of bell-shaped curve: a normal distribution curve if the changes summate until they reach a threshold level for speciation, or a lognormal curve if individual changes multiply together and the threshold is attained more quickly.

To their amazement, neither of the expected curves matched the data. It turned out that the distribution of branch lengths from 78 per cent of the trees exhibited an exponential curve, while only 8 per cent of the data sets were lognormal and not one was normally distributed (Figure 9.2).[14] Mathematicians are familiar with exponential curves, and such curves have a straightforward explanation. They are the pattern you obtain when you wait for some single, infrequent event to happen, for example, the time between the ejection of gamma particles from radioactive elements or the intervals between supernova in the Andromeda Galaxy. A more prosaic example, and Chris Venditti's favourite, is the distance you have to travel between roadkills on a motorway. But the finding that phylogram branch lengths describe an exponential curve has profound and unsettling implications for biology. It implies that new species emerge from single events, each rare but individually sufficient to cause speciation. As Pagel told me, 'It isn't the accumulation of many small events that causes speciation, it's the result of a single random event falling, as it were, from the sky.' Nevertheless, their findings were so at odds with conventional wisdom that they spent the next two-and-

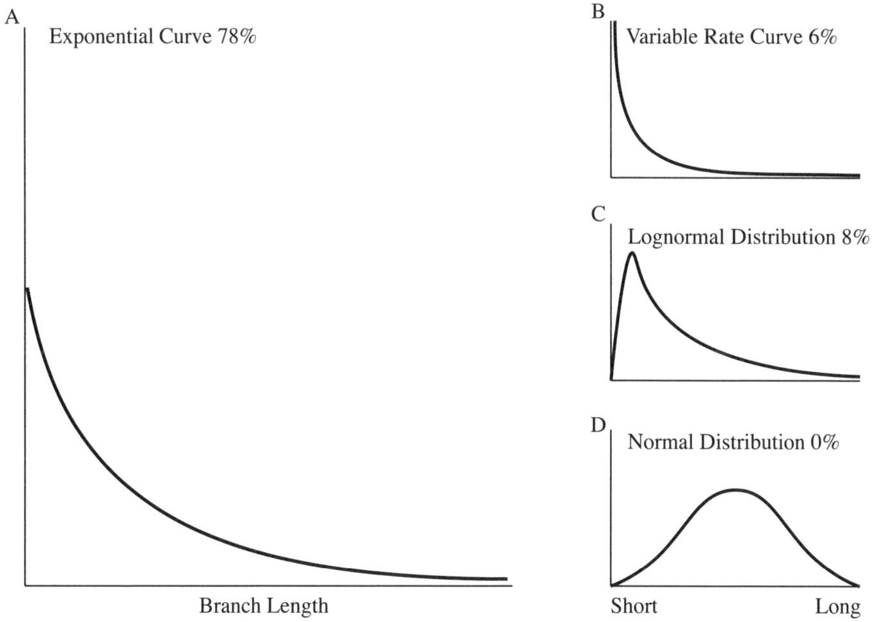

Figure 9.2 Mechanisms of speciation deduced from phylogram branch lengths, showing the percentage of branch-length distributions fitting each curve type. (A) Exponential curve reflecting a single random and rare speciation event. (B) Variable rate curve produced when bursts of speciation occur. (C) Lognormal curve, indicating that many changes have multiplied together to reach a threshold for speciation. (D) Normal distribution bell curve, reflecting the slow addition of many small changes. Adapted from Venditti *et al.* (2010).[14]

a-half years trying to 'make the exponential curves go away'. But, no matter how they reanalysed their data; the results were always the same. Even the inclusion of biases that were expected to favour the prevailing view produced little effect: the exponential curves could not be made to disappear.

Another type of curve, the variable rate curve, provided the best fit for 6 per cent of phylograms and indicates that speciation events have occurred in bursts. Textbook examples include the adaptive radiations of Darwin's finches and Hawaiian honeycreepers, clades whose phylograms possess lots of branches at irregular intervals, reflecting the availability of many unfilled ecological niches. However, such rapid diversifications are not dependent on rare and random events but result from the omnipresent pull of natural selection. A further 6 per cent of phylogram branch lengths took the form of a Weibull curve, a distribution commonly used in reliability engineering and failure analysis. However, in the context of Pagel's study, a Weibull curve implies that the probability of speciation is time-dependent. In other words, as the branch length of a tree increases so does the likelihood of speciation.

The key message from Pagel's statistical approach is that evolution is unpredictable and that all species are subject to rare random events that may lead to reproductive isolation. Whatever form this takes – alterations in mating preferences for plumage colour or song, or genetic changes that result in incompatibility – the results are the same. The affected individuals become isolated from other family members and, in doing so, evolve into new species. Of course, natural selection still has a major role to play, as Darwin envisaged, but one restricted to the moulding and shaping of any new species to the particular conditions it experiences.

Evolutionary biologists have long argued whether rewinding the tape of life and replaying it would produce similar results, or whether chance events dictate the outcome and force evolution down novel and unpredictable paths. These two opposing views yield very different scenarios for the history of life. For example, some scientists argue that you can rerun the tape of life as many times as you like, and the outcome will always be very much the same. The leading advocate, the Cambridge Professor of Palaeobiology Simon Conway Morris, even believes that human-like, self-conscious intelligence is an inevitable product of evolution, rather than a historical accident or a fluke as neo-Darwinians typically assume. In contrast, supporters of the late Stephen Jay Gould, who popularised the tape-of-life metaphor, argue that if the clock were rewound, evolution would not repeat itself. Instead, the world would look quite different, with an absence of familiar life forms, including humans.

The results of Pagel's branch-length analyses favour Gould's model and add weight to the central role of contingency in evolutionary transformation. While not all biologists fully embrace this idea, no one has yet been able to find a flaw in Pagel's scientific approach. Furthermore, in 2015, an ambitious project, involving the construction of a 'timetree of life' based on data from 2,274 separate studies and representing over 50,000 species, came to the same conclusion: 'that speciation and diversification are processes dominated by random events and that adaptive change is largely a separate process.'[15]

As readers may have guessed, the exponential curve provides the best fit for the branch lengths from the buzzards' phylogram. If Pagel and his team are correct, then the world's buzzards, along with most other avian taxa, are the result of the unpredictable hand of fate: a reflection of the utter arbitrariness of speciation.

CHAPTER 10

The Owl's Story

NIGHTLIFE

The Afroaves radiation is taxonomically diverse and, in addition to the diurnal birds of prey, includes owls, mousebirds, the Cuckoo Roller, kingfishers, bee-eaters, rollers, todies and woodpeckers.[1] While several of these families remain tied to Africa, others, such as the woodpeckers, kingfishers and owls, dispersed across the globe. Only the owls evolved a nocturnal lifestyle, with its emphasis on acute hearing and stealth. Indeed, owls occupy one of the deepest branches of the Afroaves, a fact that probably accounts for their raptor-like phenotypes. The next two branches produced the mousebirds, a vegetarian family of poor-flying birds that 'creep' about in trees using their specialised feet, and the Cuckoo Roller of Madagascar and the Comoro Islands. It is worth noting, given their raptorial ancestry, that the mousebirds have Eocene relatives with raptor-like feet,[2] while the Cuckoo Roller feeds on chameleon prey. These observations suggest that the loss of the predatory phenotype was probably a gradual process that took place across successive divergences of the landbirds.

Based on fossil and phylogenetic evidence, owls (order Strigiformes) emerged early in the Palaeogene, around 60 million years ago, and underwent a major adaptive radiation 20 million years later, making them one of the oldest groups of non-waterfowl landbirds.[3] This chronology suggests that their evolution was accelerated by the marked rise of small mammals – voles, mice and hamsters – that occurred around the same time, many of which would have been nocturnal.[4] Most of the early owl lineages, however, were subsequently displaced by other bird orders, and only two survived: the barn owls (Tytonidae) and the typical or 'true' owls (Strigidae). At the time, the Strigidae would have resembled the Spotted Owl or the Tawny Owl, and their marked diversity of morphology and anatomy only evolved during the last 15 million years.

The oldest barn owl fossils are estimated to be at least 24 million years old, while the family's diversification that gave rise to the genera *Tyto* (barn owls and allies) and *Phodilus* (bay owls) occurred over 10 million years ago. The common ancestor of *Tyto*, like that of all owls, lived in Africa and diverged as a result of the warmer climate and tectonic upheavals that occurred during the Miocene. One population spread eastwards, reaching Asia, southeast Asia, Oceania and eventually Australia, just over a million years ago. Their dispersal may have been a consequence of the collision of the Afro-Arabian and Eurasian plates in the middle Miocene and the closing of the Tethys Sea. Resultant species include the Eastern Barn Owl, found throughout Australasia, and the stunning Greater Sooty Owl, confined to New Guinea and southeast Australia. In contrast, the route taken by barn owls to reach the Americas is uncertain. One possibility is that American barn owls may have used the North Atlantic land bridge as a corridor to the continent during the warmest period of the early Pliocene. An alternative scenario is that an early representative of the eastern branch crossed Beringia and then spread all the way from Alaska to Patagonia.[5] The status of the extant American Barn Owl was resolved in 2017, after the International Ornithological Committee accepted it as a full species. In the future, the identification of Tytonidae may become even more confused, as European settlers have introduced Western Barn Owls in many countries and the genetic makeup of local populations will be influenced by hybridisation between native and introduced birds.

Fossil evidence reveals that giant barn owls, about twice the size of extant species, once thrived throughout the Caribbean during the Quaternary, between 30,000 and 10,000 years ago.[6] The largest of these evolved on Cuba, where they preyed on nocturnal hutias (*Geocapromys*), plump, rabbit-sized rodents. Later, during one of the many glacial periods when sea levels fell, both predator and prey made it across to the Bahamas, where they appeared to thrive. Indeed, the deposits of one Bahamian fossil site consist almost entirely of tightly packed bones of hutias that must have accumulated over many generations of owls. Early colonisations also involved Hispaniola and the Lesser Antilles, while giant barn owls reached Jamaica relatively recently, around 11,000 years ago. The extinction of all these giant owls was linked to the disappearance of their mammalian prey, since midden deposits across the West Indies show that rodents were a favoured food item of the islands' colonisers. Hunting pressure increased further with the arrival of European settlers in the sixteenth century, while the introduction of predators, such as cats, dogs and pigs, would have been detrimental to the owls and their prey. There is some evidence to support the idea that large owls persisted as late as the eighteenth century. The ornithologists Wetmore and Swales noted that in 1788 three residents of Haiti climbed a local mountain and for five hours heard 'hollow cries imitating the human voice that they attributed to some nocturnal bird, as they had seen feathers resembling those of a swan at the edge of a sort of den or cavern'. Equally intriguing is the folklore from one of the islands of the Bahamas of a large mysterious nocturnal creature called the *chickcharnie* that lived in old-growth pine forest – a tale that has been interpreted as referring to a giant barn owl in historical times.[7]

A further example of insular gigantism is provided by the prehistoric barn owls of Monte Gargano in Italy, an area that was an island during the Neogene when sea levels were much higher than today. Indeed, one species, *Tyto gigantea*, grew to around 80 centimetres in length and may have reached twice the size and weight of the Eurasian Eagle-Owl. Like their Caribbean cousins, the Italian 'giants' became extinct following the loss of their primary food source, the hairy hedgehog (*Deinogalerix*), which coincided with the arrival of humans and their associated mammalian predators.

The true owls (Strigidae) are more complicated than the barn owls in their evolution and biogeography, given that there are 223 extant species on six continents, compared to only 20 Tytonidae. Nevertheless, the Old and New World species cluster in separate monophyletic clades, which shared a common ancestry in Africa.[8] One of the earliest offshoots was the Australasian genus *Ninox*, which includes the boobooks, although little is known about the timing of their dispersals. In contrast, the New World screech owls (genus *Megascops*) reached Central America approximately 20 million years ago. Following the separation of the Puerto Rican Screech Owl and the Flammulated Owl, the core screech owl population colonised South America, including the Andes and the Atlantic Forest, by 7 million years ago. The uplift of the Andes had a significant role in the speciation of much Neotropical avifauna by producing a series of isolated habitats, where populations could evolve independently. The diversification of screech owls is consistent with this concept, and their speciation also supports the hypothesis that the Andean uplift occurred from south to north, since the southern taxa tend to be older than the northernmost species.[9]

While morphology is not always sufficient for the identification of Strigidae – most are barred and spotted in various shades of brown with white – their distinctive calls, which are inherited and not learned, are helpful. Indeed, several new species have been identified recently solely by their unique repertoire of vocalisations. In 2003, for example, two independent teams noted that the scops owl on Lombok in the Lesser Sunda Islands of Indonesia had a clean whistling hoot, unlike any other species. Despite the owl being known to science for over a century, detailed analysis of its call revealed it to be a new species, the Rinjani Scops Owl.[10]

Nocturnal raptors

Given that owls are closely related to raptors and mousebirds, it is likely that the earliest species were diurnal and relied primarily on sight to obtain their prey. As a result, all owls have large eyes that are very efficient at capturing and processing light. They are not spherical organs, as in most birds, but elongated structures, held in place by bony protuberances known as sclerotic rings. Such large, forward-facing eyes provide owls with an improved depth perception, but they are essentially fixed and cannot be moved within their sockets. To compensate, all owls can rotate their heads through 270 degrees in either direction without having to move their bodies. For us, this would mean turning our head to the left so that we end up looking over our right shoulder. Not a manoeuvre I suggest you try, as your delicate blood vessels would be severely damaged, resulting in a major stroke or even death. So how do owls manage it?

To find out, a team from Johns Hopkins University School of Medicine, led by medical illustrator Fabian de Kok-Mercado, injected radio-opaque dye into the bloodstream of dead owls and took a series of CT scans while the head was twisted.[11] The results revealed a number of previously unknown and unique adaptations. Firstly, the owl's cervical vertebrae contain transverse foramina ('holes' for the vertebral arteries and veins) that are much larger than those of other species, including humans. In our neck, the foramina are roughly the same size as the blood vessels, whereas in the owl they are almost 10 times the diameter of the arteries, with the extra space acting as air sacs that cushion the vessel during any twisting motion of the head. Furthermore, the last few cervical vertebrae lack transverse foramina, an anatomical adaptation that provides the cord-like arteries with some slack when the head rotates. Owls also possess connections or anastomoses between their carotid and vertebral arteries, a feature that ensures a constant blood flow to both sides of the brain even if the vessels on one side of the neck are occluded. Finally, the main arteries – carotid, mandibular and maxillary – contain contractile reservoirs that are hypothesised to ensure an uninterrupted blood supply to the brain and eyes, even during the most extreme of head movements. It seems that the 'wise old' owl has not just evolved one answer to the neck-twisting problem; it has come up with a whole raft of solutions.

While 40 per cent of extant species are commonly active in daylight, especially around dawn and dusk, only two are truly diurnal, a fact largely dictated by prey preference. The Northern Pygmy Owl, for example, has a predilection for songbirds that are mostly active during the day, while the Northern Hawk Owl feeds on insects and birds as well as small mammals.

However, most owls have evolved to occupy the vacant ecological niche of a nocturnal raptor: a lifestyle that required several adaptations, including the development of a sophisticated acoustic system.

It had long been assumed that the owl's heart-shaped facial disc was evidence that auditory clues played a role in finding prey. It consists of a concave collection of feathers that form a circular paraboloid – a shape that enables sound waves to be collected and funnelled towards the ears. The disc's efficiency is further enhanced by a ridge of feathers located between the eyes that splits the face into two dishes, each directing sound to its respective ear opening, much the same way that satellite dishes collect TV signals. (The 'ear tufts' visible on the top of some species' heads are not ears at all but merely display feathers.) Furthermore, the bill is pointed downwards, increasing the surface area for sound collection, while the focal length of the discs can be adjusted by special facial muscles, allowing the owl to focus at varying distances to pinpoint its prey.

Proof that nocturnal owls rely on sound to locate prey came in 1971 when Roger Payne from the Rockefeller University in New York published the results of a series of simple but ingenious experiments.[12] His subject was a trained Barn Owl which he kept in a free-flight room while the light levels were gradually reduced to zero over several days. By observing his subject with infrared light (invisible to birds), Payne showed that the bird was able to catch mice in the dark as they scuttled beneath a floor-covering of dry leaves. He also demonstrated that the owl would successfully strike a mouse-sized wad of paper that was dragged across the leaves. These simple observations excluded visual and body heat (infrared) detection, and, since the object did not smell like a mouse, Payne concluded that the owl must be using auditory cues. His suspicion was confirmed when he found that if he made the mouse walk on foam rubber while towing a rustling leaf several inches behind its tail, the owl would strike the leaf and not the mouse. The owl's accuracy was impressive, as it could locate a sound-emitting target with an error of between one and three degrees in both vertical and horizontal planes. However, if the bird's hearing were impaired, by blocking each ear in turn, it would land just short of its target. Also, the owl had to be thoroughly familiar with the room's layout, since it was reluctant to hunt in the dark if moved to a different room. This observation indicates that prior knowledge of the surrounding topography is important, and could explain why many nocturnal owls occupy the same territory for most of their lives.

In theory, however, a single sound source should be hard for an owl to locate in the dark, since two ears are insufficient to define a point in three dimensions. So how does the Barn Owl do it?

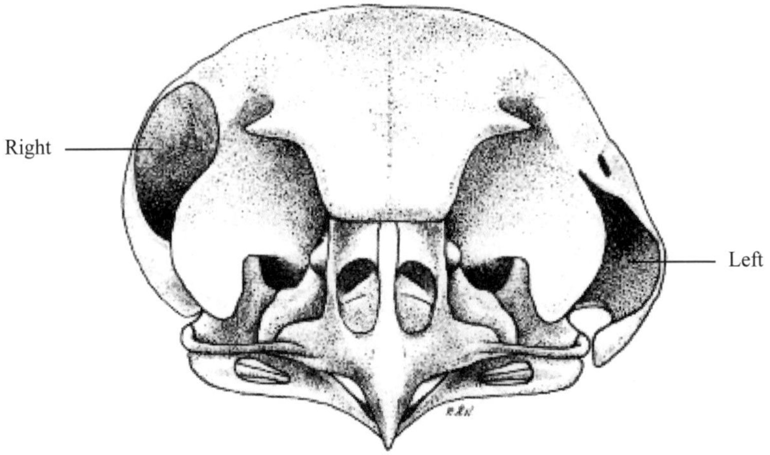

Figure 10.1 The skull of a Boreal (Tengmalm's) Owl appears misshapen because of the unequal positioning of its ear openings. The right ear is placed higher and further forward, while the left is placed lower and further back. Ear asymmetry evolved to improve the ability of owls to locate prey by sound alone. Modified and published with permission from *Ecology and Conservation of Owls*, by Ian Newton, Rodney Kavanagh, Jerry Olsen and Iain Taylor. Published by CSIRO Publishing, 2002.[14]

To hunt in the dark, owls must be able to accurately localise both the horizontal angle (azimuth) and the elevation of a sound source, a task made more complicated by the fact that the sound usually forms a shallow angle with the ground. As a result, any angle of error in the vertical plane results in a greater displacement of the sound source than the same error in the horizontal plane. This crucial factor placed the ancestral nocturnal owls under intense selection pressure to improve their ability to localise sound, and led over millions of years to the evolution of a unique anatomical trait: vertical ear asymmetry.

The positioning of the ears at different levels appears to have evolved independently in at least five owl lineages, since each one has a unique anatomical adaptation.[13] In some species, the asymmetry results from differences in soft tissue. For example, the Eurasian Eagle-Owl has a skin opening for the right ear that is larger than that of the left ear. Barn owls possess opercula (ear flaps) that lie directly in front of the ear canal openings – but, crucially, the centre of the left flap lies slightly above the line of the eyes and is directed downward, while the right one is slightly below and directed upward. In contrast, four owl species – the Ural, Great Grey, Boreal and Northern Saw-whet Owls – possess asymmetrical skull bones that result in dramatically different ear shapes and positions and give rise to a decidedly lopsided appearance (Figure 10.1).

It is the asymmetry in the placement of the ear opening, irrespective of the underlying anatomy, which enables owls to determine how far the sound source is above or below the horizontal plane. In the case of the Barn Owl, in which the left ear opening is higher than the right, any sound coming from below its line of sight will reach the right ear first, while sounds from above will reach the left ear first. This time delay results in small differences in the intensity level of sound that reaches each ear, known as the interaural level difference (ILD). In many owls, the ILD for high-frequency sounds (above 4 kHz) is the principal cue for locating sound elevation. In contrast, the lateral localisation of sound is dependent on the minute time difference between the sound reaching the left and right ears, termed the interaural time difference (ITD). By turning their heads so that the sound arrives simultaneously in both ears, owls can determine when the prey is straight ahead. But to do so, they need to be able to detect a right–left time difference of as little as 0.00003 seconds, or 30 millionths of a second.[15] Each auditory cue (ITD and ILD) is computed and processed in parallel nerve pathways that converge in the owl's midbrain to produce a two-dimensional map of auditory space. Remarkably, the acoustic map is an emergent property of higher-order neurones, in contrast to the direct projections of the sensory surface that occur in the visual and somatosensory systems.[16]

Symmetrically eared owls, such as the Great Horned Owl and the Burrowing Owl, possess smaller auditory brainstem nuclei with far fewer neuronal cells. These observations support the principle of proper mass, which states that the size of a neural structure reflects its processing capacity and the complexity of the behaviours that it subserves.[17] Symmetrically eared owls, however, still possess nerve pathways for ILD that provide additional cues to detect sound in azimuth. Since these are not essential, it is likely that these additional ILD neuronal pathways were co-opted by asymmetrically eared owls to detect differences in elevation. In other words, the bi-coordinate detection of sound, which allowed the nocturnal raptor niche to be occupied, did not depend on the evolution of novel neural circuitry, but rather the harnessing of traits already present in the ancestral owl's auditory makeup.[18] As will repeatedly be stressed, survival adaptation rarely starts from scratch but typically involves the modification of existing blueprints that have already been honed over millions of generations.

Acoustic stealth

Any selective advantage of facial discs and asymmetric ears would be lost if the birds' prey could hear them coming and escape. So to avoid detection, owls have evolved a range of specialised wing features that collectively dampen air turbulence and contribute to their acoustic stealth. The leading wing edge, for example, possesses a comb of stiff serrations on the outermost primary, with hooks and bows, which breaks up the local airflow into smaller currents, resulting in less noise.[19] In effect, the serrations act as vortex generators, disrupting the boundary layer of air and reducing the overall turbulence during flight. The disrupted airflow then moves across a soft downy layer, one with barbed fibres that interlock with other feathers to create a canopy that acts as a buffer between the rough wing surface below and the air flowing over it. Finally, the trailing edge possesses a flexible, porous fringe that mutes the sound of the airflow as it comes off the back of the wing. By combining these three adaptations with low flight speed, owls can almost eliminate the higher sound frequencies – above 2 kHz – that fall within the hearing range of their prey. Of course, it is also vital that the owl's self-generated noise is kept low to avoid any interference with its own bi-aural sound detection system.

Nature's solution for noise reduction has not gone unnoticed by mechanical engineers, who believe that 'silent owl technology' will assist in the future design of wind turbines, computer fans, and even aircraft. Since wind turbines are heavily braked to minimise noise pollution, any reduction of turbulence would allow higher operating speeds and greater energy production. By adding small serrations to the blade's leading edge, the noise can be dampened 10-fold such that an average-sized wind farm could generate several additional megawatts of electricity. It is also hoped that 'owl technology' could reduce the noise of commercial aircraft so that residents living within several kilometres of airports will be less disturbed.

CHAPTER 11

The Oilbird's Story

EVOLUTIONARY DISTINCTIVENESS

In 1799, the great German explorer Alexander von Humboldt was taken to an extensive limestone cave in the Caripe area of northeastern Venezuela and shown a colony of breeding Oilbirds, a species unknown to science at the time. In Spanish they are known as *Guácharos*, from an old Castilian word for one who shrieks or cries, because of their characteristic calls. Humboldt testified to the aptness of the species' name when he wrote: 'It is difficult to form an idea of the horrible noise occasioned by thousands of these birds. The shrill and piercing cries of the guácharoes strike upon the vaults of the rocks, and are repeated by the echo in the depth of the cavern.'[1] La Cueva del Guácharo (the Cave of the Guácharo) remains the largest Oilbird colony in the world, although smaller breeding populations are known from Peru, Colombia, Ecuador, Bolivia and Trinidad.

The Oilbird is one of the most peculiar of all living birds (Plate 18). Not only do they spend the daylight hours in caves, but they are also the only nocturnal fruit-eating bird on the planet. Each night the adults travel up to 150 kilometres to gorge themselves on fruits of palms (Palmaceae), tropical laurels (Lauraceae) and incense (Burseraceae), although they will eat more than 80 different kinds of fruit. When suitable trees are found, the birds use their unusually long wings to hover, while plucking off the fruit, swallowing them whole, and regurgitating most of the seeds. Their chicks are fed exclusively on semi-digested fruit pulp, and they rapidly gain weight to become nearly 50 per cent heavier than their parents within 8–10 weeks. It is only after its flight feathers have developed that the fledgling's weight falls to match that of the adult. Not surprisingly, the Venezuelan locals used to harvest the young birds for their oil content, which was used for cooking, and as fuel for lamps and torches. Only experienced villagers were trusted with the important task, as the odourless oil was a much-needed resource that lasted all year round. Humboldt gave an account of one such killing spree:

> The Indians enter the Cueva del Guacharo armed with poles, by means of which they destroy the greater part of the nests. The young, which fall to the ground, are opened on the spot. Their peritoneum is extremely loaded with fat, and a layer of fat reaches from the abdomen to the anus, forming a kind of cushion between the legs of the bird.[1]

ED and EDGE scores

Oilbirds are placed alongside frogmouths, potoos, nightjars and owlet-nightjars in the order Caprimulgiformes, a sister group to the Apodiformes (swifts and hummingbirds). Although the exact relationship of these families remains controversial, the phylogenetic tree published by Gerald Mayr, based on an analysis of nearly 70 different morphological features, is the most recent and the most robust (see Figure 11.2).[2] Oilbirds, unlike other Caprimulgiformes, forage for fruit and use echolocation and vocal communication to find their way in the dark (see below). As a result of these unusual traits, the Oilbird is placed in a family and suborder of its own, and there has even been a proposal to elevate it to the rank of order.[3] But why is the Oilbird so different?

A clue to this puzzle emerged from an international study, led by Walter Jetz from Yale University, that used a concept known as the evolutionary distinctiveness (ED) score.[4] Phylogenetic trees can reveal more than simply the evolutionary relationships of a group of species. As we encountered in *The Buzzard's Story*, the analysis of branch lengths may provide hints about the underlying mechanisms of speciation. Also, biologists have recognised that time-calibrated phylograms offer pointers to a species' evolutionary originality or evolutionary distinctiveness. It has long been understood, for example, that species that evolved earlier in history or those that lack a close living relative are more genetically isolated than those that developed more recently or have many family members. By studying time-calibrated phylograms, it is possible to quantify the degree of genetic isolation by calculating a species' ED score.

How ED scores are derived is worth explaining, as they are not only relevant to the Oilbird's story, but are also of value to conservationists to help prioritise and target their efforts. The ED score was initially devised by Nick Isaac and his colleagues at the Institute of Zoology in London to help identify mammals at the greatest risk of extinction, although such scores can be applied to any group of organisms.[5] ED scores are calculated by allocating each branch of a phylogeny a value that is equal to its length in millions of years, divided

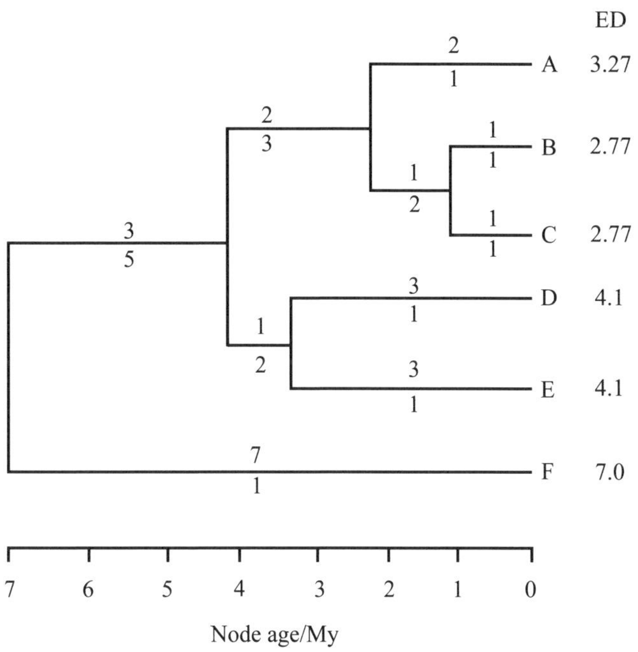

Figure 11.1 Hypothetical phylogeny of six species (A–F) with evolutionary distinctiveness (ED) scores. Numbers above each branch indicate its length in millions of years (My); numbers below indicate the number of descendant species from that branch. Species F has the highest ED score and is the most genetically isolated species. Modified from Isaac *et al.* (2007).[5]

by the total number of species that derive from it. The score for a given species is then derived by adding up the values for all the branches from which that species is descended, from its terminal branch to the root of the tree. For those interested, the next paragraph, in conjunction with Figure 11.1, explains how ED scores are calculated for a hypothetical phylogram of six species (A–F).[5]

The ED score for species A, for example, is calculated by adding the ED scores for each of the three branches between A and the phylogram's root. The terminal branch, which possesses just one species (A), is 2 million years (My) long and, as a result, is given a score of 2. The next branch is also 2 My long but gives rise to three species (A, B, and C), so it has a score of 2/3. The deepest branch that is ancestral to species A is 3 My long and is shared among five species (A–E), so its score is 3/5. The total ED score for species A is the summation of all the branch scores (2/1 + 2/3 + 3/5), making a total of 3.27. Similarly, the score for B is 2.77 (1/1 + 1/2 + 2/3 + 3/5). Since species C is sister to species B, it must have an equal score. Using the same methodology, D and E have scores of 4.1 (3/1 + 1/2 + 3/5). Note that species F has the highest score (7.0), as it is the sole taxon on a branch that lasts 7 million years. We will return to species F later.

ED scores only provide a measure of a species' evolutionary isolation. But if ED scores are combined with a measure of how globally endangered a taxon is (global endangerment, or GE score), then the result is an estimate of a taxon's expected loss of evolutionary history per unit time. The GE score, or extinction risk, is taken directly from the International Union for the Conservation of Nature (IUCN) Red List rankings. The resultant EDGE score (evolutionary distinctiveness and globally endangered) is important for biologists, as it incorporates a species value, in terms of originality, or irreplaceability, weighted by the urgency of action, or risk of extinction. Species with the highest EDGE scores are those that should be prioritised for conservation efforts, given the limited availability of resources. The two metrics (ED and EDGE) convey different information. The Duck-billed Platypus (*Ornithorhynchus anatinus*), for example, although being the sole member of its family and having the highest ED score in Isaac's analysis, is not a conservation concern as defined by its EDGE score. In contrast, the Yangtze River Dolphin (*Lipotes vexillifer*) or *baiji* had the highest EDGE score, despite belonging to a family of around 40 members, and was declared officially extinct several weeks after the paper's publication.

Let us now return to the Oilbird, and Walter Jetz's study. By analysing over 10,000 phylogenies, the scientists were able to construct an evolutionary tree that contained all the bird species recognised at the time. The complexity and difficulty of such a study are reflected in the words of co-author Arne Mooers: 'We didn't want to have any graduate student doing their PhDs on it, because we didn't know if we could do it when we started. Nobody had attempted anything as big, nobody had built a tree this big before.'[6] Overall, the task took nearly seven years and involved calculating the total amount of time evolutionary processes 'worked' to create all the world's species: over 77 billion years. Each species was then ranked according to how much of that work it accounted for.[6]

Most taxa were shown to be relatively young, with little evolutionary distinctiveness: the ED scores were markedly skewed to the right. In fact, the average avian taxon harbours only 6.2 million years of evolutionary distinctiveness. Species with the highest ED scores include several that we have already met, such as the Magpie Goose, Hoatzin and Cuckoo Roller (Table 11.1), all of which have terminal branches that link them to the rest of the avian tree before the K–Pg boundary. Other species in the top 50 include lesser-known and range-restricted taxa such as the New Caledonian Owlet-nightjar (possibly extinct), Solomons Frogmouth and Kagu, while only three passerines are represented: Sapayoa, Przevalski's Finch and Palmchat. The absence of other members of ancient passerine lineages, such as lyrebirds, scrubbirds and New Zealand wrens, results from their ED scores being reduced as a result of partial redundancy with close sister species. For example, if one of the two extant scrubbirds were to become extinct, then the remaining species would immediately jump to a much higher ranking.

As readers must have guessed by now, the Oilbird was ranked number one in the ED list and, as a result, takes the prize for the world's most genetically distinct species. In fact, the Oilbird has not shared its genes with any other taxon for nearly 73 million years (equivalent to F in Figure 11.1, as the only species to arise from its branch). Such a prolonged genetic isolation probably underpins the Oilbird's ecological specialisation, since it is the only cave-dwelling, fruit-eating, echolocating species among over 10,000 extant taxa. Should the Oilbird become extinct, a disproportionate amount of unique evolutionary history would be irretrievably lost, because there are no remotely similar species left on Earth.

Despite its high ED score, the Oilbird does not feature in the top-100 list for EDGE scores, as its conservation status is only of Least Concern. In fact, it is not ranked in the top 300. It is the Giant Ibis, a critically endangered wading bird from Cambodia, with a population of fewer than 100 pairs, that has the 'dubious' honour of heading the EDGE rankings. While Table 11.1 lists the top five species for ED and EDGE scores, readers are encouraged to visit the OneZoom Tree of Life Explorer website (www.onezoom.org/EDGE_birds.htm), where similar data are available for every avian species on Earth.

Table 11.1 Top five bird species for evolutionary distinctiveness (ED) and evolutionary distinctiveness and globally endangered (EDGE) scores. Adapted from Jetz *et al.* (2014)[4] and the EDGE website.[7]

Top ED scores	**Top EDGE scores**
1. Oilbird	1. Giant Ibis
2. Cuckoo Roller	2. New Caledonian Owlet-nightjar
3. Hoatzin	3. California Condor
4. Magpie Goose	4. Kakapo
5. Secretarybird	5. Kagu

Nocturnal frugivores

At some point during its 73 million years of genetic isolation, the Oilbird evolved to become the only nocturnal frugivorous bird on Earth. But how this came about remains unclear, given that some clade members (swifts and hummingbirds) are diurnal while others (nightjars, potoos, frogmouths and owlet-nightjars), although nocturnal or crepuscular, are insectivores.

The most parsimonious scenario is that nocturnal or dark-activity arose just once in the insectivorous stem lineage, and that a reversal to a daytime lifestyle occurred in the common ancestor of hummingbirds and swifts (Figure 11.2A). While simple theories are preferable to complicated ones, the idea of a single emergence of nocturnal behaviour within the Caprimulgiformes presents several problems. For example, there are no anatomical or physiological features to suggest that early swifts or hummingbirds were nocturnal. Furthermore, the evolution of Apodiformes from night birds would require the loss and reacquisition of visual pigments, as well as significant morphological adjustments to the retina, changes which seem highly improbable. There are also anatomical and genetic differences between the various nocturnal families that support the coevolution of night-activity. Firstly, the retinal anatomy of Oilbirds and nightjars differs, with Oilbirds having far more rods and lacking a tapetum lucidum, a layer of tissue lying behind the eye that reflects light back through the retina.[8] Secondly, the nucleotide sequence of *Aanat*, a gene involved in melatonin synthesis and nocturnal activity, varies significantly between the owlet-nightjar, frogmouth and nightjar lineages.[9] Although such changes are not proof of an independent origin of night-activity, it is not easy to account for their presence if the stem species was nocturnal.

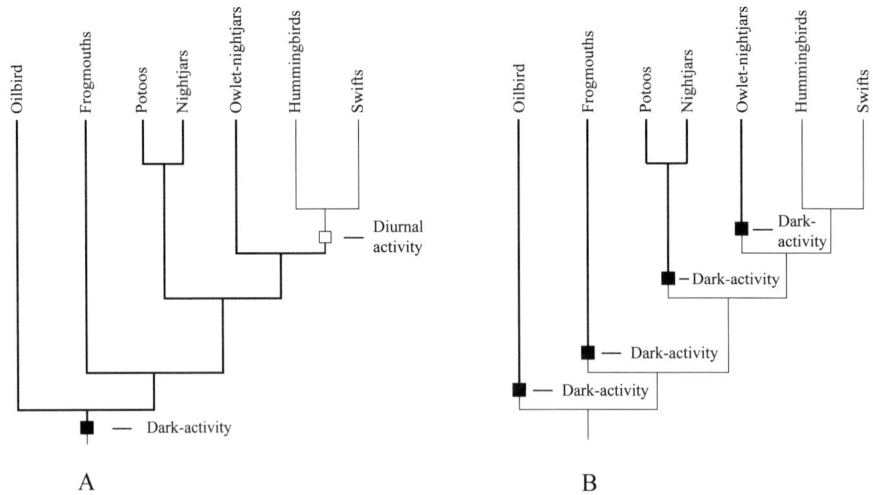

Figure 11.2 Two hypotheses for the origin of nocturnal or dark-activity (thick lines) in the clade of Apodiformes and Caprimulgiformes. (A) Single origin in the stem lineage of the whole clade and a reversal to diurnal activity in the stem lineage of Apodiformes (swifts and hummingbirds). (B) Fourfold origin in the stem lineage of the Oilbird, frogmouths, potoos and nightjars, and owlet-nightjars. Modified from Mayr (2010).[2]

It may also be relevant that no extant family of Caprimulgiformes includes a secondary diurnal representative. Indeed, the evolutionary biologist David Holyoak believes that such a switch may have been prevented by the combined effects of loss of colour vision, susceptibility to raptor predation, and potential competition from the many diurnal insectivorous species.[10] Furthermore, Gerald Mayr argues that there would have been a strong evolutionary incentive for Oilbirds and their allies to become nocturnal, given the dramatic radiation of night-flying insects that occurred during the early Palaeogene.[2] Although some critics argue that an evolutionary switch to night-activity also involves considerable physiological barriers, this may not always be the case. Nocturnal lifestyles, for example, are well represented among modern birds, being common in the procellariiform and charadriiform orders, as well as occurring in a few predominantly diurnal clades, such as the Night Parrot and Kakapo within the Psittaciformes and the kiwis among the Palaeognathae. Therefore, despite being a less parsimonious scenario, it seems likely that a fourfold origin of a nocturnal lifestyle occurred, once in each of the stem lineages of oilbirds, frogmouths, nightjars and potoos, and owlet-nightjars (Figure 11.2B).

In 1987, Storrs Olson published details of the first diagnostic fossils of an Oilbird-like species and concluded that the Oilbird's frugivorous lifestyle had already evolved by the early Eocene. The nearly complete skeleton, from the Green River Formation of Wyoming, was placed in a new subfamily Preficinae, as it was smaller and more primitive than extant Oilbirds, with a wing structure that would not have supported hovering.[11] Importantly, its mandible, which differs from that of all other Caprimulgiformes, suggests that the transition to fruit-eating had already evolved by 50 million years ago: a conclusion strengthened by the existence of appropriate food sources. Indeed, fossil remains of practically all the major fruiting plants favoured by living Oilbirds have been found in the Green River Formation, including palms, laurels and torchwoods. This suggests that the lineage leading to the Oilbird has been feeding on essentially the same specialised diet throughout its entire evolution. Such a finding has significant botanical implications.

During the late Cretaceous, angiosperms were not common, and those that occurred were mainly shrubs and small trees with correspondingly small fruits and seeds. By the early Palaeogene, however, angiosperms had become the dominant species in the forests owing to the development of bigger fruits and the evolution of birds and mammals that acted as seed dispersers. Recently, GPS tracking has shown that Oilbirds are one of the most important long-distance seed dispersers in the Neotropical forests.[12] On average, birds only return to their caves every third night, and instead roost mainly in trees located some distance from their foraging sites. Since the high lipid content of a frugivorous diet requires lengthy transit times, the fruit's seeds are deposited widely across the bird's feeding range. It is likely, therefore, that the earliest Oilbirds were part of the rapid coevolution of frugivorous dispersers and fruit-bearing angiosperms in the Palaeocene and contributed to the dramatic change in the epoch's flora.

While *Prefica* was smaller than the extant Oilbird, its head was the same size, implying that the evolution of a larger body size was not an adaptation for swallowing larger fruit but rather a development to allow an increased amount of food to be carried over greater distances. Because the young of the fruit-eating Oilbirds develop slowly – taking 110–120 days to fledge – safer nesting sites would have been required, which, in combination with social behaviour, favoured cliff edges or caves. Since caves are sparsely distributed throughout their breeding range, the ability to carry more food supported the subsequent development of their cavernicolous lifestyle – an adaptation that led, in turn, to the evolution of echolocation.

The echo of convergence

In the 1940s, Harvard undergraduate Donald Griffin, together with fellow student Robert Galambos, began a series of detailed studies to try and determine how bats navigate in the dark. At the time, Griffin's idea that bats might use sonar was ridiculed by most scientists. Luckily, he was supported and encouraged by a senior colleague who felt that there was nothing to be lost by undertaking such an outlandish project, especially as he was young, with no reputation to tarnish. His mentor's support paid off, and, after a series of carefully conducted experiments to exclude other mechanisms, Griffin confirmed his hypothesis and coined the term 'echolocation'.[13] Contrary to conventional wisdom, the two tyros had proved that bats emit ultrasonic sounds above the acoustic range of humans, and can detect small obstacles by hearing their echoes. While most people can perceive frequencies up to 20 kHz, Griffin found that bats emit and hear sounds as high as 120 kHz. Later, while at Cornell University, he undertook further studies and showed that bats' biosonar systems were highly sensitive and could even pinpoint small insect prey.

Buoyed up by his success, Griffin turned his attention to Oilbirds. Was it possible that they too were using echolocation to navigate in the dark? As he recounts in his wonderfully lucid book *Listening in the Dark*, Griffin travelled to the same Oilbird colony that Humboldt had visited over 150 years before.[14] In the deepest part of the cave, he exposed photographic film for nearly 10 minutes and proved that Oilbirds roost and nest in complete darkness. However, unlike his previous research subjects, Oilbirds are not silent. 'Our ears were bombarded almost constantly by a variety of squeaks, screeches, clucks, clicks, and shrieks,' he wrote. However, when the birds left the cave at night to forage for fruit, Griffin realised that the character of their calls changed. Instead of ear-piercing screeches, they emitted a 'steady stream of the sharpest imaginable clicks'. With the help of his friends, Griffin captured three birds, plugged their ears with glue, and observed their behaviour in a darkened room. Although the birds clicked vigorously, they were unable to find their way around and crashed into the walls. Once the plugs were removed, the birds avoided the walls. Furthermore, if the lights were switched on, the Oilbirds not only missed the walls but also emitted far fewer clicks. Based on the results of these simple experiments and the analysis of their calls, Griffin deduced that Oilbirds rely on low-frequency echolocation in the darkness of the cave, but use mainly sight during the daytime.

These remarkable conclusions were confirmed in the 1970s by Masakazu Konishi and Eric Knudsen, who noted that the click bursts were of a very short duration, with most energy between 6 and 10 kHz. Since the Oilbird's sonar system uses long wavelengths (low frequencies), the objects they can resolve must be relatively large. This prediction was borne out by obstacle avoidance tests devised by the two scientists. According to Konishi and Knudsen, when plastic discs of various diameters were strung across the cave's passageway 'all birds hit 5- and 10-cm discs as if nothing had existed in their paths. The first signs of avoidance appeared when 20-cm discs were presented and all birds avoided the 40-cm discs.'[15] Bats, on the other hand, use very high frequencies and can detect small objects, including moths in flight. This marked difference relates to the Oilbird's biosonar system having evolved solely for use within the spatially straightforward and predictable interiors of caves. Furthermore, unlike bats, high resolution is not required for the avoidance of in-flight collisions, since their low wing loading makes for low flight speeds, and it seems that their nocturnal foraging is guided primarily by vision.

If the resolution of echolocation in Oilbirds is low, how good is their eyesight? A study by Graham Martin from the University of Birmingham has shown that Oilbirds possess a banked retina with rod photoreceptors arranged in a three-layered structure that gives them the highest density so far recorded in any vertebrate eye (approximately 1 million rods per square millimetre).[16] In contrast, their retinae possess very few cone photoreceptors (for discerning colour). Rods function in low levels of light and only require a few photons for activation, while cones need tens to hundreds of photons. The Oilbird's unique rod–cone ratio provides a high visual sensitivity (greater than owls), but a low optical acuity – features that account for their dependence on other sensory clues. Nevertheless, Oilbirds are well equipped to detect the lowest light levels above trees at night and can use vision for general orientation. Tasks that require high spatial resolution, such as the plucking of fruit, depend on the integration of other sensory inputs, including tactile cues from their prominent rictal bristles and olfaction.

During the Oilbird's 73 million years of genetic isolation, it has evolved from a diurnal insectivore to a cave-dwelling nocturnal frugivore reliant on echolocation. While it remains unclear when the individual traits evolved, I would suggest that they occurred in the following order: first, the ancestral population of insectivores became nocturnal; next came the development of a frugivorous, cave-roosting and cave-nesting lifestyle; and eventually this led to the emergence of a biosonar system. As we discussed, the driving forces for such changes included the increase in night-flying insects and fruiting angiosperms in the early Palaeogene, and the protection offered by caves for their slow-growing chicks.

The only other birds to have evolved echolocation are the smaller, cave-dwelling swiftlets. The paucity of avian biosonar systems is surprising, given the benefits of echolocation in poor light conditions, and suggests that ecological factors play a greater role in its emergence than physiological constraints and opportunities. The 33 species of swiftlet are found across the Indo-Pacific region, from the Seychelles to Tahiti, and are mainly diurnal taxa that hunt for small insects on the wing. They roost at night in nests glued to the walls of caves by their saliva, and nests of several species are collected for the profitable 'birds' nest soup' industry. Despite being monophyletic, echolocation has evolved in only half the species and occurs in two of the three genera.[17]

Avian biosonar systems, therefore, have originated on three separate occasions: once in Oilbirds and twice within the swiftlet clade. Since both groups use biosonar to gain access to their nests in caves, it is likely that this common ecological variable provided the necessary evolutionary driving force. Interestingly, an analogous connection between cave-dwelling and the use of echolocation has been reported in rousette bats (genus *Rousettus*), an Old World clade of nocturnal frugivores.[18] It seems probable that the biosonar systems of Oilbirds, swiftlets and rousette bats emerged in response to the same biological need, a phenomenon known as convergent evolution. Convergence is widespread in birds, a fact that greatly complicated taxonomic classification before the era of comparative genomics. Several examples have already been discussed in this book: the flightless wing-propelled diving of penguins and auks (*The Penguin's Story*) and the hooked beaks and talons of owls and raptors (*The Owl's Story*). Indeed, according to the English palaeontologist Simon Conway Morris, convergence is a pervasive force in evolution, and given the same environmental and physical limitations, life will eventually evolve towards similar 'optimal' solutions.[19]

CHAPTER 12

The Hummingbird's Story

A ROUTE OF EVANESCENCE

A Route of Evanescence,
With a Revolving Wheel –
A Resonance of Emerald
A Rush of Cochineal –
And every Blossom on the Bush –
Adjusts its tumbled Head –
The Mail from Tunis, probably,
An easy Morning's Ride –

Writing in 1880, the American poet Emily Dickinson encapsulates the essence of the Ruby-throated Hummingbird – speed, iridescence, and a unique relationship with flowers – in her minimalist poem *A Route of Evanescence.* In the last two lines, she ponders on the hummingbird's overall physicality and imagines it flying with ease from some foreign shore, implying that the bird is completely in harmony with nature and serenely confident of its powers. Little could Dickinson have imagined that science would one day support her poetic musings and confirm that hummingbirds did indeed arrive from some distant land. However, this was not from Tunis after 'an easy morning's ride', but from Eurasia following a dispersal that took many millions of years.

Hummingbird origins

Despite their very different lifestyles, the hummingbirds (Trochilidae), true swifts (Apodidae) and treeswifts (Hemiprocnidae) are closely related families that are placed within a single combined group, the Pan-Apodiformes. As we have seen, their nearest relatives are the owlet-nightjars, nightjars, potoos, frogmouths and the Oilbird (Figure 11.2). Evidence that hummingbirds and swifts have a common ancestry has come from a well-preserved fossil, *Eocypselus rowei*, from the early Eocene Green River Formation of southwestern Wyoming. In 2013, Daniel Ksepka, while working at North Carolina State University, found the 50.6-million-year-old specimen after it had been overlooked for several years, and named it *rowei* after John Rowe, Chairman of the Field Museum's Board of Trustees.[1] It is an exceptionally well-preserved fossil that includes most of the skeleton, as well as many complete feathers with their melanosomes, tiny cellular organelles that contain pigment. *Eocypselus rowei* was a small bird, about 12 centimetres long and weighing less than 30 grams, with a swift-like beak, long legs, and a wing structure intermediate between that of hummingbirds and swifts. A feathered head-crest may have been present, as in extant treeswifts, and its non-specialised glossy black wings lacked the necessary modifications for either soaring or hovering. Overall, the fossil's morphological features indicate that the common ancestor of hummingbirds and swifts was already small-bodied before each family evolved its characteristic flight behaviour. The swift's very short legs, therefore, must have developed after the lineage's divergence, possibly to reduce weight and enable a highly aerial lifestyle. Indeed, Common Swifts are one of the fastest-flying birds and can spend up to 10 months continuously airborne, taking 'power naps', capturing food, obtaining nest material and even mating on the wing.[2]

The next fossil in the ascent of hummingbirds, *Parargornis messelensis*, was recovered from the Messel Pit in Germany, a site that was a steep-sided volcanic lake surrounded by subtropical rainforest during the Eocene. The area is now a UNESCO World Heritage Site, one of only a few that has ever been listed exclusively for its fossil assemblage. Periodically, the vast caldera released clouds of toxic fumes that poisoned scores of creatures in the surrounding area, including *Parargornis* as it flew across the lake's noxious waters 47 million years ago. A stem hummingbird, *Parargornis* had a swift-like beak, short wings and a long tail, with feathers resembling those of the owlet-nightjars. Gerald Mayr, the German palaeontologist who described the fossil in 2003, believes that its beak shows that the Trochilidae evolved from insectivorous ancestors and that its owlet-nightjar-like feathering may well be a primitive trait of early hummingbirds.[3] Also, the fossil's peculiar wing structure – a combination of a short humerus and broad wing – has no counterpart among modern birds and reflects an early stage in the evolution of hovering flight.[4]

In the late twentieth century, the Russian palaeontologist Alexandr Karhu described two incomplete fossils from 35-million-year-old deposits of the northern Caucasus. Both specimens are now known to belong to the stem lineage of modern hummingbirds, since they share several characteristic features, including a modified 'elbow' joint and a humerus head that allows rotation of the wing during hovering flight.[5] Several years later, a more convincing hummingbird fossil was found by Mayr after he had searched through the collection drawers of Stuttgart Natural History Museum. Hidden among the many specimens, he noticed two tiny unclassified bird skeletons unearthed from the Lower Oligocene deposits in southern Germany. Remarkably, both fossils possess a combination of features unique to modern hummingbirds: small size, a short humerus adapted for hovering flight, and a long bill. The 30-million-year-old species was named *Eurotrochilus inexpectatus*, the 'unexpected European hummingbird', as it had been assumed that no modern-type hummingbirds ever existed outside the Americas. Despite the similarities to extant hummingbirds, *Eurotrochilus inexpectatus* still expresses some primitive features, suggesting that it is not a particularly close relative of any living species.[6] Three years later, another *Eurotrochilus* species, this time from the Luberon National Park in France, was reported by Antoine Louchart after he spotted the fossil in a private collection.[7] This *Eurotrochilus* skeleton is the most complete ever found and includes the skull and bill, while the surrounding rock matrix reveals the outlines of its wing and tail feathers. Collectively, these Eurasian fossils show that hummingbirds had a much wider distribution in the past, and raise the interesting questions of how hummingbirds reached the New World and why they became extinct elsewhere.

A recent molecular phylogeny has helped provide some of the answers.[8] For more than a decade, Jimmy McGuire, an evolutionary biologist at the University of California, collected DNA samples from most hummingbird taxa, as well as DNA from closely related species, including nightjars, swifts and an owlet-nightjar. Working with colleagues from the USA and Canada, McGuire sequenced six genes, four nuclear and two mitochondrial, from each species and constructed a time-calibrated phylogeny, based solely on nucleotide substitution rates. The results suggested the following scenario. Crown hummingbirds split from the swift lineage at least 48 million years ago, a date that is in general agreement with the fossil record. This divergence probably took place in Europe or Asia, given the presence of early fossil hummingbirds from several sites in Europe, and the phylogenetic diversity of swifts and treeswifts in the region. Twenty million years later, the common ancestor of modern hummingbirds reached South America by dispersing across the Bering Strait to Alaska and North America. A transatlantic route is unlikely, since hummingbirds are metabolically constrained from undertaking long overseas journeys. Why these early hummingbirds left no survivors in Eurasia and North America is unclear, but it may relate to both climatic changes and the arrival of passerine nectar specialists. It is also possible that the species-rich fauna of herbivores in the Old World added to the pressures for the limited availability of energy-rich and nutritious flowers. Once hummingbirds reached South America, around 22.4 million years ago, the founding population dramatically radiated into new ecological niches to produce the nine distinct lineages recognised today: topazes, hermits, mangoes, brilliants, coquettes, the Giant Hummingbird, mountain gems, bees and emeralds.

Speciation was especially fast in the Andes, since, although the mountains represent just 7 per cent of the land area occupied by hummingbirds, they are home to 40 per cent of the species. It seems likely that the Andean orogeny contributed directly to their dramatic diversification, since speciation was greatest when the Andes were rapidly increasing in height. But mountain environments are cold at night, and if hummingbirds stopped feeding they would cool too rapidly. Rather than consume energy trying to keep warm, high-altitude species have evolved the ability to reduce their metabolic rate by as much as 95 per cent and enter a sleep-like state known as torpor. By doing so, species such as Andean hillstars (genus *Oreotrochilus*) consume up to 50 times less energy, and reduce their core temperature to a level that is barely sufficient to maintain life.

Ten million years ago, a drought-tolerant ancestor of the mountain gem and bee clades recolonised North America, which at that time was still separated from South America by the Central American seaway. The accumulation of species in North America was slow at first, but then rapidly increased owing to multiple invasions of emeralds, coquettes, mangoes and hermits once the Panamanian isthmus had formed. The Caribbean was also invaded many times, including by the bee lineage from North America, which then recolonised South America and produced further species alongside existing lineages.

In the space of just 22 million years, hummingbirds have diversified from a single common ancestor that lived in the lowlands of South America to over 350 extant species that span the Americas, from Alaska to Tierra del Fuego and the Caribbean. And yet, according to McGuire, their speciation rate is only slowing slightly, for although some clades have saturated the available environmental spaces, other clades are still evolving into new species at an extraordinary rate. Indeed, by comparing their extinction and speciation rates, McGuire estimates that the number of hummingbird species could double before reaching equilibrium. It seems, therefore, that the ascent of the hummingbirds is far from complete.

Nectarivorous lifestyle

Hummingbirds are specialist nectar-feeders, and their ability to detect sugar-rich food sources enabled their colonisation of novel ecological niches and contributed to their dramatic rates of speciation. But how hummingbirds recognise sugars has, until recently, been unclear, since they do not possess sweet taste receptors. This scientific conundrum surfaced over 10 years ago when geneticists obtained the first complete sequence of a bird's genome, that of the domestic chicken. To their surprise, unlike that of other vertebrates, avian DNA does not contain a gene that codes for a functioning sweet receptor.

Most vertebrates perceive sweet and savoury tastes by expressing a family of receptor genes, called T1Rs. Savoury or umami flavours are detected by the heterodimer T1R1–T1R3, a receptor that is sensitive to amino acids, while the T1R2–T1R3 heterodimer functions as a sugar receptor. In 2014, Maude Baldwin, a doctoral student at Harvard University, and her colleagues used this knowledge to analyse the genomes of 10 species of birds, from chickens to flycatchers.[9] They found that seed- and insect-eating species possess savoury receptors, but not a *T1R2* gene needed for sugar detection. Since the lack of the *T1R2* gene is widespread among birds, it is likely that their carnivorous ancestors, the therapod dinosaurs, also lacked sweet receptors. According to Baldwin, ancient birds lost their *T1R2* gene because there was no need for meat-eaters to detect sugars. But this reasoning poses a problem. While the ability to detect sugars is not necessary for chickens and flycatchers, many species, including hummingbirds, live on nectar, a food source made up almost entirely of simple sugars. Indeed, hummingbirds consume more than their own body weight in nectar each day and can instantly tell the difference between a weak sugar solution and water. So how can nectarivorous species find food if they lack a gene for sweet taste?

To answer this question, Baldwin and her team cloned the taste receptors from three species of bird: the sugar-insensitive domestic chicken, Anna's Hummingbird and the closest living relative of the hummingbirds, the insectivorous Chimney Swift. After expressing all three receptors in cell lines, the scientists were able to show that the hummingbird's savoury receptor responds to sugars, unlike those of the chicken and swift. When they looked more closely, they found that at least 19 amino acids had been substituted in the hummingbird's T1R3 protein and that these changes imparted sugar sensitivity to its savoury receptor. In other words, hummingbirds have evolved the capacity for carbohydrate recognition by converting a savoury receptor into a sugar one – an event that must have occurred after their lineage diverged from its insectivorous ancestors at least 48 million years ago. Future studies on other nectar-feeding families, such as honeyeaters and sunbirds, are awaited to see if evolution has used a single strategy, or a range of different strategies, to solve the problem of sweet detection in birds.

While we cannot be sure how the change in hummingbirds' taste perception occurred, one can imagine a likely scenario. An ancestral population that lacked sugar taste could have accidentally ingested some nectar while hunting insects among flowers. Any individual with an appropriately mutated *T1R3* gene, one that enabled the detection of sugar for the first time, would have been given access to a novel energy source. If fitness were improved, then the mutated gene would increase in frequency from generation to generation. Eventually, after multiple receptor modifications, the nectar-seeking population would have gained a marked evolutionary advantage over their insect-eating ancestors.

The emergence of sugar receptors changed the course of hummingbird evolution and enabled their nectarivorous lifestyle to develop. However, to satisfy their daily energy needs, hummingbirds have to consume an extraordinary amount of nectar, equal to several times their body weight each day. This value is far higher than in any other bird species of the same size, and, as a result, their kidneys have had to evolve the ability to excrete large volumes of dilute urine. Hummingbirds also possess a unique glomerular structure and a dense nephron blood supply that allows a precise control of blood electrolyte levels, despite consuming nectars with widely different sodium and potassium concentrations.

How hummingbirds syphon nectar so quickly has, until recently, been a mystery. It had always been assumed that capillary action filled the two grooves along their tongues, in the same way that sponges and paper towels soak up water, even though such a mechanism would struggle with the volumes required. Now, according to Alejandro Rico-Guevara and Margaret Rubega at the University of Connecticut, it seems that hummingbird tongues act like miniature pumps.[10] Using slow-motion videos and transparent artificial flowers, the scientists studied 18 species from seven of the nine clades of hummingbird. In all cases, they found that when the bird's tongue extends, the grooves on each side are compressed shut by the bill, storing potential energy in their walls. But once the tongue touches the nectar, the grooves spring open, and the released energy sucks up the nectar to fill the tubes in just a few milliseconds. Each time the bird compresses its tongue to release the nectar, the pump is reset for another mouthful, a process that can occur up to 14 times a second.

But hummingbirds must also supplement their diet with occasional insects, because nectars are deficient in proteins that provide essential amino acids. To do so, they have evolved a unique means of flexing their lower mandible that involves bending the jaw in two directions simultaneously. This movement, which widens the gape and enables flying insects to be taken, is associated with a complex deformation of surrounding bone. It seems that despite hummingbirds' close evolutionary relationship with flowers, their past insectivorous lifestyle continues to have an influence on their form and function.[11] Nectar is low not only in protein but also in calcium, an element that is required in significant quantities around the time of egg production. Females of most avian species store calcium in tissue called medullary bone, but hummingbirds possess only small amounts of this substance. Instead, the birds consume all sorts of mineral-rich compounds, including wood ash, slaked lime and sand, which they obtain by hovering over the ground and flicking their long tongues in and out.[12]

To be able to hover long enough to obtain nectar, hummingbirds have evolved an insect-like flight style. Indeed, there is even a Cuban endemic named the Bee Hummingbird that, at only 5 centimetres long and 2 grams in weight, is the smallest bird in the world. Unlike most flying vertebrates, which can only produce lift when their wings flap downwards, hummingbirds can do so on the upstroke as well. By filming Ruby-throated Hummingbirds in flight, Tyson Hedrick and his team showed that this ability comes from the bird's relatively small wrist bones, which allow the wings to move through a 140-degree arc.[13] Hummingbirds are also able to beat their wings faster than any other species, up to 70–80 beats a second, and, with only slight changes in wing pitch, can fly in any direction, even upside down. To power such energy-demanding flight hummingbirds have evolved the highest metabolic rate of any vertebrate, about 30 times that of humans, and have developed flight muscles with the highest known density of energy-releasing mitochondria.[14] Their cardiovascular system is no less astonishing. Hummingbird lungs have an oxygen diffusion capacity that is 10 times greater than similar-sized vertebrates, while their hearts are proportionately twice as large, beating 250 times a minute at rest, rising to 1,200 beats a minute during flight. Furthermore, species that live at high altitudes, such as the Andean Hillstar, have evolved high-affinity haemoglobins to cope with the low oxygen levels (see *The Waterfowl's Story*).

Hummingbirds must also process visual information and respond to their environment quickly to avoid collisions, especially when hovering and fighting off intruders. To do so, a highly conserved part of the brain, the nucleus lentiformis mesencephali (nLM), is enlarged and contains neurones that are tuned to detect motion in all directions.[15] In contrast, the nLM of other birds and all four-legged vertebrates (where it is known as the nucleus of the optic tract) is proportionally smaller and primarily detects back-to-front motion. This evolutionary adaptation provides the Trochilidae with the fine motor control needed to hover and zoom quickly in every direction possible, at speeds of up to 60 kilometres an hour. Furthermore, hummingbirds have evolved a markedly enlarged hippocampal formation: an area of the brain that is responsible for memory and learning.[16] As a result, they can recall the nectar quality and content of flowers, as well as their location and distribution, so that they can forage efficiently without wasting time and energy. Spatial cognition also enabled the development of trap-lining, a feeding strategy in which some species visit the same few flowers over long distances, much as trappers check their lines of traps. In effect, trap-lining allowed the allocation of limited resources between different taxa and contributed to further hummingbird speciation.

Since the highlighted adaptations characterise most hummingbirds, they must have evolved after the lineage's divergence from swifts and before their arrival in South America 20 million years later. Collectively, they enabled hummingbirds to develop their nectarivorous lifestyle and so kick-started their remarkable divergence and speciation. But flowers do not provide hummingbirds with energy-rich nectar out of kindness. As payment, they require cross-pollination – and many plants have evolved a 'pollination syndrome', consisting of a range of 'pro-bird' and 'anti-bee' inducements. These include the provision of sucrose-rich nectar, since bees prefer fructose and glucose, and brightly coloured red flowers that lack scent, as smell is essential for insects whereas hummingbirds rely on vision. Bird-pollinated (ornithophilous) plants typically have long tubular flowers and an orientation of stamen and stigma to maximise the chances of fertilisation and prevent contamination from the wrong type of flower. Furthermore, ornithophilous species have corollas that lack a landing platform, and their petals are usually angled to prevent access to insects. Over 7,000 plants in 404 genera from 68 families are now dependent on one or more of the 353 species of hummingbird for their pollination. Nevertheless, ornithophily is thought to be a costly strategy for plants, and the condition has only evolved where there are obvious benefits, as in high-altitude ecosystems that lack insect pollinators, in dry environments, and for sparsely distributed species.

The close relationship between plant and bird has led to some remarkable morphological adaptations. The Buff-tailed Sicklebill, for example, sports a bill that arcs a full 90 degrees downwards to enable it to reach nectar located deep within the corollas of *Centropogon* flowers (Plate 19A). At the same time, the plants have evolved protruding brush-like anthers to ensure that the sicklebill inadvertently collects a dusting of pollen on its forehead while feeding. In other words, the shape of the bird's bill has coevolved with the form of the plant, as both species benefit from an exclusive nectar–pollen relationship. The Sword-billed Hummingbird, in contrast, has the longest bill of any family member and is the only species with the reach to obtain nectar, and hence pollen, from certain species of passionflower in the genus *Passiflora* (Plate 19B). Because of its bill shape, the Sword-billed Hummingbird uses its feet to preen and adopts a slightly diagonal stance, with its head pointing upwards to balance.

But which evolved first: passion flowers with 10-centimetre nectar-tubes, or hummingbirds with 11-centimetre bills? By analysing DNA from 43 species of passion flower, German scientists found that the plants with the longest nectar-tubes evolved 10.7 million years ago, shortly after the Sword-billed Hummingbird diverged from its shorter-billed relative, the Great Sapphirewing.[17] Subsequently, bird and plant evolved together, since each species would struggle to survive without the other. For this particular plant–bird relationship, the scientists made an even more instructive observation. When the Sword-billed Hummingbird population fell dramatically around 3–4 million years ago because of environmental upheavals, some passionflowers quickly evolved shorter corollas to allow pollination by other bird species, as well as bats, and ensured their survival. Biologists have also shown that some hummingbird flowers can rapidly change to bee pollination as the result of a single mutation that alters their colour from red (preferred by hummingbirds) to violet (liked by bees).[18] Evolution, it seems, is not always irreversible, and a few species can escape the perils of overspecialisation. For them, survival of the fittest is, in reality, the survival of the fastest to evolve.

The relationship between plants and insect pollinators has led to a marked speciation in flowers, as individual populations adapt to their primary pollinators and coevolve over time. However, a recent paper by Stefan Abrahamczyk and Susanne Renner from the University of Bonn has revealed that speciation has been less dramatic for ornithophilous plants.[19] One explanation might be that hummingbirds cover greater distances than insect pollinators and, by increasing gene flow between plant communities, have reduced the likelihood of the population fragmentation needed for speciation. Furthermore, the driving force for the speciation of ornithophilous plants is lessened by the fact that most hummingbirds, unlike insects, rarely restrict their food source to one taxon and will often pollinate several plant species. While the longstanding evolutionary relationship between hummingbirds and flowers is undeniable, the work by Abrahamczyk and Renner implies that the relationship may be a stagnant one for some species, since 'without the promise of greater fidelity, plants will only change so much to accommodate their partners.'[20]

Given that hummingbirds evolved in Eurasia, one might expect to find flowers in the Old World that still exhibit a 'pollinator syndrome'. In fact, botanists believe they have found such plants. Several species in Asia and Africa appear to have retained morphological features similar to those pollinated by hummingbirds in the Americas, despite growing in areas devoid of hovering avian pollinators. They include the Himalayan Lantern (*Agapetes serpens*), a beautiful shrub with tiny red hanging lantern-like flowers, and the herbaceous Canarian group of plants from west Africa (family Campanulaceae) that also have red bell-shaped flowers.[5] Both plant groups are now pollinated by other avian species, especially sunbirds. However, *Canarina canariensis* from the Canary Islands has survived by adapting to pollination by non-specialist nectar-feeders, such as the Canary Islands Chiffchaff and the local race of Spectacled Warbler.

Could it be that hummingbirds originally coevolved in parallel with the nectar-laden flowers of tropical Africa long before the emergence of nectarivorous passerines? Perhaps Emily Dickinson was right all along, and the hummingbirds did indeed originate not so far from Tunis.

CHAPTER 13

The Parrot's Story

VICARIANCE AND DISPERSAL

A short distance inland from the north coast of Puerto Rico lies a terrain of thick, ancient and undulating jungle. Most of this verdant coverage sits astride a layer of porous and soluble limestone, which in places has been eroded to form huge sinkholes and plummeting hollows of land. Hidden in one of the forest's many depressions is the Arecibo Observatory, the world's largest single-dish radio telescope. For several decades, scientists have used the research facility as part of a multi-million-dollar search for extraterrestrial life, listening for radio signals that might indicate the existence of alien species. So far, no convincing evidence has been forthcoming, although numerous exoplanets have been identified, lying many light years away, that could potentially be habitable.

How ironic it is, therefore, that in the same forest a low-budget conservation programme is also listening for radio signals, using radio-telemetry to try and save a well-known but critically endangered species, the Puerto Rican Amazon, which inhabits our very own planet. So far, this endemic parrot has narrowly escaped extinction, after reaching an all-time low of only of 13 individuals in 1975. Initial conservation efforts in the eastern mountains were thwarted by Hurricane Hugo, which reduced the recovering population to 22 birds in 1989. As a result, a second captive breeding programme was set up in the more sheltered karst region of the west, not far from the Arecibo's radio telescope. In the spring of 2017, I was privileged to observe two of these charismatic birds perched high up in the forest canopy. Both had been fitted with radio transmitters, a testament to the ongoing conservation efforts to save the species. At the time of writing, the full effects of Hurricanes Irma and Maria, the latest to strike the island, are not yet known, although at least seven individuals died in captivity as a result of stress and the high temperatures that resulted from the loss of canopy. Should the present introduction programme be successful, this rare parrot will be given a greater chance of survival and, although still critically low, the species' steadily increasing numbers provide a glimmer of hope for the future.[1]

The Puerto Rican Amazon is not the only parrot under threat; Lear's Macaw and the Kakapo number less than a few hundred, while Spix's Macaw survives only in captivity. Sadly, at least 16 species have become extinct during the last century, including the Carolina Parakeet, the Paradise Parrot and the New Caledonian Lorikeet. Indeed, scientists from the Australian National University and BirdLife International have concluded that parrots are among the most threatened group of birds, with over a quarter of species classified as globally threatened on the IUCN Red List.[2] Risk factors include small geographical distributions, especially of those species restricted to islands, large body size, long generation times and a dependency on forest habitats.

Currently, there are 381 species of parrot in the wild, one of the largest of all the non-passerine groups, with a distribution heavily biased to the southern hemisphere. Indeed, their northern limits approximate to the 30th parallel, and it was only the extinct Carolina Parakeet that extended further north, reaching southern New York and the Great Lakes. The largest number of species is found in South America, with Brazil holding the record, while the greatest diversity occurs in the Australasian region. In contrast, the whole of Africa has comparatively few species.

But what do we know of the parrot's evolutionary history? A decade ago the answer would have been very little, but recent molecular studies have clarified not only the position of parrots on the avian tree of life but also where and when they first evolved and how they spread across the globe. It is a fascinating and complex story, one that involves vicariance and multiple transoceanic dispersals.[3]

The first parrots

Parrots belong to the Australaves clade of landbirds that emerged at the time of the break-up of Gondwana.[4] As noted previously, the common ancestor of the Australaves and Afroaves was an apex predator, a view supported by the lifestyle of the seriemas that occupy the basal branch of the Australaves (Figure 13.1). The two extant species, the Red-legged and Black-legged Seriema, are large, long-legged terretsrial birds with raptor-like behaviours. Both species are well known for seizing live prey, mainly reptiles, and beating them to death with violent throws to the ground. Once it is dead, the birds rip their food into smaller pieces with a sickle claw, holding the carcass in their beaks and tearing it apart with their feet. The seriemas are the sole living relatives of an extinct group of gigantic carnivorous 'terror birds', the phorusrhacids, which resided at the top of the South American food chain during the Cenozoic.[6]

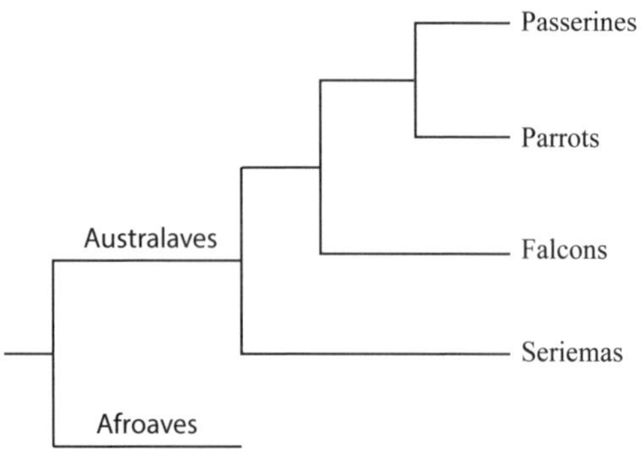

Figure 13.1 Phylogenetic relationships of parrots. Passerines ('perching birds') have parrots as their closest relatives, followed by falcons and seriemas. Modified from Hackett *et al.* (2008).[5]

The common ancestor of all parrots probably lived on Gondwana during the Cretaceous, although these early birds would have looked somewhat different from the species alive today. According to the German palaeontologist Gerald Mayr, stem parrots lacked the most familiar feature of extant species, the long and thick upper bill, or maxilla, which curves over a shorter lower bill, or mandible. Such an anatomical feature probably evolved to allow the consumption of the larger fruits and nuts that appeared during the early Cenozoic period, some time after 65 million years ago. In other words, it was the emergence of new food sources that provided the driving force for the evolution of the thick, curved bill that characterises all the species alive today.[7] Indeed, the jaws of extant parrots have been likened to Swiss Army knives, in that they can be used to crack kernels, seed husks and fruit pulp, as well as helping the birds to climb trees, manipulate food and strip wood.

Molecular phylogenies consistently show that the New Zealand parrots occupy the basal position among the group and were the first to diverge. Initially, it was accepted that the New Zealand clade became isolated after Zealandia broke away from Gondwana 82 million years ago, and that their speciation was allopatric and shaped by vicariance.[8] However, recent studies indicate that the diversification of modern parrots occurred after the K–Pg boundary, around 58 million years ago.[9] Although the revised date is incompatible with the above scenario, New Zealand parrots could still have evolved as the result of continental fragmentation. New geological data reveal that a land bridge existed between Australia and New Zealand until the early Eocene, up to 52 million years ago. Therefore, even if the initial split within the crown group occurred later than first thought, New Zealand parrots could still have been the result of vicariant evolution following the complete separation of New Zealand from Australia. Another possibility is that the ancestral population was more widely distributed in the past and that the earliest development occurred elsewhere. Indeed, the avifauna of New Zealand is composite in nature and has repeatedly experienced colonisation and extinction events. Rather than being the result of prolonged isolation, the seemingly ancient endemism of New Zealand parrots might reflect a recent expansion of range, combined with extinction elsewhere. We will return to the vicariance versus dispersal debate in the next chapter when the evolution of the earliest passerines is discussed.

Once the ancestral population reached New Zealand, it adapted to different ecological niches and evolved into two genera, *Strigops* and *Nestor*.[10] The Kakapo, the world's most genetically isolated parrot, is the only surviving member of the *Strigops* (Plate 20). It is the heaviest, reaching up to 4 kilograms, as well as the only flightless ground-dwelling parrot. Such anatomical features are typical of species that colonise oceanic islands that lack predators and offer an abundant food supply. Although nocturnal, with small eyes, it has evolved a well-developed sense of smell that allows it to track down its favourite food plants during the night. The Kakapo is a specialised foliage eater and has a large gut to allow the bulk processing of its nutritionally poor diet. Sadly, the species lies fourth in the EDGE rankings (see *The Oilbird's Story*) and, as of June 2016, there were only 154 known individuals.[11] Despite translocation to predator-free islands and intensive conservation measures, the population has been slow to increase, in part because females only breed every 2–5 years and fertility is low, with less than 50 per cent of eggs hatching.

The *Nestor* lineage diversified between 5 and 3 million years ago, during the Pliocene, as new habitats opened up after the formation of the Southern Alps. Those birds that occupied the mountain tops evolved to give rise to the omnivorous Kea, the world's only alpine parrot. This hardy species eats mainly insects but will take carrion and has been known to excavate the chicks from Sooty Shearwater burrows. Innately curious, Keas are attracted to people and are often encountered around ski slopes and mountain huts, where they will investigate rucksacks, boots, and even cars. The lowland population gave rise to three species, but only the New Zealand Kaka exists today (the Chatham Kaka became extinct in the sixteenth century, and the last Norfolk Kaka died in 1871). The New Zealand Kaka is predominantly an arboreal species, occupying the mid-to-high canopy, and is easiest to see when flying across the valleys or calling from the tops of trees. They have extra-long, slim upper bills and tongues tipped with brushy papillae that have evolved for extracting sap from trees. The two populations, on North Island and South Island, have evolved slight differences in body size and beak structure after becoming isolated by rising sea levels at the end of the Pleistocene.

Vicariance and dispersal

Once the New Zealand lineage had diverged, the remaining population split when Australia separated from Antarctica around 40 million years ago. At the beginning of the Palaeogene, Antarctica was ice-free, warmer and wetter than today and was divided into West and East Antarctica by a seaway. The common ancestor of the New World parrots evolved in the western half of Antarctica, while the common ancestor for the African subfamily, Psittacinae, evolved in East Antarctica. As the temperatures fell during the Eocene and early Oligocene, extensive ice sheets engulfed most of the continent and forced the two populations to disperse around 35 million years ago. The ancestral New World parrots established themselves in South America, while the early Psittacinae reached Africa by transoceanic dispersal, probably via the Kerguelen plateau (Figure 13.2). Although the latter is now submerged, except for a few remote islands, it once extended for over 2,000 kilometres from Antarctica towards Madagascar and rose over a kilometre above sea level. Later, multiple dispersals occurred from the Australasian population and led to the colonisation of Madagascar and Africa, Indo-Malaysia and the

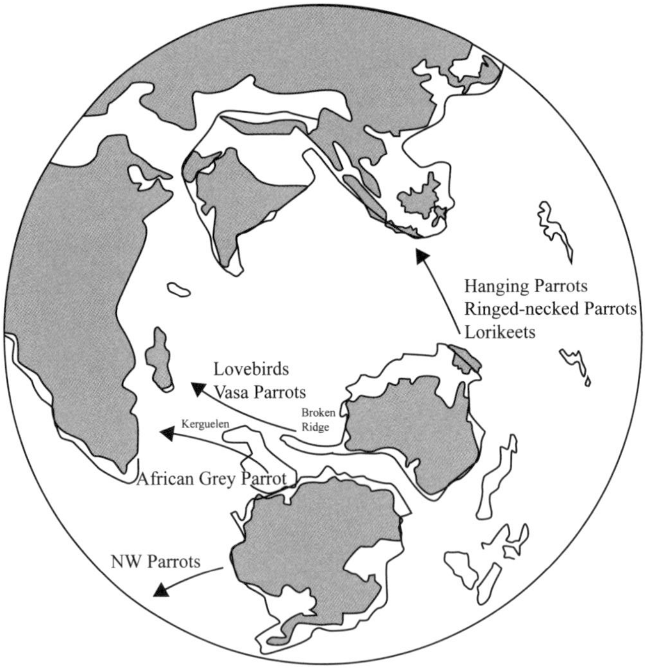

Figure 13.2 Proposed transoceanic dispersal routes of the parrots. Emergent continents above sea level today are shaded grey, and continental shelves are indicated with black lines corresponding to the middle Eocene when dispersal of the Psittacinae, lovebirds and vasa parrots could have occurred. Modified from Schweizer *et al.* (2010).[3]

Pacific islands. In other words, transoceanic dispersals from three ancestral populations – Australasia, West Antarctica and East Antarctica – account for the distribution of most of the parrots in the world today.

New World parrots

Neotropical parrots are found throughout South and Central America, including Mexico, and the Caribbean islands, and comprise some 160 species of amazons, macaws, parrots, parrotlets and parakeets. Some of the most familiar and iconic species, such as the Hyacinth Macaw, the Blue-and-yellow Macaw and Spix's Macaw, belong to this group. Indeed, the New World clade is more species-rich than any other parrot group, and it is likely that the psittacine-free continent would have presented an underutilised adaptive zone, providing multiple ecological opportunities for their speciation. Molecular phylogenies confirm that this enormous diversity all evolved from a single common ancestor that arrived from Antarctica, without any further colonisation by lineages from Africa or Australasia. Maximum speciation rates coincided with the first peak of the Andean uplift in the late Oligocene and early Miocene, suggesting that this dramatic geological event contributed to their diversification, as it did for many other South American bird groups. In contrast, lowland species were mainly influenced by Pleistocene climate changes that resulted in shifts between dry and wet forests and the creation of isolated, open savannas. Psittacine colonisation gradually spread northwards throughout the continent and reached North America around 5.5 million years ago, before the Panamanian land bridge had formed.

Sadly, the Carolina Parakeet, America's northernmost parrot, is now no longer with us. European settlers persecuted the species in their tens of thousands for food, sport and feathers, and to protect their crops, and by the early 1920s the species was extinct. A major factor contributing to their decline was their predictable flocking behaviour: the birds returned repeatedly to old haunts, enabling wholesale slaughter. Recent genetic studies, using mitochondrial DNA extracted from the toepads of museum specimens, has helped elucidate the continents' colonisation by psittacines.[12] It turns out that the Carolina Parakeet's nearest relatives are species with broad distributions in tropical South America, such as the Nanday Parakeet, the Golden-capped Parakeet and the Sun Parakeet, rather than those in nearby Mexico or the Greater Antilles, as one might have predicted. Interestingly, all four species share a distinctive blue edging to their primary and secondary feathers, suggesting that plumage patterns provide a stronger indication of the clade's kinship than their biogeographical relationships. The molecular studies also showed that Central and North America were colonised at different times by several distinct lineages of parrot, although how the Carolina Parakeet came to occupy its unique range in eastern North America remains a mystery.[13]

African parrots

The Psittacinae from East Antarctica were the first arrivals in Africa, and spread rapidly throughout the tropical and southern regions, from Senegal in the west, Ethiopia in the east, and South Africa in the south. They gave rise to two large African grey parrots (genus *Psittacus)* and nine small, stocky birds with large heads (genus *Poicephalus*, from the Greek for 'made of head') that include the Cape Parrot and the Senegal Parrot.

Ten million years later, two further dispersals from Australasia added to the African complement of parrots. The vasa parrots only reached Madagascar, but the lovebirds colonised both Madagascar and Africa. Although the routes taken are unclear, it is possible that they were facilitated by a volcanic plateau in the southern Indian Ocean, known as Broken Ridge, which served as a stepping stone between Western Australia and the Kerguelen archipelago (Figure 13.2). Interestingly, avian dispersals across the Indian Ocean from Australia were unusual, and have only been proposed for a few clades, including the blue pigeons (genus *Alectroenas*) and the cuckooshrikes (family Campephagidae).[14]

Vasa parrots are notable for their rather primitive appearance, which includes a shortened head, long neck and prominent pink beak. Furthermore, the male's cloaca can be everted into a hemipenis that becomes erect during mating. Copulation can last over an hour and involves a copulatory 'tie' facilitated by the male's genital protrusion interlocking with the female's cloaca – a sexual act that is unique among birds and may be associated with sperm competition, enabling the male to increase his chances of fertilisation. Both sexes of the Greater Vasa Parrot have multiple partners (polygynandry), and during chick-rearing it is only the female of the species that sings, holds a territory, and develops a conspicuous, orange-coloured, bald head. These unusual breeding traits may have evolved as the result of the competition among females for food provided by the males, as those with higher song rates are known to attract more males.[15]

Unlike the vasa parrots, the ancestral lovebirds went on to colonise Africa where, among the forests and savannas south of the Sahara, they diverged to produce the eight species recognised today. Evidence that the ancestral population used Madagascar as a staging post is the finding that the Grey-headed Lovebird, a Madagascan endemic, is the sister to all the African mainland species. They are social and affectionate taxa, and their anthropomorphic name derives from their strong monogamous pair-bonding and the fact that they spend extended periods of time sitting together. Furthermore, they typically feed each other to re-establish their bonding, especially after experiencing separation or stress.

Over 50 years ago, William Dilger, Professor of Neurobiology and Behaviour at Cornell University, undertook a series of classic experiments on captive lovebirds to determine if their nest-building is genetically determined or the result of learning.[16] Even though lovebirds are closely related, Dilger observed that different taxa use different strategies to obtain and transport nesting material to their pre-selected tree holes. The Fischer's Lovebird, for example, carries a single strip of tree bark in its beak, whereas the Rosy-faced Lovebird simply stuffs multiple pieces of bark and leaves into its breast and rump feathers. Such bizarre behaviour is restricted to lovebirds and their close relatives, the hanging parrots (see below). Dilger speculated that lovebird nest-building arose from fortuitous occurrences based around two psittacine activities: chewing bark to keep bills sharp and preening. Although most parrots do not build nests, some will accidentally leave bits of material in their feathers when they shift from chewing to preening. According to Dilger, such oversights are likely to have initiated the behaviour of carrying nesting material that subsequently enabled them to build nests. Since the Rosy-faced Lovebird's strategy appears less advanced than that of the Fischer's Lovebird, Dilger deduced it to be the ancestral state, a conclusion supported by recent phylogenetic studies.

The American ornithologist then went on to hybridise Rosy-faced Lovebirds with Fischer's Lovebirds to find out if nest-building is genetically programmed. Despite the hybrids being sterile, they still tried to breed and displayed nest-building behaviour intermediate between the two parental species. In other words, Dilger's simple experiments confirmed that the nest-building of lovebirds is an innate response and not a learned one, although he had no idea as to the genetic mechanism.

Just when the biogeography of African parrots seemed to have been resolved, a remarkable fossil find was reported from Siberia in 2016.[17] Nikita Zelenkov, a Russian palaeontologist, recognised an unidentified specimen in his institution's collection as being the lower leg bone of a parrot. The fossil, from the early Miocene, was found on Olkhon Island in Lake Baikal, the deepest, largest and oldest lake in the world. Although only a single specimen, the 18- to 16-million-year-old fossil raises the intriguing question of whether any African species could have arrived from Asia, rather than all dispersing directly from Australasia.

Australasian parrots

Australia's ancestral parrots diverged over 30 million years ago to produce two distinctive families – the Old World Parrots (family Psittaculidae) that spread throughout southeast Asia, the Pacific islands and Africa, and the cockatoos (family Cacatuidae) that became distributed throughout Australasia.

The Old World parrots colonised Indo-Malaysia on many separate occasions. The Guaiabero ('eater of guavas'), which is endemic to the Philippines, split from its closest Australian relatives around 28 million years ago and was the first to arrive in the area.[8] Its overseas dispersal was probably aided by a string of volcanic islands, the so-called East Philippines–Halmahera–South Caroline Arc, which approached New Guinea and the northern Australian plate at the time. Furthermore, the Australasian tectonic plate was slowly moving northwards and reached its present position in relation to Indo-Malaysia around 20–25 million years ago. It seems that all other splits between Australasian and Indo-Malayan taxa occurred after the two landmasses were in close contact, encouraged by the emerging archipelagos and the dispersal opportunities they offered. Colonisers included the racket-tailed parrots, the hanging parrots and the Afro-Asian ring-necked parakeets, a clade that also colonised Africa.

The 10 species of racket-tailed parrots (genus *Prioniturus*) are endemic to Indonesia and the Philippines and are easily distinguished by their elongated central tail feathers with a bare shaft and a spatula at the end. The hanging parrots (genus *Loriculus*) are a group of small birds with green plumage and short tails that have the unique ability among birds of sleeping upside down. Like lovebirds, hanging parrots collect nest material in their feathers, a behaviour that reflects their close evolutionary relationship. The Afro-Asian ring-necked parakeets (genus *Psittacula*) are highly gregarious, green-plumaged species with an extensive range across Africa, Asia and the islands of the Indian Ocean. They are one of a few parrot groups to have successfully adapted to living in disturbed habitats, and have withstood the onslaught of urbanisation and deforestation. The Rose-ringed Parakeet has even established feral populations in diverse urban environments from Europe to South America, where they have been able to withstand harsher climates than those in their native range. Speciation of all three genera resulted from a complex combination of island colonisations and subsequent divergences in allopatry among and within the island groups. Environmental changes in Asia, following the increased uplift of the Tibetan plateau and the onset of the Indian and east Asian monsoons during the late Miocene, also contributed to their diversification.[18]

Another member of the Psittaculidae, the Eclectus Parrot, is best known for a form of reversed sexual dichromatism (plumage colouration); a feature seen in only a few other taxa. The males of this Australasian species are bright emerald-green, whereas the females are vermillion with a vest of violet or cobalt. Indeed, for a long time, they were thought to be separate species. It turns out that the colour differences are the result of interplay of sexual and natural selection. Females remain secure inside their hollow tree nest sites for up to 11 months a year and have evolved bright red plumage to advertise that the hollow, a scarce resource in their habitat, is occupied. Males, in contrast, spend most of their time foraging in the rainforest canopy and have evolved cryptic green plumage for protection against predators. The species is also unusual in that they can control the sex of their offspring, although how and why are not yet known.[19]

Five to ten million years later, the brightly coloured lories and lorikeets split from their closest relatives, the Budgerigars, during the middle Miocene and radiated through the islands off northern Australia to colonise Sulawesi and Bali, the Philippines and several Pacific islands, as well as Australia. Given their recent divergence, they are an unexpectedly species-rich lineage, a fact that reflects a key innovation: a dietary shift from seeds to nectar.[20] This significant evolutionary event allowed an expansion into new ecological niches and led to rapid speciation through allopatric partitioning. As a consequence, lories and lorikeets underwent several anatomical modifications. Their bills became narrower and less powerful than those of other parrots, and they acquired specialised brush tongues with papillae at the tip to help mop up their food. Since nectar is easier to digest than seeds, they also evolved shorter intestines and thinner-walled, weaker gizzards, as there was no need for a grinding function.

The Budgerigar lineage remained in Australia's harsh interior, where today they roam widely, often in large flocks, breeding opportunistically when the intermittent rains produce enough grass seeds to sustain a clutch of chicks. Native birds display a light green body with pitch-black mantle markings, edged in bright yellow undulations, with a cobalt-coloured tail. Such colouration is thought to have evolved as the result of selection pressures imposed by a lifestyle of feeding on the ground and the need for cryptic plumage to help blend into the grasses and hide from predators. The first live birds were brought to Europe in 1840 by John Gould, where they soon became popular pets, owing to their playful personality, intelligence and mimicry skills. Shortly afterwards, aviculturists began to select mutations that diverged from the wild-type colouring, and today captive birds can be found in a variety of shades that include blue, grey, grey-green, violet and white.

During the mid-Miocene, the Australian plate approached and collided with the Asian plate, causing an uplift of the northern areas and a change in climate, with cooler temperatures and more arid conditions. The continent's vast rainforests became fragmented, and the vegetation changed into a mosaic of different types, including the emergence of other broad-leaf forests, eucalyptus, fire-adapted sclerophyll vegetation, grasslands and saltbush plains. The early–middle Pliocene was a significant period for migration between Australia and southeast Asia, and it is likely that the broad-tailed or platycercine parrots, as well as the cockatoos, spread and diversified into the drier habitats at this time.

The cockatoos are a distinctive family of parrots that are recognisable by their showy crests and curved bills. Their plumage is less colourful than that of other parrots, being mainly white, grey or black, although they may have colour in their crest, cheeks or tail. In 2011, Nicole White, a doctoral student at Murdoch University in Perth, compared six genes from 16 of the 21 different cockatoo species to construct a robust phylogenetic tree.[21] The results revealed that the Cockatiel is the most basal species, with the black cockatoos arising next, while the Palm Cockatoo is sister to a clade composed of the Gang-gang Cockatoo, the Galah and the white cockatoos of the genus *Cacatua*. This well-supported family tree has some interesting implications for several physical features of cockatoos. It indicates that the relatively immobile crest of the Cockatiel is an ancestral trait in the cockatoos, and that the fully erectile crest found in the other cockatoos evolved after this lineage split from the common ancestor with the Cockatiel. It also suggests that specific morphological adaptations such as plumage colour, body size, wing shape and bill structure evolved in parallel or convergently across the different lineages. For example, the large black Palm Cockatoo is more closely related to the white cockatoos than to other black cockatoos.

Unexpected kinships

Recovering the phylogenetic relationship of parrots to other avian families has not been easy. Although they are distinctive and morphologically similar, there are no apparent intermediary forms that link the clade to other well-defined groups. This limitation is reflected in the results of early morphology-based studies that variously suggested a close affinity to both woodpeckers and rollers, and even to cuckoos. But, in 2008, an international study published results that not only clarified the situation but required the avian tree of life to be rewritten. Led by Shannon Hackett, head of the bird division at Chicago's Field Museum, the research group sequenced large sections of genome-wide DNA from 169 species, representing all the major living groups.[5] At the time, it was the most complex avian phylogenetic study ever undertaken, both for the number of species involved and for the amount of DNA analysed. It turns out that parrots are the closest evolutionary relatives of the passerines, a group that includes the songbirds. Irrespective of the statistical approach Hackett's team used, the results of the analyses were the same: parrots consistently emerged as the sister group to the passerines. In other words, the closest relatives of the blackbirds and thrushes living in your back garden are the colourful macaws and lorikeets of the tropics. Despite the results closely matching a preliminary study published two years earlier by Per Ericson and colleagues, it still provoked considerable debate among the scientific and birding communities.[22] Could it be true, or was it simply an artefact?

Three years later, a team of German investigators led by Münster University graduate student Alexander Suh undertook a more detailed examination of avian evolution using a different method: retroposons, or 'jumping genes'.[23] These features are repetitive DNA fragments which can be inserted randomly anywhere in the genome after being copied, or 'reverse transcribed', from an RNA intermediary. While the original repetitive retroposon sequence is inherited like any other piece of DNA, the new insertion, together with any sequence alteration, is unique and will be inherited unchanged from the time of its insertion. In effect, retroposons are 'molecular fossils' that can be followed over evolutionary timescales, a feature used by Suh and his colleagues to obtain an improved resolution of the avian tree of life.

After evaluating thousands of retroposon insertions, Suh found seven that were unique to falcons, parrots and passerines, but absent in all other landbirds, including woodpeckers, rollers and cuckoos. The implications of this finding are twofold: first, falcons are not closely related to other birds of prey as previously thought, and, second, falcons, parrots and passerines have evolved from a common ancestor. Crucially, an additional three insertions were restricted to parrots and passerines, indicating that these two groups are each other's closest living relatives. In other words, Hackett's team was right all along.

The falcons (family Falconidae), including caracaras, forest falcons and falcons, began to diversify during the late Oligocene–early Miocene within the Neotropics. However, the species-rich genus *Falco* diversified later, around 5–7 million years ago, when climate change resulted in the expansion of grasslands and savannas, with their associated mammalian communities.[24]

Since the position of passerines on the avian tree of life has now been highlighted, it is time to move on and hear their story.

PART TWO

The Passerines

CHAPTER 14

The New Zealand Wren's Story

A NOVEL FOOT

Let's play a mind game. If you could step back to the Palaeogene and hide among the southern beech trees, conifers and podocarps at the eastern end of Gondwana, you'd have witnessed a subtle change to the local avifauna. As you watch from the edge of the forest, overlooking a glade of flowering plants, with its bees and nectar-seeking butterflies, you become aware of a small bird perched on a branch above you. You are intrigued because this species has a flexible foot unlike any other that you have seen. It differs from the familiar webbed foot of the ducks that are paddling down by the brackish lagoon and the climbing foot of the woodpecker drumming on a tree behind you. Like birds of prey, hornbills and herons, it has three small toes that point forwards and one longer toe pointing backwards, with no evidence of webbing or joining that you can discern. The hind toe, which joins the leg at the same level as the front toes, is seen to grip the branch tightly as it hops around from perch to perch in search of insects. This unknown bird also scratches itself by bringing a foot over its wing, rather than passing it underneath. What you cannot observe, however, is the presence of a tendon that runs from the underside of its toes to a muscle, located behind the leg bone, that automatically contracts when the leg bends, causing its foot to grip as it perches. Although you do not know it, you are the first person to have added a passerine, or 'perching bird', to your life list!

This fanciful scenario highlights one of the most common arrangements of toes in birds: the anisodactyl foot (Figure 14.1). Characterised by three toes pointing forwards and one – the hallux – pointing back, it was clearly a very successful trait, since the passerines (Passeriformes) have evolved to become the most diverse avian order. Currently numbering over 6,000 species, the Passeriformes are divided into at least 128 families and represent almost 60 per cent of all living birds.[1] They have adapted to a wide range of ecological niches on all continents, except Antarctica, and are highly varied in structure and behaviour. Passerines also have a tendency towards colourful plumage

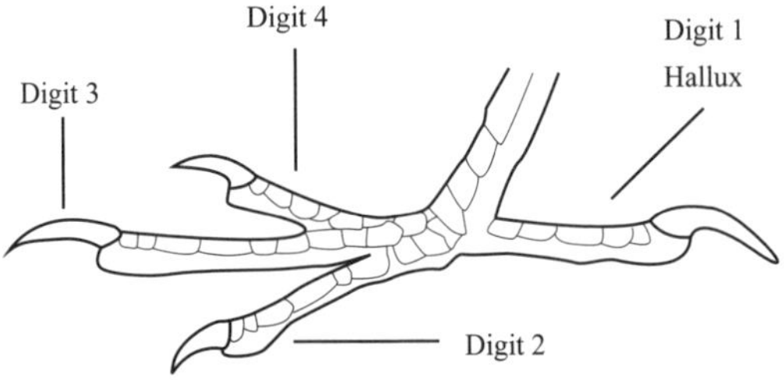

Figure 14.1 The anisodactyl foot that characterises all passerines.

and melodic song. The smallest, weighing only 4.2 grams, is the Short-tailed Pygmy Tyrant that lives in the Amazonian rainforest, while the largest, the Greenland race of Common Raven, is approximately 1.6 kilograms.

Acceptance by the scientific community that over half the world's birds, from New Zealand wrens to tanagers, have evolved from a common ancestor required data from more than just hind-toe morphology. In 1982, Robert Raikow, working at the University of Pittsburgh, reported some additional anatomical features – an unusual sperm structure, a characteristic bony palate and a distinctive wing and leg musculature – that strengthened the protagonists' argument.[2] Such conclusions have now been corroborated by biochemical and molecular research and raise two intriguing questions: Where did the passerines' common ancestor live, and what routes did their descendants take to reach the four corners of the Earth? Before addressing these issues, it is important to stress that passerines comprise two main groups, or suborders – the suboscines and oscines – a dichotomy that turns out to be crucial to the understanding of passerine evolution. So let's look at the evidence for this split.

Johannes Peter Müller, a nineteenth-century ichthyologist, herpetologist and qualified doctor, was the first to highlight that perching birds belong to one of two divisions. The son of a penurious cobbler from Koblenz, Müller was an exceptional student, excelling in both mathematics and the arts; he is known to have translated Aristotle for himself while still at school. Although destined for the church, his inquisitiveness and passion for natural history led to a brilliant career in medicine, earning the respect of his contemporaries as one of the greatest living physiologists. He never lost his initial enthusiasm for nature, and remained intrigued by how different organisms overcome the same physiological challenges. Such curiosity led to numerous seminal publications, but it was his paper of 1847, detailing the anatomy of the avian vocal organ, or syrinx, that is relevant to the present story.[3] Located at the base of a bird's trachea, or windpipe, at a point where it forks into the lungs, the syrinx produces sound without the need for the vocal folds possessed by mammals.

Müller concluded that the whole passerine assemblage could be subdivided solely by the arrangement of the voice box's musculature. The species that possessed a more complex anatomy, the 'songbirds', were termed oscines and placed in a single suborder. The remainder, with more primitive and heterogeneous syringes (plural of syrinx), were placed in several suborders, now collectively termed suboscines.

Müller's work typified the age of natural philosophy: meticulous observations coupled with precise descriptions, with no attempt at any phylogenetic or evolutionary enquiry. It is may be understandable that he became frustrated at his inability to bring order and structure to his life's work, especially since he never accepted the idea that species could evolve over time. Sadly, convinced of his scientific inadequacies, he succumbed to depression and died in 1858, less than a year before the publication of Charles Darwin's influential work.

Over a century later, Peter Ames, working at the Peabody Museum of Natural History in Yale, undertook a more comprehensive analysis of passerine syringes. He confirmed not only that the oscine's vocal organ is more complex than those of the suboscines, but that it is also anatomically uniform throughout the suborder, a finding that led him to conclude that oscines form a narrow monophyletic group.[4] Underlying Ames' interpretation is the notion that inheritable modifications, or apomorphs – the complex voice box in this case – must have evolved in an ancestral species and been passed on to all its descendants. The alternative explanation, namely convergent evolution, can be discounted, as the oscine's syrinx would have had to have evolved independently many hundreds of times.

Alan Feduccia, an American evolutionary biologist, reached the same conclusion for suboscines, after evaluating a different anatomical structure. Instead of the vocal organ, he studied the morphology of the ear ossicle. Unlike mammals, including humans, which possess three bony ossicles (malleus, incus and stapes) to transmit sound waves from the outer to the inner ear, birds perform the same function with only one, the stapes or columella. Feduccia's study was a formidable undertaking, as this bone is not only minute but exceedingly fragile and hidden away in the inner recesses of the middle ear. After several years' work, involving the dissection of bones from several hundred museum specimens, Feduccia found that all suboscines have a distinctively shaped columella, characterised by a large bulbous footplate that abuts onto the inner ear. In contrast, oscines have a much simpler bone with a flat, oval footplate. The analysis of the columella, the last of all the avian bones to be studied, was unexpectedly informative. It indicated that suboscines, like oscines, are monophyletic.[5] In other words, like the oscines, all suboscines share the same common ancestor.

The question concerning ornithologists in the 1980s was whether the idea of passerine dichotomy would hold up using the new molecular techniques – or would the traditional view vanish like mist in a new dawn? The answer was revealed theatrically by Charles Sibley, the egotistical doyen of avian evolution and systematics, at a meeting of the Nineteenth Ornithological Congress in Ottawa in the summer of 1986.[6] Surrounded by eagerly awaiting colleagues, Sibley pinned a long sheet of paper, the legendary 'tapestry', onto his allotted poster board, for all to see. Reassuringly, the long-awaited phylogenetic tree, based on the genetic study of more than 1,000 species and conducted over many years, showed the expected passerine split: suboscines and oscines are sister clades.

The tapestry, however, contained a major surprise, one that would have profound implications for our understanding of passerine evolution. Several Australian songbirds believed to be closely related to Old World species, including robins and treecreepers, were located near the base of the passerine tree. Furthermore, the enigmatic New Zealand wrens (family Acanthisittidae) were placed at the primitive end of the Old World suboscine lineage (see *The Sapayoa's Story*), suggesting a great age for these diminutive species.[7] For the first time, Sibley's 'tapestry' indicated that the widely held view of a northern origin for passerines might not be correct. Despite praise for Sibley's drive and innovation, and the universal welcome for the suboscine–oscine split, the idea that songbirds originated within the southern hemisphere was too much for most scientists to accept. Even Stephen Jay Gould and Ernst Mayr, two of the field's intellectual heavyweights, let their criticisms be known.[8] However, the tapestry's profound implications could not be ignored forever.

The emergence of passerines

A more poignant specimen to highlight the origin of the world's passerines is hard to imagine. The Stephens Island Wren (now formally known as Lyall's Wren) has languished in the Liverpool Museum for over a century, hidden from view and ignored by visitors and staff alike (Plate 21). To observe this important skin, I was kindly given access to the building's storerooms, normally the restricted habitat of the museums' curators. Tony Parker, Head of Collections, met me in the busy public gallery, beneath an enormous pterosaur skeleton, and led me down a flight of stairs to the contrasting silence of the basement. At the end of a corridor of glazed Victorian tiles, a door marked 'Vertebrate Zoology' was opened to reveal stacks of metal cabinets. Each one contained a set of trays filled with neatly arranged skins, all packaged in individual plastic bags. From one tray, a drably coloured specimen was carefully removed: a bird with a prominent beak, smallish wings and long legs. A handwritten label in Indian ink informed me that I had found my quarry: '*Traversia lyalli*, Stephens Is New Zealand'. Tony explained that all the specimens in these cabinets were part of an extensive collection of birds' skins sold to the museum in the early years of the twentieth century by the acclaimed ornithologist and collector, Henry Baker Tristram. It is appropriate that Tristram was an early acceptor of Darwinism, one of the first parson-naturalists to attempt a reconciliation of creation and evolution. Once the wren was safely back in its tray, we headed back upstairs, where I reflected on how the bird's paltry labels failed to betray the drama of its discovery. It is a tale that deserves recounting, for this flightless passerine is not only a pivotal character in our story but one that also provides a salutary lesson for ecologists and conservationists.

Stephens Island, a rugged landmass guarding the western approaches to Cook Strait, was a natural place to build a lighthouse. And so, on 29 January 1894, the assistant lighthouse-keeper David Lyall, together with his family and pregnant pet cat, apocryphally known as Tibbles, took up residence for a two-year tour of duty.[9] Within days, the cat was bringing her owner presents: carcasses of the island's fauna that included a well-preserved bird which Lyall, an amateur naturalist, had never seen before. Intrigued, he preserved it, together with at least 14 others over the course of his stay, having been encouraged by Henry Travers, a local dealer in rare species from Wellington. The skins commanded high prices, equivalent to six months' salary, and Lyall sold them all, with several being acquired by the English collectors Sir Walter Rothschild and Henry Tristram.[10] Ironically, Rothschild's idea of preserving birds meant saving their skins, nine in the case of the Stephens Island Wren, for his collection at Tring in Hertfordshire. According to Barbara and Richard Mearns in their book *The Bird Collectors*, Rothschild 'has been accused of hastening certain species into extinction. If these species had lasted a little longer perhaps some sort of captive breeding programme could have rescued them.'[11]

Sadly, the defenceless Stephens Island Wren, the world's last flightless songbird, was extinct within 12 months, exterminated by the hunting adeptness of Tibbles and her offspring. Indeed, so accomplished were their genocidal activities that an editorial in the *Christchurch Press* (1895) read:

> There is very good reason to believe that the bird is no longer found upon the island, and, as it is not known to exist anywhere else, it has apparently become quite extinct. This is probably a record performance in the way

> of extermination. The English scientific world will hear almost simultaneously of the bird's discovery and its disappearance before anything is known of its life history or its habitats.[12]

It probably didn't help that the species was mainly nocturnal and scampered about at night, very much like a mouse. Stephens Island was the wren's last refuge as, although it had once been widespread, it had been extirpated on the mainland by the Polynesian Rat (*Rattus exulans*), or *kiore*, introduced by the Māori.

Unfortunately, the Stephens Island Wren was not the only casualty of the cats' activities. It is estimated that 13 bird species were eventually eliminated from Stephens Island, including a subspecies of South Island Piopio, a bird now classified as an Old World oriole. It is tragic to think that the extinction of both species could have been prevented with a little more forethought, since the destructive potential of introduced cats was already well known. By the time the feral population had been exterminated, it was too late. Nevertheless, this episode remains a stark reminder to conservationists as they struggle to eliminate cats from other islands in an attempt to save endemics such as the Ascension Frigatebird, the Hawaiian Petrel and the Marquesan Ground Dove. Today, there are only two extant species of New Zealand wren, the Rifleman and the New Zealand Rockwren, while a third, the Bushwren, has not been seen since a sighting on Kaimohu Island in 1972.

While the New Zealand wrens are obviously passerines, they have always been regarded as taxonomic oddities, defying sub-classification. Originally, they were thought to be suboscines, based on their syringeal structure, but Feduccia argued for an oscine lineage, on account of their ear-ossicle morphology.[13] Raikow also assigned them to the oscines, claiming that their leg musculature is more characteristic of this suborder because they lack part of one of the flexor muscles.[14] Interestingly, the first molecular study had hinted at the answer. In 1990, Sibley and Ahlquist, using a crude DNA–DNA hybridisation approach, reported that 'the acanthisittids are the survivors of an ancient passerine lineage with no close living relatives. We include them in the suborder [suboscines] because they are not oscines, but it is possible that they should be assigned to a third suborder as the sister group of [them both].'[15] Writing in the journal *BioScience* in 2000, Raikow and Bledsoe's pithy comment aptly said it all: 'The New Zealand wrens seemed to disrupt the otherwise neat division of the oscines and suboscines.'[16]

A different approach was needed, and it fell to the new discipline of molecular phylogenetics to resolve the longstanding taxonomic conundrum.

As discussed in *The Buzzard's Story*, the technique assumes that the differences in base sequence between homologous DNA from two species are proportional to the amount of time since they last shared a common ancestor. The more distant they are, the more time they will have had to acquire mutations, or base changes, compared to more closely related taxa. Conversely, species that have diverged more recently will have had less time to accumulate mutations, and their DNA will be more similar. In 2002, Keith Barker and his colleagues from the American Museum of Natural History used this approach – and the results were striking.[17] They compared homologous sequences from two nuclear genes (the details don't matter, but for the record they were *RAG-1* and c-*mos*) from a range of species, which represented most of the recognised passerine families. Irrespective of the statistical approach, the Acanthisittidae separated out from all the other passerines by a long branch, indicating a considerable evolutionary distance. This finding also accords with the wrens' morphologically primitive syringes and stapes. In the same year, Per Ericson's team, based at the Swedish Museum of Natural History in Stockholm, reported identical results, despite sequencing slightly different stretches of DNA.[18] Both studies showed unequivocally that the diminutive New Zealand wrens are the most ancient of all passerines, forming a sister group at the base of the entire suboscine–oscine radiation. But just how old are they?

These days, the reliability of high-throughput sequencing and the use of sophisticated analytical programmes enable the routine construction of phylogenetic trees, or phylograms, with unprecedented precision. But such analyses can only reveal relative and not absolute time intervals. In other words, while it is possible to deduce that a given time-period is the same, or twice as long, as another, it is not possible to date accurately any of the nodes, or branching points, on a tree. Phylogeneticists have addressed this limitation by calibrating their molecular clocks against events whose dates are known, ideally one of a similar age to the root of the tree in question. In practice, this is provided by a reliable fossil, or the emergence of an oceanic island, or a major tectonic occurrence. Both Barker and Ericson chose the same tectonic event, the rifting of New Zealand from Australia–Antarctica that took place 82–85 million years ago, as the primary calibration point and used this to estimate later divergence dates in the passerine evolutionary tree. By assuming that the wren's ancestors were present on eastern Gondwana when it separated from Antarctica, it is not surprising that both groups concluded that the split of Acanthisittidae took place some time between 85 and 82 million years ago.

However, is this approach justified? It is a crucial issue, and one that deserves further discussion, since the divergence date has been widely accepted and used as a temporal framework for many subsequent phylogenetic studies. In their defence, both teams proffered several supporting arguments. They emphasised the significant basal separation of the wrens from all other passerines and the fact that no close relatives exist outside New Zealand. The latter point is important, as it suggests that the ancestral wrens were not once widespread and then died out leaving an isolated New Zealand population. Furthermore, studies using fossil-calibrated clocks had already suggested that the basal oscines and suboscines stem from the Cretaceous, a time much earlier than previously thought.[19] It is logical, therefore, that the wrens must have originated even deeper within the Cretaceous, given that they are basal to both suborders. Lastly, and most significantly, passerine phylogenetic relationships show a biogeographical pattern with a clear Gondwanan signature. In other words, the basal oscines and suboscines are all confined to continents that were once part of the southern supercontinent, the most notable being the restriction of Acanthisittidae to New Zealand. Further studies in 2004 and 2014, using updated molecular clocks, also concluded that New Zealand wrens first evolved during the Cretaceous and that their speciation coincided with the separation of Zealandia from Antarctica.[20] If these findings are correct, then the New Zealand wren's story can be summarised as follows:

Primitive passerines, probably small ground-dwelling insectivorous foragers, were once widespread throughout the Australia–Antarctic region of Gondwana during the late Cretaceous. Around 85 million years ago, a fault-line began to form in the brittle crust of eastern Gondwana, which subsequently became flooded, splitting off a continent known as Zealandia or Tasmantis. This new landmass became separated by a widening sea, the Tasman Sea, which slowly opened up like a zipper from the south. It was a vast continent, encompassing present-day New Zealand and extending as far as New Caledonia in the north and Macquarie Island to the south. The splitting process stretched and thinned Zealandia's continental crust, reducing its buoyancy and causing it to subside slowly. As a result, Zealandia had a low relief, with the highest elevations being less than 300 metres. A complete Gondwanan ecosystem was taken along for the ride, including podocarps, giant kauri trees and ferns, among which scampered the ancestors of the New Zealand wrens. These tiny passerines were poor flyers, and they remained trapped with no further opportunities for genetic exchange, having left their relatives far behind to the southwest. Over the next 50 million years, while drifting towards New Zealand's present position, they adapted to a range of ecological niches and evolved into at least six species. The absence of significant predators – for no mammals made the journey – meant that flight was an unnecessary metabolic expense, and most of the wrens lost the ability to fly. Unfortunately, flightlessness led to four species becoming extinct during the last millennium, the result of the introduction of a variety of animals, including rats, dogs and cats.[21]

Vicariance or dispersal?

Until recently, most biogeographers accepted the above scenario, and regarded Acanthisittidae as a paradigm of vicariance biogeography. But not Gerald Mayr.

I have never met Mayr, a renowned palaeontologist and Curator of Ornithology at the Senckenberg Research Institute, but I am indebted to him for his prompt and helpful responses to my many enquiries. Like Alan Feduccia before him, he challenges a Cretaceous occurrence of the modern representatives of extant bird families or orders, especially passerines. Mayr stresses that, while he accepts that the New Zealand wrens lie at the base of the passerine tree and originated somewhere in the southern hemisphere, he disputes the molecularly derived divergence dates and hence their Gondwanan origins.[22] His primary concern is the lack of supporting fossil evidence. The earliest passerine fossils have been found in Australia and date from the early Eocene, approximately 55 million years ago, while those in the northern hemisphere are significantly younger. Furthermore, the oldest fossils of Acanthisittidae, unearthed recently near Otago in New Zealand, date from the early Miocene, a mere 16–19 million years ago.[23] For Mayr and his supporters, this poses an insurmountable problem: the complete lack of fossil finds from the first 25 million years of passerine evolution. Of course, it is possible that if much of the early diversification had taken place in the southern continents, where comparatively few fossil sites are known, the missing evidence might simply not have been found. It stands to reason that the oldest fossils of a taxon set only a minimum age for any group, while the actual age is likely to be much older, by an amount that is impossible to determine. The old scientific adage that 'absence of evidence is not evidence of absence' remains germane. Indeed, Darwin emphasised this point in *The Origin of Species* to explain the lack of intermediate forms that his theory predicted.

Gerald Mayr, however, has additional concerns. A Cretaceous origin for passerines 'would imply an unprecedented evolutionary stasis for more than 80 million years in one of the most species-rich groups of endothermic vertebrates.' This is not a view, however, shared by the Swedish team led by Per Ericson. They argue that although the passerines appear to be a morphologically uniform group, this applies mainly to body parts unrelated to locomotion and feeding. In contrast, the limbs and beaks have shown substantial evolutionary plasticity.[24] Nevertheless, it is the lack of fossil evidence that underpins Mayr's anti-vicariance, pro-dispersal stance. He believes that the Acanthisittidae diverged after the break-up of Gondwana and that the wren's ancestors made landfall in New Zealand at a much later date, during the Cenozoic. In support, Alan de Queiroz, an American evolutionary biologist, has stated that most New Zealand biota 'missed the Zealandian boat as it left harbour and that colonisation of the landmass occurred only after its separation from Antarctica/Australia.'[25] In de Queiroz's view, the wrens might be Gondwanan in the sense of having come from some landmass that had once been part of the southern supercontinent, rather than survivors with an uninterrupted presence in New Zealand dating from the time of continental break-up.

Mayr also highlights, *en passant*, a potentially far more damaging criticism: the suggestion that New Zealand may have been completely submerged during the late Oligocene and early Miocene.[26] If this were true, as is the case for other Zealandian islands, including New Caledonia and the Chatham Islands,[27] then it would completely scupper the vicariance model for the wren's speciation. As I have indicated, some experts now believe that the majority of flora and fauna found on Zealandia-derived islands arrived by dispersal and replaced the ancient biota only after the islands re-emerged from the sea.[28] New Zealand's endemic land-walking bats (Mystacinidae), for example, originated from an Australian population that subsequently became extinct during the Miocene.[29] The key question, therefore, is whether New Zealand, like the Chatham Islands, became completely submerged during the Oligocene. Twenty years ago, a clue to this geological puzzle was found lurking in the most unlikely of places: the genetic code of the Stephens Island Wren.

In 1995, Alan Cooper, from the Smithsonian Institute, together with Roger Cooper, from the Institute of Geological and Nuclear Sciences in New Zealand, reported a lack of genetic diversity in the mitochondrial DNA from three species of New Zealand wren: two extant species and the recently extinct Bushwren.[30] This finding was surprising, since the degree of genetic stasis was incompatible with 80 million years of isolated evolution. Their favoured explanation was that a genetic bottleneck had occurred during the Oligocene, resulting from a drastic loss of habitat secondary to rising sea levels. As discussed, such changes took place because of continental crust thinning and the subsequent sinking of Zealandia as it rafted north-eastwards, approximately 25–30 million years ago. These genetic findings suggest that the marine environmental crisis did not completely submerge New Zealand, but probably reduced the landmass to a string of low-lying islands that reduced the wren population markedly. Cooler temperatures and the lack of significant geographical features would have further limited the range of available niches. The overall effect was to restrict the wrens' genetic pool, as many of the alleles or gene variants that were initially present in the population would have been lost.

Recently, Alan Cooper's team, now based at the Australian Centre for Ancient DNA at Adelaide University, have extended their observations by incorporating genetic data from two more extinct wrens: the Stephens Island Wren and the Stout-legged Wren.[31] Such museum specimens can only yield minute amounts of DNA, so the team had to use the kind of DNA sequencing technology that had previously been used to decipher the human genome. The results, presented at the Geogenes V meeting in Wellington during July 2014, were unexpected. The lineage that led to the Stephens Island Wren was the first to diverge, dating from well before the Oligocene marine ingressions. In other words, unlike New Caledonia, New Zealand could not have been entirely submerged.[32] The irony that such a longstanding geological puzzle could be addressed by studying the DNA from a museum skin would not have been lost, I am sure, on either Rothchild or Tristram.

Despite the highlighted caveats, the weight of available evidence still favours a link between the emergence of the Acanthisittidae, the earliest of all passerines, and the formation of Zealandia. Confirmation, however, will depend on novel scientific methods, coupled with future fossil finds. Already, an international scientific collaboration, the Avian Phylogenomics Consortium, has released data that might support Mayr's view.[33] By using whole-genome sequencing, a younger date for the divergence of passerines was obtained, a finding that came as a surprise even to some of the researchers.[34] However, it should be stressed that only a small number of passerine species were included in the study, and there remains the possibility that such an innovative approach might contain inherent and as yet poorly understood biases. The latest study, involving the detailed genetic analysis of nearly 200 species, suggests a divergence date of at least 50 million years ago.[35] Given the conflicting data, the results of the Avian Phylogenetic Consortium's next endeavour, the ambitious and expensive B10K project, in which genetic sequences for all the world's species will be generated over the next five years, is eagerly awaited.[36]

Emergence of anisodactyly

As highlighted in *The Parrot's Story*, parrots are the closest living relatives of the passerines and, as a result, both lineages must share a direct common ancestor. However, parrots have a zygodactyl foot: one characterised by a permanently reversed fourth toe and associated changes in the leg bone or tarsometatarsus (Figure 14.2). This fact raises an interesting question. Did the earliest passerines also possess a parrot-like foot? Surprisingly, palaeontologists think the answer might be yes. Recent fossil discoveries reveal that passerines had a sister clade with zygodactyl feet, the so-called Zygodactylidae that lived in North America and Europe until the middle Eocene.[37] The most parsimonious explanation for such a phylogenetic pattern is that the stem passerines were also parrot-footed, and that the fourth toe became reorientated some time before the New Zealand wrens diverged. It is now thought that zygodactyly evolved, not for improved perching as one might imagine, but for cavity-nesting and the need to cling to vertical tree trunks. Such an explanation would also account for the reversal of one of the forward-facing toes in other cavity-nesters, such as trogons and woodpeckers. It is likely, therefore, that the emergence of the anisodactyl foot in the stem passerines was the result of a transition from cavity-nesting to open-nesting. Furthermore, a relatively simple developmental mechanism could account for such an anatomical change. If several of the parrot's leg muscles had reduced in size and function, then the altered mechanical forces would have led to both the realignment of the fourth toe and the loss of the associated tarsometatarsal features.[38]

Let us leave the story of the New Zealand wrens and return to those of their ancestors that were left behind on the fragmenting southern continent. For it was these early birds that would eventually give rise to all the remaining passerines – suboscines and oscines – that today account for the majority of the world's birds.

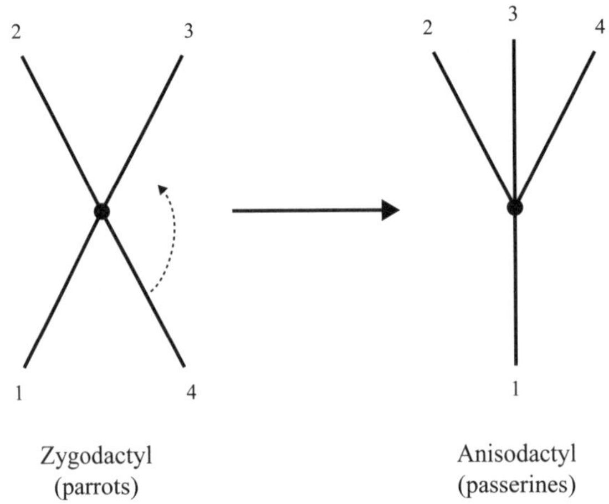

Figure 14.2 The passerine's anisodactyl foot evolved from a zygodactylous ancestor (two toes forward, two toes backward) and involved the reorientation of the fourth toe.

CHAPTER 15

The Manakin's Story

WHY SO MANY SUBOSCINES?

It is a sultry morning in mid-August. My wife and I are following our guide along a well-worn trail as it snakes around the corrugated slopes of Tobago's Main Ridge. High above us, the forest's skyline is engulfed in a swirl of cloud, dark with foreboding. Soon, the surrounding leaves rustle knowingly and, within minutes, the first cold raindrops have become one of the day's many torrential downpours. All around, spiny palms and bromeliad-adorned tree ferns sway to the water's tune, while muddy rivulets stream across the path, only to disappear into the void below. From where we stand, huddled beneath a mass of bamboo fronds, we watch as the leafcutter ants struggle with their booty and unknown creepy-crawlies seek refuge from what must have seemed like an impending tsunami. As is typical in the tropics, the clouds soon clear, the sun re-emerges, and the montane rainforest becomes alive again with avian chatter. A high-pitched, drawn-out call soon reveals a Blue-grey Tanager sitting at eye level; a subspecies (*nesophilus*) more brightly coloured than its cousins that we would soon be seeing on Trinidad. We set off once more, and it's not long before the guide stops, listens for a moment and then points excitedly to a nearby strangler fig. High up, almost hidden by leaves, is one of the key targets of our trip, a male Blue-backed Manakin. Our group peers intently at this tiny gem of a bird, with its pale blue back and scapulars, vivid red skullcap and orange-red legs, as it hops about in the dense foliage, feeding on fruit and insects. Moments later, an inconspicuous female, sporting a drab olive-green outfit, is spotted in a nearby tree. Then, as if by some prearranged signal, both take to the air and vanish for good higher up the valley.

The rationale for recounting this vignette is not that it was a memorable day's birding or that the manakin was my first sighting of a suboscine, both of which are true. Rather, it provides a snapshot of an ongoing confrontation between two major avian clades, a microcosm of a continental struggle. The Blue-grey Tanager, a foot-soldier from the newly arrived northern oscines, is competing for resources and territory with a Blue-backed Manakin, a defending suboscine from the south: oscine versus suboscine. But how did these representatives of two great avian suborders come to be facing one another on the island of Tobago?

It is a story that we will address in the remaining chapters of this book.

Ancestral suboscines

Amid the luxuriant foliage of Gondwana, the earliest passerines flourished – seeking mates, building nests, laying eggs and searching for fruit and insects to feed their young. Then, during the early Palaeogene, the ancestral population underwent a major divergence, one of the most important in the evolutionary history of birds. In eastern Gondwana, in an area that would become Australia,[1] a population began to evolve a more complex syrinx, a voice box characterised by an increased number of muscles that enabled a greater range of vocalisations. These basal songbirds or oscines may have had an evolutionary advantage, since they appear to have dominated and limited the suboscines to the western areas. Later, the ancestral suboscines underwent a pivotal divergence of their own.[2] One lineage, or infraorder, entered South America to become the New World suboscines, while the other took an alternative route north, reaching Asia and then Africa (see *The Sapayoa's Story*).

The New World suboscines rapidly colonised South America, by way of a broad continental shelf that linked the continent's southern tip to Antarctica. These early passerines encountered a vast fertile continent, an eighth of the world's landmass, inhabited by ancestral toucans, motmots, jacamars and trogons. Despite competition from these well-established families, it would be the momentous climatic and geological upheavals over the next 50 million years that would shape their evolution. So let us look at these processes and see how they led to such a diversity of species.

Initially, South America's climate was distinctly warmer and more equable than today's, with warm-temperate zones extending from the tropical equatorial forests towards the poles, where cooler conditions predominated.[3] The Andes, which converted Patagonia into a desert by shielding it from the moist Pacific air, would not rise for another 30 million years. At the end of the Palaeocene, the Earth entered a 'greenhouse' phase, with temperatures reaching the highest levels experienced at any time during the past 65 million years. Polar seas reached 23 °C; the Earth became free of snow and ice; crocodiles and hippopotamuses wallowed in Arctic waters, and early primate-like mammals and palm trees flourished in the American West.

This episode, which began 55 million years ago, is known as the Palaeocene–Eocene Thermal Maximum (PETM) and is likely to have resulted from a sudden release of methane, a potent greenhouse gas.[4] The trigger was a rise in the Earth's temperature that resulted from high atmospheric carbon dioxide levels, caused by volcanic eruptions from tectonic rifting. Methane, which is produced continually by decomposing microbes, is trapped in ice-like structures, called clathrates, under the ocean floor. The addition of large amounts of methane gas to the atmosphere would have caused further increases in the Earth's temperature and further clathrate destabilisation. In effect, a positive feedback mechanism occurred that led to a runaway increase in temperature. Overall, some 2,500 gigatonnes of carbon were released into the atmosphere and ocean, resulting in a 'greenhouse' Earth within a very short time, estimated at between 2,000 and 3,000 years.[5] Climatologists are interested in these gas levels, as they are comparable to those predicted from gross anthropogenic emissions by the end of the twenty-first century.[6]

The ecological consequences of PETM were complex. Deep-water organisms, known as foraminifera, were reduced by 50 per cent, possibly because of the increasing oceanic acidity. Their reduction would have contributed to the global warming, since foraminifera capture atmospheric carbon, in the form of carbon dioxide, and lock it up in their shells. In contrast, many new species of tree and plant evolved, despite a high extinction rate, while insects and mammals thrived – although the latter became much smaller, almost dwarf-like.

The early New World suboscines adapted to the tropical environment and spread throughout the extensive forests of South America, filling many vacant niches. At the height of the greenhouse period, suboscines divided into two independent lineages or parvorders.[2] One, the Furnariida, produced the woodcreepers, ovenbirds, tapaculos and antbirds, while the other, the Tyrannida, gave rise to the manakins, cotingas and tyrant flycatchers. Before most of these crown splits had occurred, however, the Earth was subjected to yet another dramatic climatic upheaval.

By the middle Eocene, 40 million years ago, South America had separated from Antarctica and a seaway, the Drake Passage, had opened up between the two continents.[7] As a result, the flora and fauna of South America was cut adrift and evolved in isolation until a land bridge was established with North America around 3–4 million years ago. To the east, Australia remained attached to Antarctica for a little longer, delaying the establishment of the Antarctic Circumpolar Current. As a result, the warm waters of the tropical Pacific and Atlantic Oceans could still reach Antarctica and keep the continent warmer than it would otherwise have been. Once

Australia broke free, and the Tasman Passage formed, the circum-Antarctic gyre developed – with dramatic consequences for Earth's climate and life's evolution. Equatorial heat transfer to the Antarctic fell dramatically, and the cooling of the surrounding seas led to the formation of the south polar ice cap and, eventually, the refrigeration of the whole planet.[8] At the same time, the rise of the Himalayas and other mountain chains contributed to the cooling, mainly due to the increased physical and chemical weathering of their rocks. Such processes expose reactive minerals that combine with carbon dioxide, so reducing the atmospheric levels of the greenhouse gas.

The marked cooling of the planet, which reached a nadir around 30 million years ago, triggered the retraction of South America's extensive tropical forests towards the equator and their replacement by savanna and desert habitats.[9] The falling temperatures led to a high biotic turnover. Megatherms – plant species that required stable high temperatures and abundant moisture – disappeared, while crocodiles and lizards retreated to the equatorial tropics. Strange hoofed herbivores evolved, with many developing extreme parallelisms to unrelated forms found on other continents. For example, notoungulates, an order of mammals that became extinct 5,000 years ago, evolved to fill the niches occupied elsewhere by rabbits and hares (among many others), while Litopterna took the place of horses and camels on the grazing lands.

The suboscines were not immune to these climatic changes. The Furnariida, however, were better able to cope, as they were primarily terrestrial and insectivorous, and indeed had already started to diversify by the middle Eocene. The earliest birds may have resembled the crescentchests (Melanopareiidae) which today thrive in the arid scrub and tropical dry forest environments of central South America. In contrast, the Tyrannida, having adapted to a forest lifestyle and a diet of mainly fruit, suffered a severe culling, with many species becoming extinct. Consequently, their diversification was delayed by approximately 15 million years,[2] although what forms the lost passerines took, we can only speculate. Robert Ricklefs, a biologist from the University of Missouri, questions whether they might have 'comprised members of groups we are familiar with today, or, like the notoungulates and many of the marsupials, were they highly unusual local products of South American splendid isolation?'[10] In the absence of future fossil finds, we may never know the answer.

A significant and permanent cooling occurred around 14 million years ago, a fact linked to both an increased production of cold Antarctic waters and a major extension of the East Antarctic ice sheet. Curiously, for reasons that are not entirely clear, the temperatures in the low latitudes remained stable and warm.[11] As a result, the north–south temperature gradient increased, and the boundaries between climatic zones strengthened. Such changes led to the Earth's flora and fauna becoming subdivided into reasonably distinct provinces defined by temperature, patterns of seasonality, and precipitation.[12] The Middle Miocene Climatic Optimum, as this period is now called, saw the New World suboscines diversify quickly, especially the spinetails and tyrant flycatchers. Most of these species-rich radiations moved into riverine or riparian habitats throughout the tropical lowlands, as well as the more open habitats to the south, before spreading northwards into the tropical Andes.[2] Riparian habitats also enabled the tyrant flycatchers to expand up into the mountains along the streams and to adapt to the newly emerging montane forest habitats.[13] Eventually, these radiations would account for approximately half of all the New World suboscines. In contrast, the diversification of antbirds, woodcreepers, manakins and some cotingas was restricted to the humid tropical rainforest habitats, predominantly in the forest understorey.

Over millions of years, the suboscines colonised all areas of South America. Then, during the late Pliocene, following the formation of the Panamanian isthmus, they invaded Central America, Mexico, Middle America and the Caribbean. Furthermore, several species of tyrant flycatcher began to migrate further north in summer, with the Alder Flycatcher now regularly reaching central Alaska. The payoff for these energy-sapping migrations is that the northern latitudes offer a greater number of territories, more food and fewer nest-robbing predators than their winter quarters in the tropics.

Suboscine diversity

A birder's first visit to South America can be a daunting experience, given the continent's unparalleled avian diversity. Peru, for example, has the highest concentration of bird life on Earth, with over 1,800 species. Many bird tours see over 500 species in two weeks, while one company has recorded a staggering 1,000 species in less than a month in Colombia. Such diversity is reflected in the New World suboscines, a radiation that has given rise to 1,289 extant species, comprising predominantly tyrant flycatchers (435 species), ovenbirds and woodcreepers (314 species) and antbirds (235 species).[14] To put this in perspective, the number of species in each one of these families greatly exceeds the total number of species that regularly breed in the United Kingdom. But how is it that so many suboscine species have evolved in South America, and how can their local distributions be explained? It is certainly not just a consequence of the continent's large area. Sub-Saharan Africa has only half the number of species, despite being of equal size. The answer to these questions, which have intrigued evolutionary biologists for over a century, turns out to be much more complicated than one might first imagine. As we will highlight, suboscine biodiversity results from the interplay of many vicariant events, including riverine barriers, marine transgressions, vegetative shifts and Andean uplift. Also, their adaptive radiations were facilitated by the length of time available for speciation, their anatomical and behavioural plasticity, and the emergence of many novel ecological niches.

Alfred Russel Wallace, an intrepid adventurer, collector and co-author of the theory of evolution, was the first to consider the role of geographical barriers, or vicariance, to explain Neotropical biodiversity. During his travels in South America, Wallace was struck by how often different species inhabited the opposite sides of rivers. He wrote:

> During my residence in the Amazon district I took every opportunity of determining the limits of species, and I soon found that the Amazon, the Rio Negro, and the Madeira formed the limits beyond which certain species never passed.[15]

Importantly, he added: 'I have only referred to the monkeys, but the same phenomena occur both with birds and insects, as I have observed in many instances.' It was an impressive piece of detective work, as his observations were rather patchy, and he had to rely on the reports of several other naturalists. A year later, in 1853, Wallace ventured a theory to explain his conclusions in a book entitled *A Narrative of Travels on the Amazon and Rio Negro*.[16] He imagined that continuous populations of species inhabiting the low-lying plains of South America had become suddenly partitioned by the development of the area's three main rivers: the Amazon, the Rio Negro and the Rio Madeira. Eventually, these rivers would have formed natural dividing lines, beyond which certain species never passed. But he also observed that near their narrower sources, the rivers ceased to be boundaries and the ranges of species overlapped.[17]

Wallace's astute deductions laid the foundations for what is now termed the 'river-barrier' hypothesis of speciation. This idea supposes that previously widespread ancestral species became separated when the Amazon and its major tributaries formed during the late Miocene. Originally, the western area of Amazonia was a million-square-kilometre marshland – the Pebas mega-wetland – consisting of shallow lakes and swamps that drained northwards towards the Caribbean (Figure 15.1). As South America drifted away from Africa, forming the Atlantic Ocean in the process, it moved across

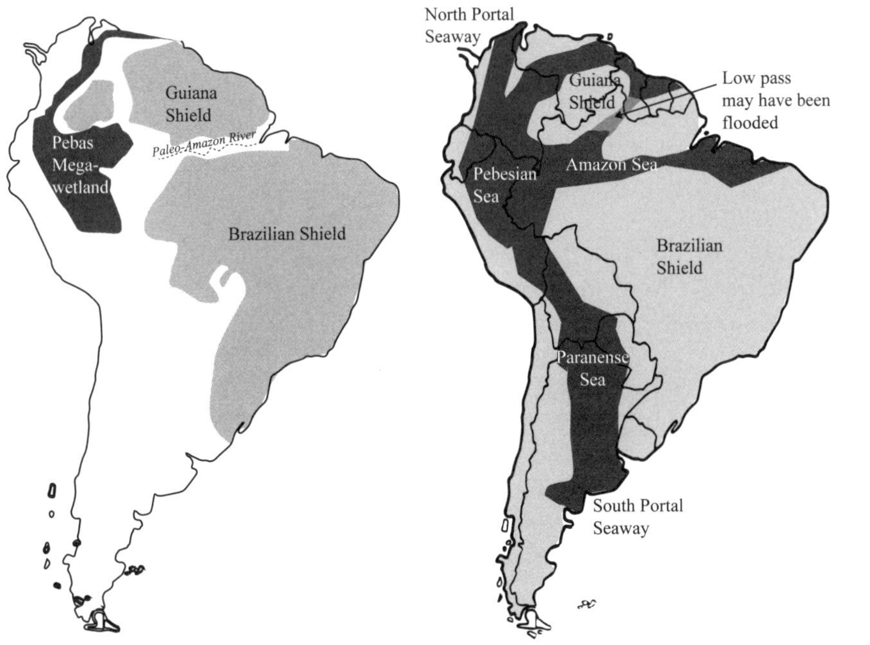

Figure 15.1 (Left) Thirteen million years ago, the Pebas mega-wetland covered over 1 million square kilometres of what is now the Amazonian basin and drained northwards into the Caribbean. (Right) Marine transgressions (dark grey) resulted in allopatric speciation of many taxa by vicariance.

a subduction zone, where slabs of the Earth's crust sink into the softened mantle. These tectonic processes resulted in an intense uplift along the continent's Pacific edge that created the Andes, the world's longest mountain range, extending over six countries. But the uplift affected more than just the plate boundary, and the whole of the western half of South America became elevated. At the same time, the continent's northeastern areas subsided by as much as 400 metres. The net effect was that South America tilted like a giant seesaw, and the resultant 'slide' redirected the waters of western Amazonian over 6,400 kilometres into the Atlantic Ocean.[18] The newly formed river systems then fragmented the early bird populations, so that many became isolated and subjected to different selection pressures, genetic drift and mutations which eventually led to the evolution of new species.

The late Ernst Mayr, a doyen of neo-Darwinism, was the first to extend Wallace's deliberations, concluding that the Amazonian rivers could have 'initiated the first steps of speciation.'[19] Biologists had long known that the distribution of many suboscine taxa appears to coincide with the course of the Amazon and its tributaries. White-breasted Antbirds, for example, never cross the Tapajós. In northeast Peru, the Golden-headed Manakin is restricted to the forest understorey to the north of the river while the Red-headed

Manakin replaces it to the south. Such riverine delineations can also apply to the ranges of subspecies. The red-crowned race of the Blue-backed Manakin abuts the Amazon's north bank, while the yellow-crowned subspecies resides only on the south side. These and other field observations imply that the ancestors of many antbirds and manakins must have inhabited the area before the river was formed.[20] Recently, Angelo Capparella documented marked genetic differences between morphologically identical populations of Blue-crowned Manakin that inhabit the opposite banks of the Amazon. This finding confirmed for the first time the river's potential effectiveness as an impediment to gene flow.[21] Interestingly, this effect appears unique to the Amazon and its major tributaries; the Amazon's headwaters and the continent's second-largest river, the Paraquay–Paraná system, fail to provide adequate barriers.

In 1985, Joel Cracraft, a biologist working at the Field Museum of Natural History in Chicago, proposed the existence of 33 areas of avian endemism in South America.[22] This seminal work, based on the distribution charts of species, provided indirect evidence for the river-barrier hypothesis. In essence, Cracraft superimposed the range maps for most of the continent's endemics and theorised that the areas of maximal overlap were major sites

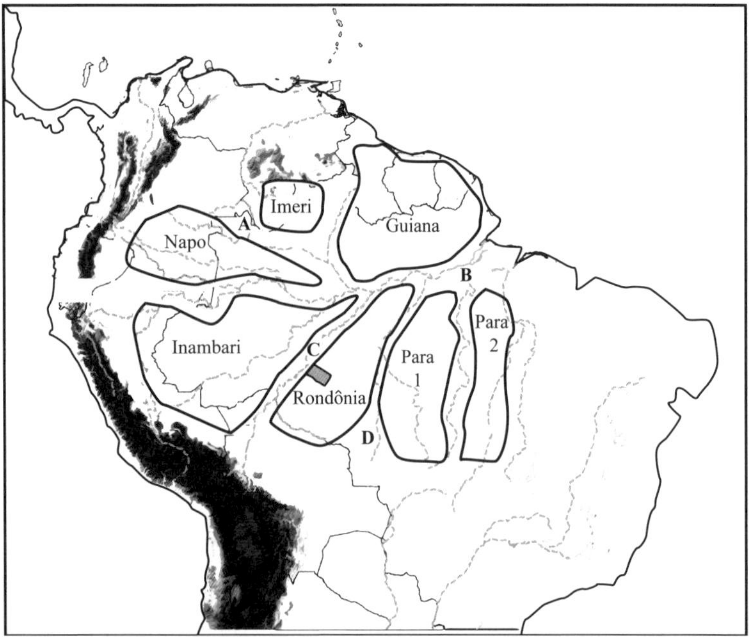

Figure 15.2 Location of the Amazonian areas of avian endemism, based on Cracraft (1985).[22] Major rivers: A, Negro; B, Amazon; C, Madeira; D, Tapajós. The 'mini-interfluvial' area (grey block) inhabited by the Rondonia Warbling Antbird lies between the Aripuaña and Jiparaná rivers.

of speciation. As shown in Figure 15.2, all the endemic centres he identified within Amazonia are bordered by major rivers; for example, the Rondônia area lies between the Madeira and Tapajós, while the Imeri is between the Rio Negro and the Rio Branco.

Intriguingly, smaller areas of endemism are now thought to exist within some of these larger zones. Mario Cohn-Haft, a scientist working at the National Institute for Amazonian Research (INPA), reported that several species in the Rondônia area have ranges delineated by the Madeira's smaller tributaries. Indeed, the pattern was so apparent that his research team referred to the areas as 'mini-interfluves'.[23] A typical example is that of the warbling antbird complex, a group that has been split recently into six distinct populations. One of these, the Rondonia Warbling Antbird, is confined to a small area between the Aripuaña and Jiparaná rivers (see Figure 15.2).[24] Mini-interfluvial or fine-scale endemism has also been described for two members of the Tyrannidae that inhabit the Rondôna zone: the Sucunduri Flycatcher and Chico's Tyrannulet.[25]

As in the manakins, genetic differences have also been documented between morphologically indistinguishable populations of tody-tyrants, in particular, the Snethlage's Tody-Tyrants that reside on the opposite banks of the Aripuaña and Jiparaná rivers.[26] A more detailed analysis of three typical understorey suboscines, the Wedge-billed Woodcreeper, the Spot-backed Antbird and the Chestnut-tailed Antbird, produced similar results.[27] Collectively, the above findings raise two important issues. First, cryptic endemism appears to be much more widespread than previously appreciated. Indeed, half the 16 lineages of Wedge-billed Woodcreeper described by Alexandre Fernandes and his team are morphologically indistinguishable. Four of these inhabit the Madeira–Tapajós interfluvium, where they replace each other on opposite banks of the Madeira's tributaries.[28] Fine-scale endemism also has significant conservation implications, because many of the critical mini-interfluvial areas have been earmarked for agricultural development. Sadly, it may be too late for the Rondonia Warbling Antbird, as rapid deforestation has already destroyed most of its habitat. Second, the highlighted studies seem to imply that the smaller Aripuaña and Jiparaná rivers have been more effective barriers to gene flow than the main river, the Madeira. This paradox is resolved if the direction and flow rates of the tributaries changed markedly in the past – events that are now thought to have occurred.

Many rivers provide only 'leaky' barriers, and most species will eventually find a way across, either in their narrower headwaters or by using islands as stepping stones. Nevertheless, the principal rivers of Amazonia seem to represent formidable barriers for many understorey inhabitants,[29] especially manakins, antbirds and woodcreepers. Why this should be so remains unclear. It is possible that such species have adapted over millennia to low levels of light and, as a result, avoid crossing large open spaces. In the case of antbirds, there is an additional factor, one suggested by foraging theory, a model that helps predict how species behave when searching for food. For the 'professional' antbird, meals are a smörgåsbord of grasshoppers, cockroaches, praying mantises, scorpions and centipedes, all flushed by army ants on the move. To locate such an event and to compete with the hordes of other ant-followers takes both time and effort. The more successful an individual bird is, in terms of the energy expended, the greater its fitness and the greater the likelihood of it passing on its genes to the next generation. For an obligate antbird, it is more cost-effective to follow insects on one side of the river than to cross over and search for new swarms on the opposite bank. Swarms are uncommon, and antbirds will not know the whereabouts of ants' bivouacs in unfamiliar territory, and the chances of encountering a swarm by accident are slight.[30]

The influence of rivers on suboscine diversity, however, extends beyond vicariant speciation. In Amazonia, rivers have created a range of unique habitats adjacent to *terra firma*, a term biogeographers use for forested areas that never flood. These include sandbar scrubs, river-edge forests and *várzea*, or seasonally flooded forests – each of which supports its specialist taxa.[31] For example, the Scaled Spinetail is endemic to *várzea* bordering the east Amazon, while the poorly known Klages's Antwren is restricted to a few isolated areas of lowland *várzea*. In contrast, the widespread River Tyrannulet, a small grey flycatcher, is confined to the sandbar scrubs of the Orinoco and Amazon watersheds. Intriguingly, such riverine endemism appears unique to Amazonia.

Rivers are not the only means by which water can isolate or segregate avian populations. Manuel Nores, an Argentinian biogeographer, has noted that the Amazon basin was flooded 10–15 million years ago when sea levels rose by up to 100 metres. By studying relief maps of South America, he concluded that two broad marine incursions would have formed via the Amazon and Orinoco rivers, as well as a narrower one from the smaller Branco. Most of southern Amazonia would have been flooded with brackish water, but two major areas were spared: one located in Venezuela and another straddling the borders of Guyana and northern Brazil (see Figure 15.1). Interestingly, these two 'islands' correspond roughly to the Imeri and Guiana areas of endemism identified previously by Cracraft (see Figure 15.2). According to the 'marine transgression' hypothesis, as it is now called, many species with continuous distributions became separated and experienced an interruption of genetic exchange, an event that led to allopatric speciation by vicariance. The striking Crimson Fruitcrow, a member of the cotinga family, and the Spot-backed Antwren, a diminutive canopy specialist, are thought to have evolved in this way.[32]

Pleistocene refugia

A popular view during the mid-twentieth century was that the rich Amazonian biodiversity was merely the consequence of ecological stability. Species never became extinct, but merely accumulated in a static environment. However, this 'museum' model was challenged in 1969 by a remarkable amateur ornithologist, Jürgen Haffer. Born in Berlin in 1932, Haffer became fascinated by birds as a youngster, having found a dead bird that had been ringed. After taking it to the local Zoological Institute, he was introduced to Erwin Stresemann, one of the pioneers of modern ornithology. Haffer was inspired by this chance meeting and later studied at the Institute, although he soon realised that zoology wouldn't pay the bills and he switched to geology and palaeontology. Once qualified, he worked for Mobil Oil as a field geologist and spent the next eight years surveying the rainforests of remote northern Colombia. It was while out in the field, often involving arduous expeditions by mule and canoe, that Haffer became intrigued by the variety and distribution of Neotropical birds. He collected an impressive array of skins and drew up detailed range maps that highlighted where distinct and unique sets of birds were found. He also puzzled over the existence of hybrids that appeared to be limited to narrow contact zones often located far away from any contemporary barrier. Seeking an explanation, he realised that Pleistocene climatic fluctuations could be the missing link, an idea that culminated in his landmark paper published in the journal *Science*, and entitled 'Speciation in Amazonian forest birds'.[33]

Haffer's novel hypothesis envisaged the Amazonian rainforest to have cycled through several periods of expansion and contraction during the Pleistocene and post-Pleistocene epochs. Glacially driven, colder, drier intervals allowed the forests to contract and fragment into numerous smaller, isolated units or refugia. Birds stranded within these forest tracts were separated by impassable areas of open grassland and underwent allopatric speciation. As the temperature and rainfall increased again during the interglacials, the forest patchwork merged back to its original state, but now with a high level of species diversity and endemism. Since the alternating glacial advances and retreats occurred many times, with each cycle lasting up to 80,000 years, there would have been plenty of scope for new species to evolve. In summary, Haffer was proposing a species pump driven by glacial fluctuations.

The 'Pleistocene refugia' hypothesis, a synthesis of palaeoclimatology, biogeography and the concept of allopatric speciation by vicariance, gained widespread acceptance. The model's elegance encouraged many other workers to propose similar explanations for the speciation of frogs, butterflies, lizards and even plants. It was also used to explain population patterns in other parts of the world, including Africa, Eurasia and America. But the idea had its limitations: the precise locations of the putative refugia were unknown, and there remained the difficulty of discriminating between alternative hypotheses. Nevertheless, the concept of Pleistocene refugia was readily adopted by the scientific community and went untested for several decades. Untested, that is, until the ecologist Paul Colinvaux took an interest.

Colinvaux, a Professor Emeritus at Ohio State University, predicted that if Haffer's theory were correct, then it should be relatively easy to prove, because grass pollen extracted from lake cores laid down during the glacial periods should be increased. This seemingly straightforward project took Colinvaux and his team nearly 10 years to complete, because informative historical lakes, ones that had existed in the lowlands with sediments undisturbed by rivers, were extremely hard to find. Eventually, a number of likely candidates were identified, and the results were a surprise: ice-age deposits failed to show an increase in grass pollen.[34] In other words, Colinvaux's findings did not support the fragmentation of Amazonia, the keystone of Haffer's widely accepted theory.

In the past decade, scientists from the Deep Ocean Drilling Project have come to the same conclusion after studying core samples of ocean sediment obtained from the mouth of the Amazon.[35] This river system acts as a vast pollen trap and carries its cargo to the Atlantic coast, where it is deposited into the ocean. Analysis of pollen counts, or their surrogate markers, shows no alteration throughout the Pleistocene. It seems that the Amazonian rainforest, far from being ephemeral, is quite resilient, and has adapted well to past climate changes. Colinvaux's uncompromising conclusion, to paraphrase Thomas Huxley, was that the refugia hypothesis was just another 'beautiful theory … destroyed by an ugly fact'. However, his response may have been an over-reaction. Pleistocene climate changes are widely accepted to have increased savanna at the expense of rainforest in tropical Africa. Furthermore, the results from a recent genetic study of South American leafcutter ants favours a combination of refugia and marine incursions to explain their biodiversity.[36]

Scientists have also considered other possible interpretations of Haffer's hypothesis. Could the colder glacial periods, with their reduced carbon dioxide concentrations, have produced significant changes in the composition of forests, leading to the isolation of species even in the absence of refugia? Or could the Pleistocene climatic fluctuations have resulted in effective refugia, not by the formation of savanna during the colder times, but by producing wider rivers during the warmer periods. In other words, could the increased interglacial rainfall have altered not just the size and discharge of rivers, but also their courses, resulting in enhanced barriers to genetic transfer during the wetter periods?

While Haffer's hypothesis may now be thought suspect by many, his legacy has been an acceptance that past climatic fluctuations significantly influenced Amazonian biodiversity, and that this helped shape present avian distributions.

The impact of the Andean orogeny on avian biodiversity also needs consideration. The familiar snow-daubed sentinels, separated by high plateaus, that form a scoliotic spine down the continent's western rim were not always a feature of South American geography. Rather, they are the ongoing progeny of a relentless and tumultuous slide of one tectonic plate beneath another. Each peak, from the lofty Aconcagua to the plethora of lesser mountains, is the result of the Nazca oceanic plate sinking beneath the lighter South American plate. Despite the initial uplift beginning before the Gondwanan break-up, by the mid-Miocene they were only 1,200–1,800 metres high, still covered with tropical vegetation and populated by lowland birds. It is only during the last 10 million years that the central Andes have risen to heights that are likely to have divided ancestral populations and provided new ecological niches. In support of this view, d'Horta and colleagues have shown that between 5 and 6 million years ago a species of leaftosser became separated on either side of the Andean uplift. As a result, the Black-tailed Leaftosser, the darkest and dullest of the genus, is found to the west while the Scaly-throated Leaftosser is restricted to the east.[37] However, most species-level diversity appears to have occurred after the major Andean orogeny in the Neogene, which suggests that dispersal into the mountains from the lowlands, rather than vicariance, may have been the predominant driver for speciation.[38]

Focusing on single vicariant mechanisms may be overly simplistic. A team from Illinois State University, led by Angelo Capparella, undertook a detailed study of the Blue-crowned Manakin and found its past evolutionary history to be highly complex. Indeed, the geographical distributions of its many subspecies were best explained by the combined effect of three events: Andean uplift, river barriers and climate-induced shifts in vegetation.[39]

While these and other hypotheses fail to explain all of South America's suboscine speciation events, they do help clarify why the immense biodiversity and high endemism in Amazonia are not randomly distributed. Lowland Amazonia is a mosaic of large endemic areas, each of which has its own uniform avifauna, but which differ from other apparently similar interfluvial zones. World birders seeking big lists will be well aware of this, since to see most of Amazonia's birds it is necessary to visit many different areas of seemingly comparable habitat.

Evolutionary shifts in behaviour can serve as an ecological release and facilitate diversification into new habitats. For example, speciation in the ovenbird–woodcreeper assemblage, in particular the spinetails and their allies (genus *Synallaxini*), was aided by a switch from building 'closed' cavity nests to the construction of vegetative nests.[40] It is likely that closed nests were the ancestral, or plesiomorphic, condition within the whole parvorder. Cavities, however, cannot be readily concealed, and numerous predators exploit their vulnerability. Also, natural holes or ground areas suitable for excavating tunnels are often limited in open landscapes, and the ability to build a vegetative nest would have provided a competitive advantage in such environments. Despite shifts in nest construction being rare events, it appears to have happened on at least three occasions during the early evolution of the Furnariida, involving the horneros, spinetails and foliage-gleaners. Indeed, the spinetails, which are adapted to dry habitats and build exposed vegetative nests, show the highest diversification rate of all the major clades.[41] The move to building vegetative nests is likely to have been a gradual process, as many basal furnariid species still adopt a combined approach. For example, the Sharp-tailed Streamcreeper and the bartails (genus *Premnoplex*) build domed mossy nests within underground cavities, while the cup-shaped nests of rayaditos (genus *Aphrastura*) partially cover the walls of their cavities. Although such an energy-demanding 'double approach' may have evolved to reduce underground nest humidity, it undoubtedly facilitated the subsequent diversification into more open habitats.

Ecological factors

The importance of ecological factors as drivers of suboscine biodiversity should not be underestimated, especially the role of epiphytes, palms, bamboos, vines, dead leaves and ants. In contrast to the forests of temperate climates, these Neotropical features offer a range of 'all-year-round' food resources. As a result, many suboscines evolved specialist foraging strategies unique to the New World that, in turn, led to further speciation.

Up to half of all plants in the Neotropics are epiphytes, species that include mosses, ferns, liverworts, orchids and bromeliads. Epiphytes grow non-parasitically on other plants and rely on specialised aerial root systems to absorb water and nutrients directly from the air. Most are found carpeting tree branches or rooting in pockets of humus and rotting leaves, and provide rainforests with an extra dimension of biodiversity. Canopy soil and detritus, collectively known as crown humus, and non-vascular epiphytes (bryophytes and lichens) are an important food source for birds as they support a diverse invertebrate community.[42] Amazingly, epiphytes account for 40 per cent of the rainforest's entire biomass, so it is not surprising that many birds have evolved to exploit their microhabitats. Indeed, a study conducted in Costa Rica documented that nearly 200 species regularly make use of their offerings for nesting material, water and food (including fruit, nectar and invertebrates). Unlike their hosts, epiphytes can provide resources throughout the year, a fact exploited by several insectivorous birds that have become epiphyte specialists. The Spot-crowned Woodcreeper, for example, favours foliose lichens, while the Ruddy Treerunner and the Buffy Tuftedcheek specialise in bryophytes and arboreal bromeliads respectively.[43]

Another unique Amazonian niche is provided by the swamp palm, notably the elegant Moriche Palm (*Maurita flexuosa*). This tall species favours wet areas, either in savannas or deep within forests, where it can reach high densities, known locally as *aguajale*. Although patchily distributed, such groves provide a home for some species, including two suboscines: the Point-tailed Palmcreeper, a member of the ovenbird family (Furnariidae), and the Sulphury Flycatcher, a tyrant flycatcher (Tyrannidae). The Point-tailed Palmcreeper, as its name suggests, has evolved to become the ultimate palm specialist, spending its whole life among the tree's fronds, where it can remain frustratingly inconspicuous. Characteristically, it rummages about in the basal pleats of the palm's fan-shaped leaves, where it is often found hanging upside down feeding acrobatically on insects. The rather noisy and excitable Sulphury Flycatcher, while not an obligate specialist, rarely strays far from the palms, where it pursues its sallying lifestyle.

Bamboo contributes to the region's biodiversity by supporting a taxonomically and ecologically diverse suite of birds. Indeed, the American ornithologist Ted Parker listed over 100 species of Neotropical birds that are associated with bamboo microhabitats.[44] The synchronous production of seeds by many bamboos results in an abundant and nutritious food source for granivores, while their fast growth, hollow stems and densely tangled habitat support a wealth of arthropods that provide subsistence for insectivorous birds. The latter forage among both live and dead bamboo shoots, where they prise open the internodes or probe in holes, as well as capturing insects in flight between the stands.[45] Although bamboo clumps can be many kilometres apart, a seminal study recorded 19 bamboo-specialists at just one site in western Amazonia, including foliage-gleaners, antbirds and pygmy tyrants.[46] The contribution of bamboo to avian diversity may be even greater than previously appreciated. According to Kristina Cockle and Juan Areta, 'in the past decade, new species of bamboo-specialist birds have been described, known species have been identified as bamboo-specialists, and bamboo species have been studied in greater detail, revealing relationships with specific species of bamboo.'[45]

Canopy and mid-storey liana tangles are a resource that takes on considerable importance in the Neotropics. Indeed, ecologists regard the presence of large vines as an indicator of pristine, undisturbed rainforest. Lianas, or climbing vines, are rooted in the ground and use trees to climb to the canopy to gain access to well-lit areas of forest. They provide year-round cover and foraging habitats for a variety of vine-gleaning and vine-inhabiting specialists including foliage-gleaners, treerunners and some antshrikes.

Dead leaves are also a rich food source and are typically found trapped among the vine tangles or hanging *in situ* from branches. They harbour more insects than green leaves but are much less abundant and more patchily distributed. Extreme dead-leaf specialists, therefore, spend more time and energy searching and handling prey, as extracting insects from curled dead leaves requires probing and acrobatic skills not necessary when gleaning green leaves.[47] Most of the dead-leaf specialists are suboscines: foliage-gleaners and antwrens, which are restricted to the Neotropics where leaves are replaced at a low but relatively constant rate throughout the year. A typical example is the Checker-throated Antwren,[48] a species which spends all its time foraging dead leaves caught in dense vine tangles for arthropods, many of which are unique to aerial leaf litter.[49] Interestingly, phylogenetic studies indicate that all such specialist antwrens are related, and that they evolved separately from other antwrens after diverging around 9 million years ago. It seems that dead-leaf foraging is a primitive trait within the antwren group and appeared well before the radiation of modern species.[50]

Following army ants is a foraging strategy unique to the tropical regions, with no ecological counterpart in temperate zones. Contrary to popular opinion, antbirds do not eat ants. Instead, the ants increase the birds' foraging efficiency by acting as 'beaters' to drive out hidden prey from beneath the leaf litter. Such activity is parasitic since the birds significantly reduce the ants' success rate in capturing prey.[51] All antbirds favour a single species of ant: the diurnal, swarm-raiding *Eciton burchelli*, found from southern Mexico to southern Peru and Brazil. A second species, *Labidus praedator*, is also

Plate 1. The family Picathartidae consists of two very unusual species: (A) the Yellow-headed Picathartes from west Africa, and (B) the Grey-necked Picathartes, restricted to Cameroon and Gabon. (A, Willie de Vries; B, Markus Lilje)

Plate 2. Giant Haast's Eagle attacking moas. The ancestral moa flew from Gondwana to New Zealand before undergoing an adaptive radiation that produced at least nine flightless species. (John Megahan)

Plate 3. The Grey Tinamou is a flying palaeognath that inhabits Amazonia. Tinamous lie deeply within the flightless ratite 'tree of life', a finding that disproved the continental drift theory for palaeognath biogeography. (Lars Petersson)

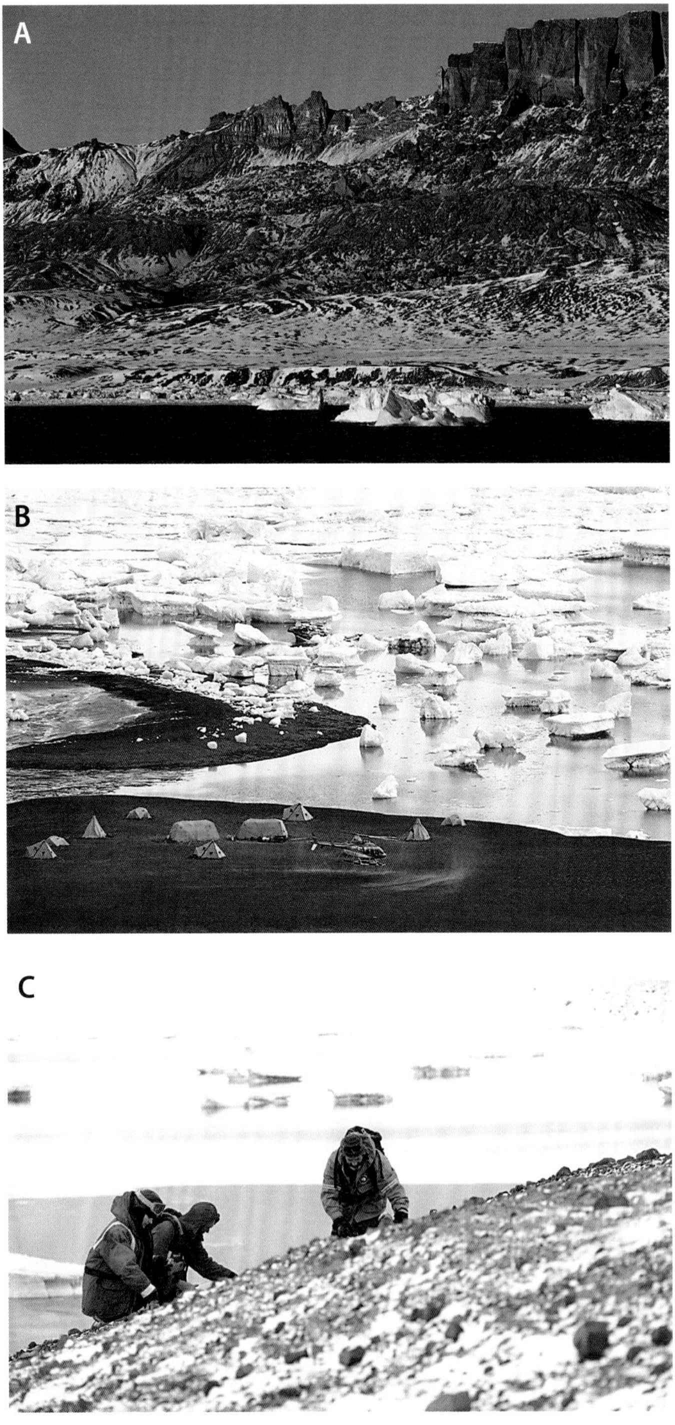

Plate 4. Fossil hunting on Vega Island, Antarctica, during the austral summer of 2016: (A) base camp on the shoreline of Vega Island; (B) the expedition's helicopter was a new and welcome addition to the scientists' toolkit; (C) palaeontologists scouring the frozen ground for Cretaceous fossils. (Jin Meng)

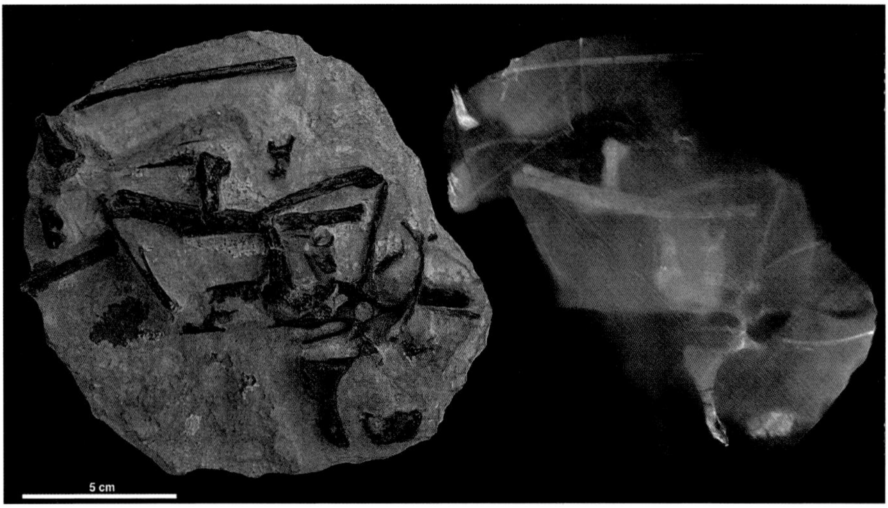

Plate 5. *Vegavis iaai* fossil (left) and computerised tomographic (CT) scan of same rock concretion (right). This Magpie Goose-like Cretaceous bird was discovered on Vega Island, western Antarctica, in 1992. At least five modern bird lineages diverged before the K–Pg boundary, based on the inferred placement of *Vegavis*: palaeognaths, chickens, screamers, the Magpie Goose and ducks. (Julia Clarke)

Plate 6. *Vegavis* flying above a mid-sized dinosaur amid a *Nothofagus* forest on the shoreline of Vega Island, Antarctica. The discovery of its fossilised syrinx suggests that the avian voice box originated after modern birds diverged from dinosaurs. (Nicole Fuller)

Plate 7. The flightless Chubut Steamer Duck of Patagonia evolved 15,000 years ago, as the result of glacial melting and rising sea levels. (John Reilly)

Plate 8. (A) The Coscoroba Swan and (B) the Cape Barren Goose diverged from the rest of the Anserinae around 23.5 million years ago in the southern hemisphere, and well before the swan–goose split in the northern hemisphere. (John Reilly)

Plate 9. Lucy Hawkes, a physiological ecologist at the University of Exeter, fitted several Bar-headed Geese in Mongolia with devices to monitor their altitude, wingbeat frequency and heart rates. The birds were caught during the brief period of wing moult when they couldn't fly. (Lucy Hawkes)

Plate 10. The 42.5-centimetre-long everted phallus of the Lake Duck is not only longer than the duck itself but also longer than the penis of any other species of bird. (Kevin McCracken)

Plate 11. The prehistoric-looking Hoatzin reached South America by crossing the Atlantic Ocean on floating vegetation. (Lars Petersson)

Plate 12. Reconstruction of the 61.6 million-year-old penguin *Waimanu manningeri* that lived along the east coast of South Island, New Zealand, soon after the K–Pg extinction event. *Waimanu* was already flightless, like all modern penguins. It swam, loon-like, on the surface using its feet, and dived using its modified wings for locomotion. (Chris Gaskin, Geology Museum, University of Otago)

Plate 13. The King Penguin (illustrated), like the larger Emperor Penguin, belongs to the genus *Aptenodytes*, a lineage that is basal to all other living penguins. The 'great penguins', as they are collectively known, evolved from an ancestor that lived 40 million years ago. (John Reilly)

Plate 14. An Emperor Penguin huddle or 'turtle' during the Antarctic winter. Periodic shuffling ensures that each penguin takes its turn at the centre of the huddle: a remarkable adaptive strategy to conserve heat. (Stefan Christmann)

Plate 15. Luis Monteiro (1962–1999), a Portuguese scientist whose field and laboratory investigations led to the discovery of the sympatric speciation of storm petrels. (Robert Furness)

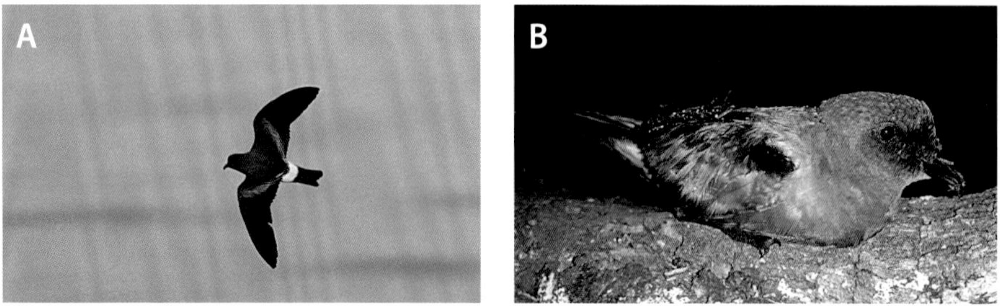

Plate 16. (A) Monteiro's Storm Petrel off the Azores: the brownness of the bird's feathering, created by bleaching, suggests a hot-season breeder. (B) Monteiro's Storm Petrel on Praia islet. Remarkably, these birds may use the same nest burrows as the Band-rumped Storm Petrel, which breeds in the cool season. (A, Peter Alfrey; B, Nuno Oliveira, Portuguese Society for the Study of Birds – SPEA)

Plate 17. The Bar-tailed Godwit is a record holder among migrating birds: a staggering non-stop journey of 11,500 kilometres from Alaska to New Zealand that lasts nine days. (Shaun Templeton, Elm Wildlife Tours)

Plate 18. An adult Oilbird at its roost in Humboldt's Cave, Venezuela. The Oilbird possesses the highest evolutionary distinctiveness (ED) score of any bird and has not shared its genes with any other taxa for over 70 million years. (Walter Jetz)

Plate 19. The coevolution of plant and hummingbird has led to some remarkable morphological adaptations. (A) The Buff-tailed Sicklebill has a bill that arcs a full 90 degrees downwards to enable it to reach nectar from *Centropogon* flowers; (B) the Sword-billed Hummingbird possesses a beak longer than its body and can obtain nectar from the elongated corollas of passion flowers. (A, Christopher C. Witt; B, Rolf Nussbaumer)

Plate 20. The Kakapo, the world's most genetically isolated parrot, is the only surviving member of the genus *Strigops*. (Dylan van Winkel)

Plate 21. Skin of the extinct Stephens Island Wren sold to the Liverpool Museum by the ornithologist and collector Henry Baker Tristram. The flightless basal passerine may have had the smallest natural range of any known bird. (National Museums Liverpool, John Reilly)

Plate 22. The enigmatic Sapayoa split from the broadbills 50 million years ago and crossed Beringia to reach Central America. (Petra Rank)

Plate 23. Superb Lyrebirds, most noted for their excellent mimicry, are basal songbirds restricted to southeast Australia. (Ian Montgomery)

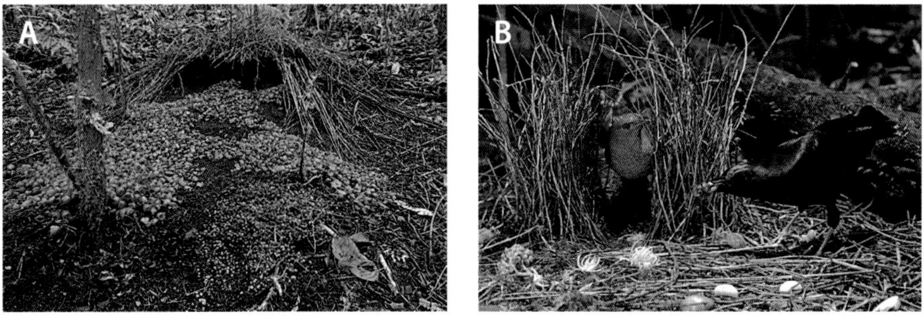

Plate 24. Extended phenotypes. (A) The maypole bower of male Vogelkop Bowerbird, with his decorations. (B) A male Satin Bowerbird holds a cicada case in his bill and displays to a female who has entered his 'avenue' bower. (Tim Laman)

Plate 25. A Hawaiian Crow will carefully choose and shape a stick to snag its prey. (Minden Pictures)

Plate 26. (A) The Wilson's Bird-of-Paradise and (B) the Magnificent Bird-of-Paradise evolved by vicariance after geological forces split their ancestral population. (Tim Laman)

Plate 27. Two adult male Emperor Birds-of-Paradise displaying to a nearby female. The species evolved after becoming isolated in the Huon Peninsula. (Lars Petersson)

Plate 28. Sexual selection underpins the absurd ornamental head-wires and bouncing displays of the male King Saxony Bird-of-Paradise. (Tim Laman)

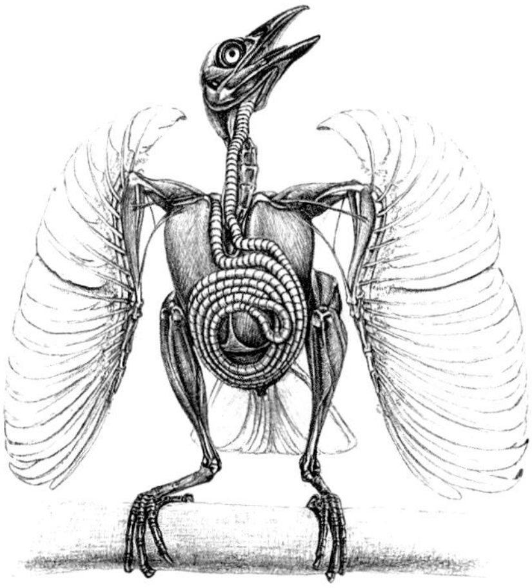

Plate 29. The extraordinary loud and powerful vocalisations of the male Trumpet Manucode are the result of an extremely long trachea: one that has six concentric loops between the skin and breast muscles. Females lack a coiled trachea, indicating that the male's windpipe has evolved as the result of millions of years of sexual selection. (Katrina van Grouw, *The Unfeathered Bird*, 2012)

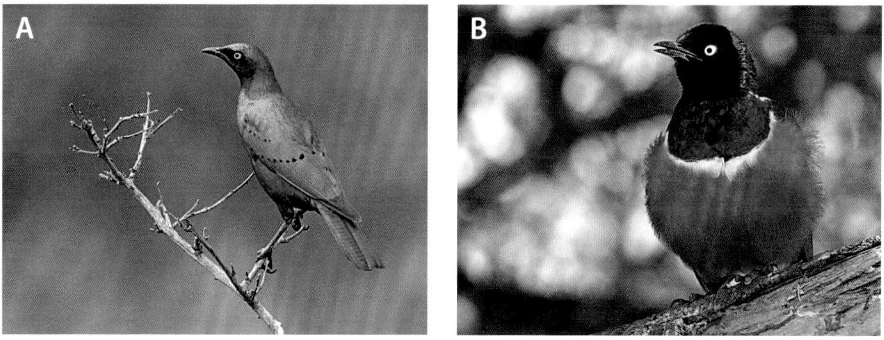

Plate 30. Four types of melanosome are responsible for the structural colours of African starlings: an evolutionary development that underpinned the clade's rapid speciation. (A) Greater Blue-eared Starling; (B) Superb Starling. (Lars Petersson)

Plate 31. The evolution and speciation of the Common Blackbird involved two sweepstake dispersals across the Atlantic Ocean. (Andreas Trepte)

Plate 32. (A) The Italian Sparrow arose 8,000 years ago as the result of hybridisation between (B) the House Sparrow and (C) the Spanish Sparrow. (A, alamy.com; B, Peter Garrity; C, Steve Mills)

Plate 33. Study of Zebra Finches has contributed to our understanding of the genetics and evolution of birdsong. (pixabay.com)

Plate 34. The Louisiade White-eye, like other insular white-eyes, exhibits behavioural or psychological flightlessness. (Lars Petersson)

Plate 35. Cassia Crossbill, a species that has coevolved for the past 6,000 years with the Rocky Mountain Lodgepole Pine. (A) A close-up of a male's beak; (B) a female employs her beak to prise open a Lodgepole Pine cone. (Craig Benkman)

Plate 36. The diversification of beak morphology facilitated the speciation of Darwin's finches by allowing a greater range of food resources. The insectivorous Grey Warbler-Finch (A) has a small pointed beak; the Common Cactus Finch (B) has a large pointed beak for medium-sized seeds, while the Large Ground Finch (C) has a large blunt beak for large seeds. (B, Rosemary Grant)

Plate 37. The Masked Flowerpiercer from South America has evolved a long, hooked bill and a grooved tongue for extracting nectar from flowers. (Lars Petersson)

followed, but they are less predictable and tend to swarm only after heavy rains.[52] Obligate army-ant followers have evolved a variety of specialised traits not seen in other species. They can sit 'sideways' on small vertical branches above the swarm, snapping up prey from their perches. They do not defend exclusive territories, and they undertake 'bivouac-checking', the regular sampling of army-ant activity: a behaviour that reflects the unpredictability of ant swarm activity. But how and when did this specialised foraging strategy evolve?

Robb Brumfield at the American Museum of Natural History thinks it likely that antbirds underwent an evolutionary progression from the least to most the specialised.[53] The ancestral antbirds only foraged around swarms as they moved through their territories. This behaviour then progressed, via a stage of regularly attending swarms outside territories, to the obligate ant-following strategy observed today. Although both *Eciton burchelli* and *Labidus praedator* have existed for around 20–40 million years,[54] it was not until the late Miocene, some 6 million years ago, that the ant-following behaviour of birds evolved.[53] Then, intense competition, a characteristic of extant obligate antbirds, would have encouraged their early diversification. This trait, coupled with the hierarchical division of advancing ant swarms into a series of concentric feeding zones, would have favoured the evolution of differently sized antbirds. Today, the largest ant-followers are found at a swarm's productive leading edge, while the smallest species occupy the trailing edge.

The disarmingly simple question of why there are so many South American passerines turns out to be an unexpectedly complex one. As we have highlighted, the answer is in part the result of the combined effects of prolonged geographical isolation, geological upheavals, climate change, vicariance, and the presence of many unusual ecological niches. Also, as the gap between South and North America reduced during the late Miocene and Pliocene, the continent's avian diversity dramatically increased due to an invasion of oscines from the north: a collection of songbirds that included jays, wrens, thrushes, wood-warblers, vireos, finches and tanagers.

Life for the suboscines, including the manakins of Tobago, would never be the same again, for they now faced competition from these new arrivals. But before we discuss how songbirds came to dominate the avian world, there is one further New World suboscine worthy of mention: the Sapayoa. For its evolutionary tale is as strange as that of any bird.

CHAPTER 16

The Sapayoa's Story

ODD ONE OUT

A soft musical trill emanates from deep within the nearby foliage, only to be followed by a long and frustrating silence. We are nevertheless excited, as this brief call indicates the presence of a Sapayoa, an elusive species restricted to the humid rainforests of southern Panama and western Colombia. Thankfully, the bird is quickly located, a uniformly olive manakin-like species, perched on a horizontal branch amid the forest's lower canopy. Over the next few minutes, we watch keenly as the Sapayoa scans the surrounding leaves, sallies forth in pursuit of some unsuspecting insect and returns to its favoured vantage point. For the Sapayoa is the New World's most enigmatic suboscine, a newly arrived interloper from distant shores, an evolutionary conundrum that bears the apposite epithet, *Sapayoa aenigma* (Plate 22). To fully appreciate its story, we need to return to Antarctica (western Gondwana) and follow the second wave of suboscine dispersal, one that would eventually give rise to the Old World suboscines (Figure 16.1).

Old World suboscines

Around 64 million years ago, amid the relatively mild climate of the Palaeocene, the ancestral population of Gondwanan suboscines split into two distinct groups.[2] As we have discussed in *The Manakin's Story*, the larger clade entered South America by way of a broad continental shelf and gave rise to the species-rich Furnariida and Tyrannida. In contrast, the second clade (Eurylaimides) dispersed northwards to Asia and diversified from about 55 million years ago to produce two relatively small families, the pittas (Pittidae) and the broadbills (Eurylaimidae), each with distinctive morphology, diet and breeding behaviour.[3]

The 42 species of pitta live in the tropical forests of Asia and Australia, with only two species occurring in Africa. The name 'pitta' probably stems from the Andhra Pradesh district of India, where it is a commonly used

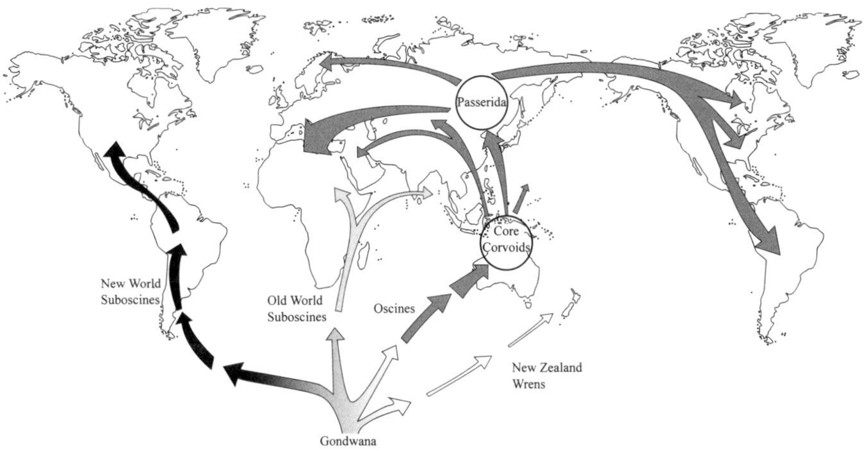

Figure 16.1 Proposed dispersal routes of major passerine groups from Gondwana: New Zealand wrens (Chapter 14), New World suboscines (Chapter 15), Old World suboscines (Chapter 16) and oscines, including core corvoids and Passerida (Chapters 17–27).

word for any small bird.[4] Pittas vary considerably in length, although they all appear stocky, almost tailless, with long, sturdy legs that reflect their terrestrial lifestyle. They have retained their ancestral dietary habits, feeding mainly on invertebrates, including ants, termites and beetles, obtained by rummaging through the leaves and detritus on the forest floor. Despite their brilliant plumage – they were once called 'jewel-thrushes' – pittas are masters of disguise and extraordinarily shy and retiring. They often stand motionless, and their colours can be rendered invisible amid the darkness of the forest. In shafts of sunlight, however, they are transformed from evanescent shadows to objects of exquisite beauty, a fact emphasised by many of their names: Garnet Pitta, Superb Pitta, Fairy Pitta, Elegant Pitta, Rainbow Pitta. It is the unique combination of being hard to locate and their gem-like appearance that makes the Pittidae such a highly sought-after family for globe-trotting birders. Indeed, the pittas' lure can become an obsession for some. Chris Gooddie, for example, resigned from his job and spent his life savings to track down all the world's pittas in a single year.[5]

The 19 species of broadbill are also confined to the tropical forests, occurring in sub-Saharan Africa, as well as extending from the eastern Himalayas to Indonesia and the Philippines. As their name suggests, they tend to have broad, heavy bills, with that of the Dusky Broadbill being the widest and most grotesque of any passerine. Such large gapes enable the capture of small animals, including insects, snails, frogs and lizards. Three species in southeast Asia, however, eat primarily fruit and, as a consequence, have evolved narrower bills. Several species have very restricted ranges. Hose's Broadbill and Whitehead's Broadbill, for example, are confined to a few mountains in northern Borneo, while Grauer's Broadbill is limited to a narrow belt of montane forest in central Africa. Walter Rothschild, who first described the latter from a skin collected by the ornithologist afrer whom it is named, considered it to be a flycatcher. He named it *Pseudocalyptomena* ('false *Calyptomena*'), on account of its superficial similarity to Asian broadbills of the genus *Calyptomena*. Today, Grauer's Broadbill is regarded as a true broadbill, one of only four African representatives of a primarily Asian family.

Broadbills from the genus *Smithornis*, including the African Broadbill, Rufous-sided Broadbill and Grey-headed Broadbill, have evolved the unusual ability to 'sing' with their wings. Males of all three species produce a distinctive, klaxon-like 'breeeeet' while making a short circular flight around their perches. Recently, experiments by Richard Prum and his colleagues from the University of California have revealed that the sound is produced during the down-beat of the wings and results from the vibration or 'aeroelastic flutter' of two outer primary feathers.[6]

The Club-winged Manakin, a New World suboscine from the cloud forests of Colombia and Ecuador, has also evolved a 'wing-song', although by a different mechanism. The male possesses a pair of club-shaped wing feathers that are knocked together over 100 times per second, the fastest limb movement known for any vertebrate.[7] Intriguingly, this mechanism is similar to that used by the Castanet moth (*Hecatesia exultans*) and highlights an arthropod–vertebrate evolutionary convergence, probably brought about in each case by choosy females. Why broadbills and manakins evolved new ways of producing sound to replace their vocal songs, despite the possession of functional syringes, is unclear. The most likely explanation, according to Prum's team, is that the novel sounds developed as an arbitrary replacement for a vocal song as the result of sexual selection, for no other reason than that it was preferred by the females. But for the male Club-winged Manakin, female choice has come at a cost. To maximise the sound, males evolved solid, bulky wing bones, with high calcium concentrations, a unique combination within the passerine lineage that makes flight far more demanding.

Our understanding of the evolution of Old World suboscines has been helped by recent phylogenetic studies of the asities of Madagascar and the Sapayoa of Central and Southern America. The four asities are forest dwellers with diets consisting of fruit and nectar. They exhibit sexual dimorphism, with the colours of the males' wattles, or caruncles, being structural in nature (see *The Starling's Story*). Indeed, the vivid blues and greens, expressed predominantly during the breeding season, derive from an arrangement of collagen fibres previously unknown in the animal world.[8] Since each species is coloured differently, it is likely that their structural basis has evolved continuously following their diversification around 40 million years ago.[3] Asities resemble broadbills morphologically, as they have the same number of tail feathers, albeit on shorter tails, and possess similar syringes and hindlimb musculature.[9] A detailed morphological study has confirmed the relatedness of the two lineages and suggested that the asities are a subfamily of broadbills.[10]

The Sapayoa has been a taxonomic challenge ever since the German ornithologist Ernst Hartert described the species in 1903. The only skin sent to Hartert was assumed to be from a female since, according to the accompanying notes, it was taken from a nest with two eggs near the Rio Sapayoa (now known as the Rio Zapallo Grande) in northwest Ecuador.[11] After a 'careful comparison', Hartert concluded that the 'remarkable' species was a New World suboscine and best placed among the manakin family.[12] Indeed, he thought it closest to the genus *Neopelma* (tyrant-manakins), which has similar colours, bill shape and yellow coronal patch. However, as noted by Jon Fjeldså, 'parallel evolution of a broad flycatcher bill is seen in several lineages of perching birds, and the coronal patch is a kind of "flash colour" that has evolved independently to enhance short-distance communication in dense foliage.'[13] Others stressed the Sapayoa's similarity to the New World tyrant flycatchers and placed it among the large Tyrannidae family.

In 1990, Sibley and Ahlquist included a sample of Sapayoa DNA in their celebrated hybridisation studies and obtained a completely different result: the species was most closely related to the Old World suboscines.[14] In other words, the Sapayoa was not a member of the manakin or tyrant flycatcher family as previously thought but related instead to the pittas and broadbills of Asia and Africa. These rather crude genetic experiments did not allow a precise classification, and the Sapayoa was placed initially in a taxonomic category *incertae sedis* (Latin for 'of uncertain placement') in front of all the Old World suboscines. It was not until 2006 that the enigmatic species was given a permanent taxonomic home. Using DNA extracted from museum collections, Martin Irestedt and his colleagues at the Swedish Museum of Natural History proved that the Sapayoa belonged to a deep branch within the broadbill clade.[15] This conclusion has been strengthened by a recent field study conducted by Cornell University. The American team found that many of the Sapayoa's behavioural traits – diet, nest structure, breeding system and mode of parental care – are also found in broadbills but not in any other New World suboscine.[16]

Given these findings, how can the Sapayoa's present distribution be explained? As discussed previously, the Eurylaimides' stem emerged 64 million years ago somewhere on the southern continent. Subsequent phylogenetic reconstructions have revealed that this ancestral population then gave rise to the pittas and broadbills (including the asities and the Sapayoa), but only after they reached mainland Asia.[2] Such conclusions pose a problem, because by this time the African and Indian landmasses had already separated from Antarctica. In fact, Africa had started to split from South America during the early Cretaceous,[17] while India had separated from East Gondwana and then from Madagascar by 100 million years ago.[18] Furthermore, there were no possible stepping stones, as the remnant land connections between East Antarctica and Madagascar–Seychelles–India had already disappeared. It now seems likely that the Old World suboscines, in contrast to their New World cousins, flew across the emerging Southern Ocean and then hitched a lift on the Indian tectonic plate as it rafted towards Asia. The architects of this hypothesis have stressed that, although the broadbills and pittas both contain African species, it is less parsimonious to infer an African origin. Interestingly, several families of amphibians may also have used the 'Indian conduit' to Asia, including some frogs and the limbless caecilians.[19]

Once India had crashed into the Eurasian continent during the Palaeocene, the insectivorous suboscines entered Laurasia. There then followed a series of major radiations during a 10-million-year period that led to the Eurylaimides becoming widespread, reaching southeast Asia, Africa, Madagascar and Australia.[3] The broadbill lineage started to diverge around 46 million years ago, and while most retained their ancestors' insectivorous lifestyle, several lineages became frugivorous. Five million years later, the asities diverged from the broadbills and evolved to become fruit and nectar specialists. They reached Madagascar, not from Africa as originally thought but by way of exposed land that existed between India, the Seychelles and Madagascar during the Eocene.[3]

The pittas were the last to diversify, around 23 million years ago, and spread out to reach Africa and Australia 16 and 9 million years ago respectively. Some pitta species are migratory, and they are widely distributed, occurring in Africa and Australia. Since such birds belong to a single clade, one that also contains many sedentary species, it is likely that the migratory habit was ancestral, or plesiomorphic, within the group.[15] Furthermore, reconstruction of the pittas' phylogenetic tree suggests that it was a migratory ancestor that gave rise to the taxa found on remote islands including Halmahera, Manus and the Solomons.

However, it is the biogeographical story of the Sapayoa that is the most unusual and unexpected. Fifty million years ago, at the time of the Sapayoa's divergence from the broadbills, the climate was much milder, with evergreen and seasonal forests extending to high latitudes. Alligators, turtles, flying lemurs and primitive hippopotamuses thrived in the Canadian Arctic only 500 miles from the North Pole.[20] Such favourable climatic conditions enabled the ancestral Eurylaimides to spread much further north than at present, a conclusion supported by fossil finds from France and Germany.[21] Crucially, there also existed land connections between Laurasia and the New World, areas that could have supported insectivorous passerines. Given these facts, Robert Moyle and colleagues at the American Museum of Natural History believe that the Sapayoa's ancestors then spread northwards and gained access to the Americas via either the North Atlantic or the Beringian land bridge.[2] Once established in the New World, their distribution would have mirrored the extent of available habitat, which contracted relentlessly towards the equator throughout the Palaeogene and Neogene.

Many millions of years later, deep within Central and South America's extensive rainforest, the manakins and tyrant flycatchers met their long-lost cousin from the Old World. It was a significant meeting, as the diminutive and unobtrusive Sapayoa was the only suboscine from over 1,200 species to have approached the New World from the north, rather than from the south. Ernst Hartert's binomial *Sapayoa aenigma* turns out to be far more appropriate than he could have ever imagined.

The Old World suboscines, however, present a further evolutionary conundrum. Why are there so few species, despite the lineage having dispersed across half the world and having had the same time to evolve as their New World cousins? In other words, why do the extant Old World suboscines occupy only a fraction of the available ecological niches? The answer to this puzzle is to be found in the emergence of the most successful avian lineage of all, the oscines or 'true songbirds'.

CHAPTER 17

The Scrubbird's Story

WHERE SONG BEGAN

On Christmas Eve, 1961, the Australian naturalist Vincent Serventy was at home making his last-minute festive preparations, when the phone rang. It was a reporter from the *West Australian* newspaper who sought his views on the reported rediscovery of the Noisy Scrubbird. Serventy's reply – 'If it's true, it's the most exciting find of the century' – encapsulates one of the most dramatic episodes in Australian ornithology.[1] Indeed, the finding of the scrubbird (family Atrichornithidae) so excited the general public that its loud call was aired on radio stations throughout the world.

The story of this Lazarus species began in May 1838, when the English ornithologist and renowned artist John Gould travelled to the continent to collect material for his monumental tome *Birds of Australia*. Among his entourage was the naturalist, explorer and taxidermist John Gilbert, whom he sent out to remote areas to seek new botanical and zoological specimens. On 3 November 1842, while working at Drakesbrook on the southwestern tip of Australia, Gilbert first heard and eventually collected a bird unknown to science: the Noisy Scrubbird (or *jee-mul-uk* according to the local Aborigines). The skins of four individuals, together with many other species, were forwarded to Gould the following year, accompanied by a letter that included the following account of the new find:

> This is without exception the loudest of the Western Australian songsters I have yet heard. It inhabits the densest and rankest vegetation, on the sides of hills and thick swamp grass on the banks of small running streams or swamps, and is of all birds, the most difficult I have yet had to procure … as it runs with the utmost rapidity on the ground, sheltered from view by the overhanging vegetation which renders it almost impossible to get a shot at.[2]

Gould showed keen interest in the skins of the 'noisy bird' and noted, 'few of the novelties received from Australia are more interesting than the species to which I have given the generic name of *Atrichia*.'[3] Gould's English name, Noisy Scrubbird, has remained, although his original generic name was later found to be already in use, so it was subsequently altered to its present spelling: *Atrichornis*. Surprisingly, the usually meticulous artist-entrepreneur paid scant regard to Gilbert's accurate description, and his published illustration portrays a pair of birds perched on leaf-adorned branches, rather than on the ground.[3] Twenty years later, in the 1860s, the energetic Australian George Masters, an excellent shot who caught venomous snakes with his bare hands, obtained seven further specimens, and then another collector took eight birds in the 1880s.

The Noisy Scrubbird was last seen in 1889 and remained lost, but not forgotten, until it was rediscovered in 1961. In December of that year, Harley Webster, headmaster of Albany High School and a keen birdwatcher, was attracted by an unfamiliar birdsong at Two People's Bay, south of Perth:

> I came away that evening with impressions of a brown bird with a call that really made my ears ring and with the knowledge that it was almost certainly the Noisy scrub-bird.[4]

After extensive field work, Webster found several birds, although a nest eluded him. Like most birders who have seen the scrubbird, I can vouch for the difficulties faced by Gilbert, Masters and Webster. While I was staying at Cheynes Beach near Albany, the unmistakable song of a male scrubbird was heard coming from an area of dense scrub near the shoreline. Despite a patient wait of nearly an hour, the bird remained hidden, even though it could not have been more than 5 metres away. Then, without warning, a rat-like apparition scuttled across the track and vanished into the thick undergrowth on the other side, where it proceeded to tease me once again with its ear-piercing call. Although I never managed another sighting, I nevertheless cherish my 3-second, full-frame view of this notoriously difficult bird. Today, only around 1,000 adults remain, and despite an intensive conservation programme, including fire protection, habitat management and translocation to new sites, the species remains threatened.[5]

In 1866, a second member of the Atrichornithidae, the Rufous Scrubbird, was described by the Australian zoologist Edward Ramsay. It is also rare, and inhabits a few areas of rainforest on the eastern slopes of the Great Dividing Range along Australia's east coast, some 3,000 kilometres from its cousins. Ramsay was struck by the bird's ventriloquial powers and noted that 'they must be heard to be believed.'[6] Like the Noisy Scrubbird, the Rufous Scrubbird inhabits impenetrable undergrowth, creeps and runs along the ground like a rat, and feeds on invertebrates found among the leaf litter.

Lyrebirds (family Menuridae) are also exclusively Australian, and are renowned for their remarkable displays and use of mimicry (see *The Zebra Finch's Story*). Specimens were much easier to obtain than those of scrubbirds, and several skins had already reached European naturalists by the end of the eighteenth century. In 1800, a British Army officer, Thomas Davies, presented the first description of Menuridae – the Superb Lyrebird – to a scientific meeting of the Linnean Society in London (Plate 23). The English name 'lyrebird' reflects the male's spectacular tail, consisting of 16 highly modified feathers that were originally thought to resemble a lyre. Such a misunderstanding occurred when an early specimen was prepared for display at the British Museum by a taxidermist who had never seen the bird in the wild. As a result, he incorrectly placed the tail's feathers in an upright position, similar to that of a peacock during a courtship display. Later, John Gould reinforced this view when he painted the lyrebird in a similar pose, using the museum's exhibit as his model. In fact, the male lyrebird inverts the tail over his head and neck during courtship displays and fans the feathers to form a silvery-white canopy.

The only other species of lyrebird, the Albert's Lyrebird (named in honour of Prince Albert), is smaller and rarer than the Superb Lyrebird and was not described until 1860. It has one of the smallest distributions of any bird

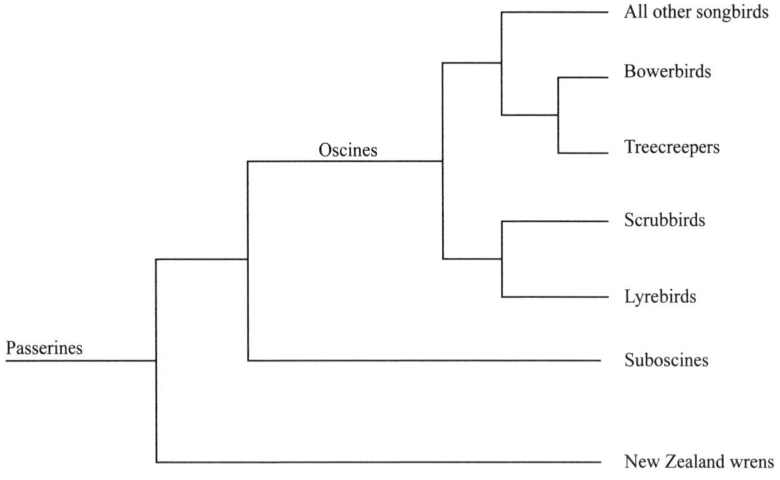

Figure 17.1 The basal passerines. After the New Zealand wrens diverged, the passerines split to produce the suboscines and oscines (songbirds).

on the Australian continent, being confined to small scattered areas within southern Queensland and northeast New South Wales. Its total population is estimated at only 3,000–4,000 breeding birds, and the species remains under threat since its fragmented distribution and sedentary nature renders gene flow between populations unlikely. Any environmental disaster, such as a severe regional drought, has the potential to affect every individual.

What the early collectors could not know was that scrubbirds and lyrebirds are closely related, and that both families lie at the base of the entire phylogenetic tree of the songbirds (Figure 17.1). They are ancient sister groups to all the other oscines. As we will discuss, this finding turns out to be crucial in the search for the origin of the world's songbirds, or, as Tim Low poetically puts it, the land 'where song began'.[7]

Erosion of northern-centric certainty

The Australian passerines were discovered many years after European ornithologists had classified most of the world's species. It is not surprising, therefore, that when Victorian adventurers and species hunters dispatched their large collections of skins back home, taxonomists were content to accept that many belonged to families already well defined, including wrens, thrushes, treecreepers and flycatchers. Novel species from down-under were seen as 'the poor last gasp of radiations that had been accumulating over eons in the Old and New Worlds … relatively recent derivatives of corresponding groups better known and already described from Eurasia and the Americas.'[8] Personality and emotion play a surprisingly large role in science, and the unwavering support for the 'northern-centric' view by Ernst Mayr, a tower of twentieth-century evolutionary biology, influenced many scientists of his generation. In an article entitled 'Timor and the colonization of Australia by birds' (1944), Mayr expounded with typical authority that Australia had received its birds in waves from Asia.[9] The first to arrive were the early emus, lyrebirds and honeyeaters, to be followed later by the ancestral robins and treecreepers. This northern-centric view even withstood the challenge offered by Australasia's many endemic families: scrubbirds, lyrebirds, bowerbirds, birds-of-paradise and logrunners. As Michael Heads emphasises, New Guinea's avifauna was seen traditionally as 'a sink, made up of waifs and strays that arrived from elsewhere in recent Neogene times.'[10] In other words, for over 150 years the ancestral home of all songbirds was thought to be located in the northern hemisphere. It is against this backdrop that the full impact of the latest data can be appreciated. For recent molecular studies and fossil finds have literally, and metaphorically, turned the world of oscine evolution upside down.

As we have discussed, it was the brilliant and mercurial Charles Sibley who first challenged conventional wisdom when he compared species' DNA rather than their morphological traits. Such novel approaches to systematic biology relied on the unique binding behaviour of DNA: single-stranded molecules bind, or hybridise, more tightly to themselves than to less similar strands. Sibley realised that genetic material would bind best between individuals of the same species, less well between closely related species, and least of all between species from different families. By determining the strength of DNA binding between any two species, a phylogenetic tree can be constructed whose branching points mirror the recency of common ancestry. Working with his colleague Jon Ahlquist between 1975 and 1986, Sibley analysed 26,000 DNA–DNA interactions from 1,700 hundred species, representing all the avian orders and most of the recognised families. The results were startling: the endemic passerine families of Australia and New Guinea (scrubbirds, bowerbirds, lyrebirds and Australasian treecreepers) were more closely related to each other than to the passerines of Eurasia. Sibley and Ahlquist concluded that such endemics must have 'arisen within the area and were not the products of a series of invasions from Asia.'[11] Their findings suggested, for the first time, that Australia may have had a significant role in songbird evolution.

Several years later, a young Melbourne graduate, Les Christidis, teamed up with fellow Australian Richard Schodde to check Sibley's findings. They adopted a slightly different approach. Rather than compare species' genetic material directly, they studied the encoded proteins, or enzymes, extracted from the songbirds' liver, heart and muscle. By analysing the behaviour of these proteins in electrophoretic gels, they were able to construct phylogenetic trees: the greater the protein separation, the greater the difference in gene sequence, and the greater the evolutionary distance. What they found contrasted with the prevailing dogma and required the entire oscine family tree to be redrawn. In contrast to Sibley's study, the protein approach showed the Old World songbirds to be nested entirely within the Australo-Papuan songbirds. This finding implied that all Eurasian passerines, including our familiar starlings, thrushes and sparrows, must have evolved from ancestors that inhabited the southern hemisphere. Furthermore, the two lyrebirds occupied the most basal position and therefore had the longest evolutionary history – likely remnants of an ancestral population from which all the other oscines arose.[12]

At that time, a southern origin of songbirds was not accepted by the scientific community, and the authors had difficulty getting their results published. After several years of rejection, their paper finally appeared in the British Ornithologists' Union's journal *Ibis,* although with an imposed watered-down conclusion to avoid controversy.[13] The final paragraph, while understandably vague, was nevertheless profound in its implications:

> It is not beyond reason to draw attention to the possibility of a Gondwanan origin for the order … Although purely speculative at present, this hypothesis does warrant testing.[12]

The Australians' conclusion is a classic example of how science can pan out in unexpected ways: a 'paradigm shift' in biological thought that lends credence to Thomas Kuhn's view of scientific revolutions.[14] Except, in this case, the paradigm shift was overlooked for nearly 10 years, until the publication of confirmatory evidence.

The above outcome seems odd in hindsight, since a detailed morphological study published much earlier had hinted at the same conclusion. In 1982, Alan Feduccia and Storrs Olson noted that while lyrebirds are oscines, their bone structure, including that of the stapes, is more primitive than that of other songbirds, including the bowerbirds and the birds-of-paradise. In fact, the bones appeared most similar to a family of South American suboscines: the tapaculos. Their prescient conclusion – that the passerines 'probably arose in the southern hemisphere' – was ignored, and their findings languished in the recondite tomes of the *Smithsonian Contributions to Zoology.*[15]

It was not until 2002 that the scientific argument was settled once and for all. Two independent groups, one led by Keith Barker at the American Museum of Natural History and a second headed by the Swede Per Ericson, compared gene sequences from a selection of oscines.[16] Despite using different genes, the results were the same: lyrebirds lie at the base of the oscine tree and, as a result, are the most primitive of all songbirds. However, the phylogenetic position of the scrubbirds remained unresolved, as neither Barker nor Ericson had suitable DNA for study. One might have predicted, given field observations and Sibley's early DNA–DNA hybridisation studies, that scrubbirds and lyrebirds would turn out to be closely related and that their ancestry would be similar. Both families, for example, are excellent mimics, have the same volume of song, and possess a similar syrinx structure. Scrubbirds and lyrebirds are highly secretive and rare within their respective ranges and are primarily ground-dwellers, inhabiting similar ecological niches. Furthermore, the male scrubbirds' display is like that of immature lyrebirds, with elongated and fanned tails, lowered wings, and a torso that quivers from the effort of their sustained, loud and melodious song. In 2007, an Australian study by Terry Chesser and José ten Have confirmed what was expected: scrubbirds are sister to the lyrebirds, and both genera are sister to all other songbirds.[17]

To some, the construction of phylogenetic trees, whether based on morphological traits or DNA structure, may seem a little dry and theoretical. But it is not! In the words of Keith Barker, 'the inescapable conclusion is that songbirds had their origin on the Australian continental plate.'[18] He then added 'the only alternative is to postulate a previously widespread oscine lineage that invaded Australia many times and then went extinct with the exception of the Passerida.' We will return to the Passerida in *The Starling's Story*, but the evidence is irrefragable: songbirds first evolved in Australia.

CHAPTER 18

The Bowerbird's Story

EXTENDED PHENOTYPES

The next oscine branch to evolve after the scrubbirds and lyrebirds gave rise to the Australasian treecreepers (Climacteridae) and the bowerbirds (Ptilonorhynchidae), approximately 45 million years ago (Figure 17.1).[1] While the Australasian treecreepers are very similar in appearance and behaviour to their northern namesakes (family Certhiidae), they are a much older lineage, and any likenesses are the result of convergent evolution due to the occupation of similar ecological niches. The Australasian family's ancient origin is suggested by their early design of syrinx, which reveals a 'striking departure from the basic muscle pattern that prevails throughout the rest of the oscines.'[2] Also, the oldest Climacteridae fossils are known from the early Miocene, and were extracted from deposits at Riversleigh in northwestern Queensland.[3] This important site, which has also produced the oldest example of lyrebird, is unusual because the fossils are located in soft freshwater limestone which has not been compressed.[4] The resultant well-preserved fossils, together with recent phylogenetic studies, provide yet further evidence that the world's songbirds originated in eastern Gondwana – present-day Australasia.

The Climacteridae is a small family of only seven species; six of them live in Australia, while one, the Papuan Treecreeper, inhabits the mountains of New Guinea. All are medium-sized, scansorial species, with greyish-brown to black plumage and streaked undersides, which forage for arthropods on the bark of eucalyptus and other trees. Unlike their northern counterparts, Australasian treecreepers do not use their tails for support while climbing but only employ their feet. They climb upwards, placing one foot ahead of the other, circling up the trunk and outwards onto the main branches, using their long, slightly curved beaks to seek out prey. Because they always keep one foot attached to the bark, they can climb upside down beneath branches, making them the only birds in Australia to access this niche. In contrast to other passerines, Australasian treecreepers possess a hind toe that lacks ligaments and an extensor system that allows the rotation of this toe to produce a firmer grip on the toughened bark.[5] Australasian treecreepers are cooperative breeders, as are many of the ancient branches of oscines, and obtain help, often from the young males of previous broods, to feed the incubating female and defend the young.

Despite molecular studies placing bowerbirds on the same ancient branch as the Australasian treecreepers, their skeletal anatomy is very different. Indeed, Walter Bock, Professor of Evolutionary Biology at Columbia University, noted that bowerbirds possess 'some of the most peculiar and unique cranial features in the entire order of perching birds.'[6] In particular, their lacrymal bone is unusually large and unlike that of any other songbird, except lyrebirds. Given the marked anatomical, as well as behavioural, differences between treecreepers and bowerbirds, Tim Low has questioned in his book *Where Song Began* whether both families can belong to the same evolutionary branch. Could it be that the genetic evidence is misleading: the result of a phenomenon known as 'long-branch attraction'? This recognised limitation of phylogenetics is the tendency of DNA sequences from lineages with long terminal branches to group together, regardless of their actual relationships. In part, the phenomenon relates to the limited number of possible states that rapidly evolving sites can change to (20 amino acids and four nucleotides). Since the branches of bowerbirds and treecreepers are long, sequences could contain many such changes, leading to the development of spurious similarities. In effect, these false resemblances override the real phylogenetic signal, allowing the sequences to group, or 'attract' to each other. However, most phylogeneticists, including Professor Christidis, co-author of *Systematics and Taxonomy of Australian Birds*, believe that both families do belong to the same branch: the second one after the scrubbirds and lyrebirds.[7]

The precise evolutionary relationship between treecreepers and bowerbirds has little bearing on our present story, however. What is important is that readers appreciate that phylograms based on gene sequences, including all those used to illustrate this book, are merely the most likely result from a number of possible options. In truth, despite the use of sophisticated statistical manipulations, phylogenetic trees are just scientific hypotheses, subject to falsification by the addition of further data or the use of alternative analytical techniques. This fact should come as no surprise, as genetic sequences constitute 'noisy' data: the result of millions of years of genetic recombination, horizontal gene transfer and hybridisation, not to mention the confounding problems of conserved sequences and long-branch attraction. Nevertheless, the analyses of known phylogenies, such as viral populations that have evolved in the laboratory, indicate that the widely used statistical approaches (e.g. bootstrapping) give, in general, a reliable measure of phylogenetic accuracy (values of 70 per cent or higher, supporting reliable groupings).[8]

Now let us return to the bowerbird's story, one that explores the concept of the extended phenotype and its role in sexual selection.

The bower

Most male traits that catch the female eye have a clear benefit for the species. Many female insects, for example, require a 'nuptial gift' from their suitors. This is usually something edible, such as another bug, which can supply nutrients and so directly increase the chances of successful mating when food supplies are limited. The presence of butterflies dancing in the sunlight around a drying mud puddle is usually the result of males collecting salt crystals to present to potential mates. Birds are no different. In the United States, Blue Jays pass tasty morsels from beak to beak, while the majority of roadrunner matings involve the transmission of food. Birds also select partners by their nest-constructing abilities. The male Eurasian Wren, for example, offers his mate a choice of three or four nests that he has built. This approach makes sense, as a well located and optimally built nest will increase the offspring's survival rate. For other species, including Marsh Wrens, European Robins and Red-winged Blackbirds, it is the male's ability to defend a large territory that wins out, as bigger areas are more likely to provide sufficient resources for rearing young. Most seabirds, although they breed in colonies and do not hold territories, still need to show prospective mates that they have real estate of their own.

In contrast, some avian species favour traits that appear, at first sight, to be somewhat arbitrary. Why, for example, would a female select a mate by his song, mimic ability, plumage decoration or prowess on the dance floor? Although the genetic advantages of these seemingly 'whimsical' criteria are now partially understood (discussed in subsequent chapters), it is the biological benefit of the bowerbird's bizarre displays, especially the decorated bower, which has been the most challenging to explain. In the words of the American ornithologist Thomas Gilliard, bowerbirds 'raise difficult questions, questions that penetrate to the very foundation of our biological theories.'[9] But before we attempt to explore such issues, we need to describe the bower itself, the 'extended phenotype' and central character of the bowerbird's story.

Of all the structures built by animals, there can be none as strange and magical as the bowerbirds' 'love-shacks'. Indeed, the first Europeans to encounter these remarkable chambers were convinced that they were the handiwork of local women, playthings constructed for the amusement of local children. Bowers can be one of two types, each built by species that appear to be phylogenetically closer to each other than to the builders of the alternative configuration. The first form, known as a 'maypole' bower, consists of a central supporting vine or sapling, around which are added woven sticks that radiate outwards in all directions, in the form of a tepee or Christmas tree (Plate 24A). The Vogelkop Bowerbird and the Streaked Bowerbird, however, add massive roofed, hut-like structures up to 2 metres across, often with a double entrance and a covered porch.

The second form, or 'avenue' bower, consists of parallel vertical walls of grass stems and sticks, planted into a base of vegetation. Each wall curves gently outwards, with one end of the bower exiting onto a display area where most of the decorations are arranged (Plate 24B). The Satin Bowerbird's avenue consists of just two rows of sticks, while that of the Yellow-breasted Bowerbird contains a double avenue, with two parallel paths on a raised platform.

Paradoxically, not all bowerbirds build bowers. Catbirds, as well as the Tooth-billed and Archbold's Bowerbird, merely clear areas of forest floor for their displays, although Archbold's Bowerbird covers his with a thick mat of ferns. The construction of bowers and their absence in several family members raise some interesting questions: how, for example, did the behavioural trait evolve, and what is the function of the bower?

To address the first question, Rab Kusmierski and colleagues from the University of Maryland constructed a molecular phylogeny, one that revealed all bowerbirds to have evolved from a single ancestral species.[10] By studying individual relationships, the team concluded that the common ancestor of all extant species (currently 27) was probably a monogamous, dull-plumaged bird that lacked bower-building and court-clearing skills. Indeed, this ancestral state characterises the catbirds, the first bowerbird genus to evolve, around 20 million years ago.[11] Catbirds are found in eastern Australia and the mountains of New Guinea and are mainly green, nondescript birds. Although they are known to attract mates by displaying food or colourful objects in their beaks, they do not build bowers. Furthermore, catbirds are monogamous, and once a mate is chosen they remain together for life since, unlike in other bowerbirds, both parents are required for raising a brood. It was only during the last 20 million years, following the divergence of catbirds, that the family's characteristic court-clearing and elaborate decorating behaviours evolved. Such traits, coupled with the transition to polygyny (males mate with more than one female), provided the preadaptations that enabled the evolution of bower-building to occur. According to the leader of the research group, Gerald Borgia, the first bowers were probably simple structures consisting of a display court that surrounded an isolated sapling or vine. As the bower-building lineage evolved, a second major divergence occurred, one that produced the avenue- and maypole-building clades. Later, two members of the maypole clade, the Tooth-billed and Archbold's Bowerbird, independently lost their bower-building abilities. Such dramatic shifts in behaviour must have evolved quite rapidly, given that both species have close phylogenetic relationships to the maypole-building Vogelkop Bowerbirds.

Learning is likely to play a significant role in bower construction. Young male Satin and MacGregor's Bowerbirds, for example, are inept at building bowers and often adorn them with inappropriate objects, often of the wrong colour. Youngsters also spend extended periods of time watching old males at work and take up to 4–7 years to develop the necessary architectural skills, a fact that may account for bowerbirds having the longest life expectancy of any passerine. Similarly, females go around in groups to visit bowers, so that the younger females can learn the species' style preferences from their older kin.

The wide variation in bower design between the different species may have evolved as a result of changes in female preference or choice, a view first proposed by Jared Diamond.[12] While undertaking field work in West Papua (the Indonesian half of the island of New Guinea) in the 1980s, Diamond discovered an isolated population of Vogelkop Bowerbirds that, although only 100 kilometres away from other populations in the same mountain range, built remarkably different bowers with distinctive decorations. Instead of large, hut-like bowers, copiously decorated with colourful objects, the new population constructed maypole bowers without a roof, and decorated exclusively with drably coloured objects – stones, acorns and moss – even though bright objects were available. Importantly, these differences have evolved despite the males in this isolated group having access to the same range of environmental materials as other Vogelkop Bowerbirds. Diamond concluded that ongoing sexual selection was the most likely explanation for an individual male's preference. In support of this idea, variations in female choice are believed to give rise to local, or fine-scale 'cultural' differences in the bower design of the Spotted Bowerbird.[13]

Albert Uy and Gerald Borgia from the University of Maryland decided to conduct an experiment to see if Diamond was right.[14] They placed a collection of differently coloured small plastic tiles just beyond the males' court and filmed the birds to see which objects were incorporated into their displays. Although both the new population and males from elsewhere readily made use of the novel items, their choices were very different. The traditional Vogelkop Bowerbirds used the red and blue tiles and ignored the dull ones, while the new population preferred the drab ones and disregarded the coloured tiles. Also, females preferred the display traits, including the bower shape, size and decoration, of males from their own population. Given these findings, the researchers concluded that the different displays are the result of rapidly evolving female preference or choice. Since the two populations are allopatric, with minimal genetic differentiation, it is likely that they are incipient species. In other words, if the trend continues, female preference will lead eventually to two different species of bowerbird. These findings are significant because they provided the first direct evidence for the 'speciation by sexual selection' hypothesis in birds: a model in which the reproductive isolation of a population results from the relatively few genetic changes associated with female preferences and male display traits. But what could account for a change in female preference? Uy wonders whether 'visual conditions, such as the play of light and shadow on foliage, could select for signals that are most effective in those particular environments. Or are these changes arbitrary relative to habitat, analogous to fads we see in our own society?'[15]

Whatever the cause, the highlighted field study suggests that female choice over many millions of years may have contributed to bowerbird speciation. It is a mechanism that is likely to have acted in conjunction with other processes, especially vicariant evolution. For example, population fragmentation, due to Pleistocene glaciations and the uplift of New Guinea's central mountains, is known to have underpinned the speciation of catbirds.[11] Furthermore, New Guinea's orogeny has contributed to the speciation of the genus *Amblyornis*, which includes a central cordilleran form (MacGregor's Bowerbird), the Bird's Head form (Vogelkop Bowerbird), and a North Coastal Range form (Golden-fronted Bowerbird).

Bowers are obviously important to their owners, since males spend hours constructing, decorating and maintaining them. In fact, no male bowerbird has ever been noted to have mated successfully without first having built one. Surprisingly, this unique behavioural trait was only made possible by the availability of fruit. For bowerbirds thrive on the region's variety of abundant, high-energy fruit and, as a result, spend little time or effort foraging. The resultant release from the daily pressures of finding food opened up new evolutionary possibilities and allowed the development of bower-building and polygyny. Indeed, the 'many-female' breeding system is only possible if females have enough spare time to observe courtships displays, build nests, incubate eggs and provide for their chicks. Males, on the other hand, must be able to devote time and energy to tending their bowers and wooing females without the competing need to search for food. As a result, successful male

Satin Bowerbirds may mate with up to 25 different females, with the record being held by one male who inseminated 53 individuals within 12 months. Interestingly, the non-bower-building Tooth-billed Bowerbird is the most committed leaf-eater, having evolved specialised serrated jaws, with cusps and notches, to facilitate its folivorous lifestyle. Without fruit, therefore, bower-building and the polygynous breeding system of bowerbirds could never have evolved.

But why are bowers and their associated decorations so important for female bowerbirds? In other words, what is the female assessing when she selects a mate? To try and answer these questions, biologists have undertaken numerous field observations, using remote cameras, and structurally altering bowers and their decorations, to determine the effect on male mating success.

Collectively, these experiments indicate that bowers may reduce the threat of sexual coercion, physical molestation and forced copulations, and allow females to observe court decorations from close range. Indeed, for females to profit from mate choice they need to be able to visit and watch the courtships of as many different males as possible and reject those deemed unsuitable. At the same time, a male has to carefully assess a female's readiness to mate while performing his dynamic vocal and dancing displays. If the signals appear favourable, then the male approaches slowly behind her to copulate. Any female with second thoughts can easily escape while the male negotiates his way around the intervening bower wall or maypole. In support of the threat-reduction hypothesis, experimental tinkering with the structure of bowers has shown that female Spotted Bowerbirds prefer to assess males through an intact straw wall, only to choose the most vigorously displaying male. If the wall is experimentally removed, then courting males lessen the intensity of their displays – an observation that implies that threat reduction has influenced the evolution of the species' physical and behavioural displays.[16]

Further evidence for a linkage between the bower and female sexual autonomy is provided by studies of the Satin Bowerbird, a species in which the male modulates his display intensity according to the female's body language. Gail Patricelli, a graduate student working with Gerry Borgia, created a remote-controlled robotic female bowerbird, dubbed a 'fembot', to determine the finer points of the species' sexual interactions. The highly engineered, feathered construction was able to rotate her head, fluff her wings, tilt her beak and assume a mating posture, to fully mimic the female's range of reactions. After carefully placing the fembot inside a bower, Patricelli operated the controls from a hide and, once a male was in attendance, sent four different female courtship signals, including consent. By studying the male's response, she was able to confirm that those individuals who adjust their display intensity to keep females more at ease and relaxed were the most sexually successful.[17]

Species that do not construct bowers but possess display courts have evolved alternative ways of reducing the risk of unwanted male attention. The female Tooth-billed Bowerbird selects a suitable male before arriving on a court, so there is little need for physical protection. In contrast, the male Archbold's Bowerbird has modified his display court such that aggressive behaviour is prevented. Orchid vines are draped on numerous overhanging branches to produce a series of curtains that criss-cross the court, almost reaching the ground. Should a female inspect his handiwork, the male flies down and maintains a crouched position below the vine curtains. If interest is shown, he approaches in a non-threatening way by keeping his body low to the court floor. In effect, the low-hanging vegetation reduces any opportunity of taking the female by surprise.

The reduction of threat by male bowerbirds may help our understanding of the display adaptations of other species. The Blue Bird-of-Paradise, for example, hangs upside down when displaying, a strategy that may have enabled the species' highly intense courtships to evolve, while at the same time reducing the threat to females. Since many species, especially waterfowl, gain reproductive success through aggressive sex (see *The Waterfowl's Story*), why would male bowerbirds build structures that thwart such behaviour? A likely explanation, according to Borgia, is that bowers encourage more females to attend on a greater number of occasions, outcomes that more than compensate for any loss of forced copulations.[18]

Cognitive ability

Mating success, however, depends on much more than just the offering of a functional or 'protective' bower. Females also take note of the overall quality of the male's building efforts. For instance, the mating success of Satin Bowerbirds depends on the degree of bower symmetry, as well as the density and type of sticks used – thick curved walls, made from fine tightly packed sticks, being the winners.[19] The critical importance of a carefully sculptured structure is reflected in the extraordinary lengths males will go to repair any damaged or asymmetrical bower. Indeed, Satin Bowerbirds have evolved a unique mechanism of rebuilding known as 'templating', in which they use the standing wall as a template to measure the size and position of every stick to be placed in the new wall. This process is repeated along the whole length of the bower and ensures that the new wall is a mirror image of the original standing wall.[20]

The courtship of Satin Bowerbirds has received particular attention, with the activity of individual birds being monitored for many years using automatically triggered cameras. Typically, once a female Satin Bowerbird finds an 'aesthetically' pleasing bower, she steps into the avenue of parallel sticks and surveys the decorated court from a vantage point predetermined by the male. Once she is inside, the male becomes very excited and runs to his display area to pick up his most prized possession and shows it off to his potential mate. He then carries out a ritualised display of exaggerated movements and sounds, dubbed the 'buzz/wing-flip'. This frenetic performance involves strutting and bowing, with outstretched and quivering wings, while simultaneously issuing a variety of mechanical-sounding vocalisations, such as buzzing, hissing and chattering. Finally, he makes a more intimate appeal, coming almost beak to beak, and softly mimics a variety of other Australian bird calls. If she is impressed, the female signals her willingness to mate by adopting a low, submissive crouching posture, and after copulation leaves to raise the next generation on her own, while the male readies himself for courting more prospective females. The relevance of this account is that it highlights two additional behavioural traits that influence female choice, or sexual selection: court decoration and mimicry.

The arrangement and quantity of displayed objects appear paramount, and this explains why males spend many hours improving their courts, even if it means stealing objects from rival males. Each species has its particular taste in colour, size and alignment of ornamentation. Satin Bowerbirds, for example, prefer bright blue items, Great Bowerbirds like green and white, Streaked Bowerbirds favour yellow and red, while Fawn-breasted Bowerbirds opt for green. 'Novelty' items, rather than objects of any particular colour, are preferred by Vogelkop Bowerbirds, while Regent Bowerbirds coat their bowers with pea-green saliva paint, sometimes using a leaf as a paintbrush. Objects may not be confined to natural debris and can include human rubbish, such as broken glass, bottle tops, snack-food wrappers, ballpoint pens, plastic straws and even children's toys. In general, the more decorations on display, the more successful will be the male owner: a finding underscored when researchers demonstrated that experimentally removing objects reduces a male's attractiveness to females.

Laura Kelly and John Endler from Deakin University have shown that male Great Bowerbirds use visual tricks to influence female judgement, by arranging objects so that they increase in size as the distance from the bower increases.[21] The resulting size–distance gradient creates a forced perspective and results in false perceptions of the bower's geometry when seen from inside the bower. If the displays are experimentally rearranged, by placing the larger objects closest to the bower and the smaller ones furthest away, the males set about restoring the original pattern. This happens very quickly – in all instances, the forced perspective is recreated within three days. The overall geometrical effect is that females may discern the court to be smaller than it is, while the male appears to be bigger, an illusion that seems necessary for successful mating.[22]

According to the biologist Seth Coleman, mimicry also plays a significant role in the final decision of mate choice by female Satin Bowerbirds. By analysing vocalisations from thousands of hours of recordings, Coleman showed that the best mimics, those with the broadest repertoires, were the most successful in the mating game. The top males accurately mimicked four to five species, including kookaburras, honeyeaters and cockatoos, while the inferior males could only imitate poorly one or two species.[23] The best mimics also possessed the largest number of court decorations and built the better-quality bowers. Although mimicry is widespread within the bowerbird family, only Satin Bowerbirds use it during courtship, suggesting that it must have been a pre-existing trait that was later adopted for courtship displays by the Australian species.

Despite Thomas Gilliard's view that understanding bowerbird courtship is a scientific challenge, progress has been made. Field experiments have shown that it is the female's assessment of male cognitive ability that underpins their elaborate courtship routines and accounts for the importance of bower design, court displays and mimicry. Jason Keagy and his colleagues, for example, gave male Satin Bowerbirds a set of problems to solve as an indicator of their cognitive ability. Since males have a strong aversion to anything red on their courts, the team created situations in which added red objects were increasingly difficult to remove. One test involved placing a transparent container over the offending items so that the males needed to work out how to tip the container and remove the objects. In a more challenging test, the red objects were fixed to the ground so that they could not be eliminated. Only the 'brightest' birds realised the nature of the problem and concealed the objects with leaf litter. When the team looked at the mating success of the birds, they discovered that the males with the best performances were also the most sexually attractive and mated with the most females.[24]

Females, it seems, may be using a combination of display traits as a sort of sexually selected intelligence test of potential mates. If there are fitness advantages to better cognitive ability and these are inheritable, then females choosing the 'cleverest' males are likely to have offspring with the same benefits. The higher cognitive ability of the descendants may enable them to live longer by avoiding predation, find more food, establish better territories, and avoid parasitic infection by having better immune systems. According to Gerald Borgia and Jason Keagy, females can be seen as sophisticated decision-makers able to make complex fitness-enhancing mating decisions, while males can be perceived as using their cognitive processes to construct highly effective new displays.[20]

Extended phenotypes

I have not yet fully explained the rationale for the present chapter's subtitle, 'extended phenotypes'. Before the seminal work of the evolutionary biologist Richard Dawkins, the phenotypic influence of a gene was thought to be limited to an individual's body. Even so, the outcome of a single genetic change can be wide-ranging and affect far more than a single facet of an organism's phenotype.

As an example, let us return to one of the key adaptations of the Bar-headed Goose, namely its high-affinity haemoglobin (discussed in *The Waterfowl's Story*). A mutation in one of the haemoglobin genes enabled the species to gain a significant advantage by being able to fly longer and higher for the same amount of expended energy. Those individuals that possessed the adaptation would have survived longer and been more likely to pass the mutated gene to subsequent generations. Over millions of years, the 'improved' section of DNA has come to characterise the species, rather than the alternative forms of haemoglobin gene present in other geese. The modified gene and its resultant protein have had a cascade of consequences for the species: a structurally different haemoglobin within red cells, increased oxygen binding by blood, enhanced oxygen delivery to heart and brain, and, ultimately, the ability to undertake the highest-altitude sustained migration of any species. This impressive range of consequences, from red cell to migratory behaviour, still falls within the classical view of a gene's phenotypic effect.

However, in his influential 1982 book *The Extended Phenotype*, Dawkins stresses that such a cascade of events should not be limited to the physical characteristics of an individual, but should include 'all effects of a gene upon the world.'[25] Each stick in a beaver's dam and each heavy stone carried by the male Black Wheatear during the breeding season are expressions of the individual's genotype, and therefore belong to the organism's phenotype.[26] So it is with bowers: female bowerbirds can read a male's genetic quality not only from his appearance and behaviour but also from the artefacts that he produces. To paraphrase Dawkins, although the bowerbird's phenotype may be made of sticks and bottle tops rather than living cells (hence an *extended* phenotype), it is no less a true phenotype.[27]

Since bowers constitute an extended phenotype, then, by definition, there must be genes controlling their construction and decoration. Such an argument holds true even though there is a major learning component. But how can we be so sure, given that no genetic research has yet been undertaken? Firstly, males reared in isolation still construct bowers, albeit rudimentary ones, suggesting that such behaviour is innate. Secondly, a genetic predisposition is likely, given that most bowerbirds exhibit the trait, despite its absence in all other avian families.

Since bowerbirds have taken millions of years to evolve their unique behaviour, and each male takes many years to master the art of bower construction, it is maybe not surprising that scientists require a little longer to reveal their genetic secrets.

CHAPTER 19

The Crow's Story

COGNITIVE SKILLS

As highlighted in previous chapters, the basal oscines – scrubbirds, lyrebirds, Australasian treecreepers and bowerbirds – all stayed within present-day Australia and New Guinea. Other early songbird lineages that also remained within the region include the pardalotes (Pardalotidae) and the fairywrens, emu-wrens and grasswrens (Maluridae). In contrast, a few basal lineages, such as the honeyeaters (Meliphagidae) and Australasian warblers (Acanthizidae), spread more widely, reaching the Philippines to the north and New Zealand, Samoa and Tonga to the east. But it was the 'core corvoids', comprising more than 750 species, which gave rise to the first global radiation of songbirds – an event that followed the emergence of an arc of islands off the northern coast of Australia.

Islands and archipelagos have influenced evolutionary thinking ever since the time of Charles Darwin and Alfred Russel Wallace. Indeed, it was the subtle differences between the mockingbirds of the Galápagos Islands that eroded Darwin's faith that species were the fixed products of God's creation. In the late 1930s, the ornithologist Ernst Mayr was struck by the absence of sister species of birds on many Indo-Pacific islands, observations that convinced him of the role of geographical isolation in speciation. Later, the relationship between the number of bird species and the size of an island inspired Robert MacArthur and Edward Wilson's equilibrium theory of island ecosystems. In their classic book *The Theory of Island Biogeography*, the two ecologists recognised that island biodiversity develops as a dynamic balance between the rates of colonisation and extinction.[1]

However, MacArthur and Wilson, along with all other early biogeographers, regarded islands as biodiversity sinks and gave little thought to the possibility of upstream or 'reverse colonisation'. Such an entrenched viewpoint has been challenged recently by a series of studies that suggest that the recolonisation of continents may be more common than was previously assumed.

In 2005, Christopher Filardi and Robert Moyle from the American Museum of Natural History studied monarch flycatchers (family Monarchidae) from Australasia and the Pacific islands. They were able to show that oceanic islands are not the evolutionary 'dead ends' that they have traditionally been thought to be. In fact, the Pacific islands turn out to be a powerhouse of monarch biodiversity, with the proliferation of insular species leading eventually to the recolonisation of continental Australasia.[2] The speciation of whistlers (family Pachycephalidae) provides a further example. After an initial colonisation of Melanesia, some species returned to Australia and may have even undergone a second round of island spread.[3] Upstream colonisation has also been documented in the New World, where new evolutionary lineages of parrots, orioles and tyrant flycatchers evolved in South America after ancestors arrived from the Caribbean islands.[4] In fact, a recent study provides strong evidence that islands not only contribute to continental diversity but also provide proportionately more taxa than expected based on the number of species that inhabit them.[5]

The above case studies, however, involve a small number of species and pale into insignificance when compared to the global radiation of the 'core corvoids' that originated in the proto-Papuan archipelago during the late Eocene/Oligocene.[6] Before this time, the northern rim of the Australo-Papuan plate (present-day New Guinea) was submerged, and it only began to rise out of the sea when it approached and collided with plates to the north. The resultant uplift produced a string of volcanic islands along the plates' edges (the Sunda Arc), as well as many islands to the west of New Guinea towards Eurasia. These oceanic landmasses would have provided novel habitats and are thought to have acted as powerful drivers for speciation, as well as stepping stones for dispersal between Australo-Papua and Asia. According to Knud Jønsson, the archipelago contained islands of the right size, number and proximity to continental areas to enable them to function as engines of global songbird diversity, with many new evolutionary lineages emerging within a relatively short time span (Figure 19.1). This cauldron of speciation produced, among others, fantails (Rhipidura), woodswallows (Artamidae), shrikes (Laniidae), birds-of-paradise (Paradisaeidae) and the crows and their allies (Corvidae). Some of these families adapted to forest environments and remained within New Guinea, while others favoured open habitats and evolved morphologies and behaviours that enabled long-distance dispersal and subsequent colonisation of all continents except Antarctica.[7]

Science, however, does not proceed in an orderly fashion, as is often portrayed in popular accounts. In reality, as in many a good detective story, there are false leads, misrepresentations and wrong turns on every page. As

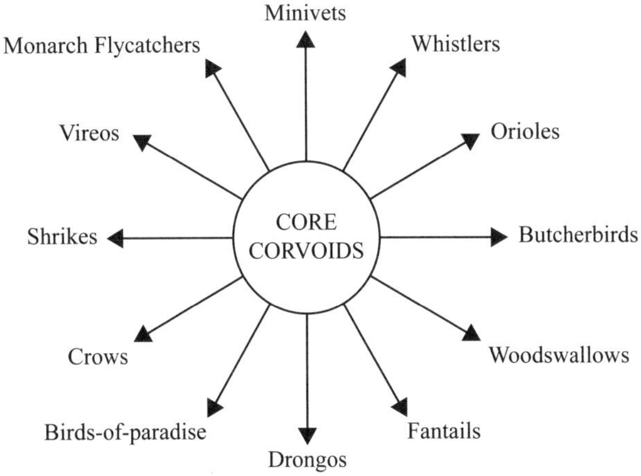

Figure 19.1 The 'core corvoids', comprising more than 750 extant species, constituted the first major radiation of songbirds from the Australian ancestral areas.

we have seen, the story of early passerine evolution is no exception, especially concerning the time calibration of basal phylogenies. As theories are published, there are always researchers eager to challenge the results. Indeed, this forms the basis of how science progresses. Given this fact, I should highlight an alternative scenario for core corvoid evolution, one recently proposed by Robert Moyle and his colleagues. This American team incorporated fossil-based time calibration and concluded that songbird diversification began in the Oligocene and accelerated in the early Miocene, a much later date than previous estimates. As a result, Moyle believes that the first major diversification of songbirds must have occurred within an isolated Australian continent, since New Guinea had already formed from the accretion of volcanic islands by the Oligocene.[8]

What is undisputed, however, is that the largest group to emerge from the core corvoids were the Corvidae, a family that today consists of 130 species, comprising jays, nutcrackers, treepies, choughs, ravens, jackdaws, rooks, magpies and crows. Jays remained within forested environments, specialising in eating nuts and acorns, and have a lifestyle that is probably similar to the corvids' common ancestor. In contrast, the magpies and crows left the forests and moved into more open spaces to become omnivores. Crows (genus *Corvus*) were the last clade to evolve, around 17 million years ago in the Palaearctic, and now make up one-third of species, currently 46. An early dispersal reached North America and the Caribbean, while Africa and Australia were not colonised until the Pliocene, around 5 million years ago.[9]

The corvids' social learning skills, flexible behaviour and problem-solving abilities have contributed to their success and allowed them to occupy almost every ecological niche throughout the world. But how intelligent are they, and how did their cognitive skills evolve?

Crows and ravens have long been viewed as clever by humans, and they feature prominently in our mythology and folklore. Some Native Americans, like the Haida people, believed that the raven was the ultimate spirit that created both the Earth and the Milky Way. Siberian folklore tells of ravens teaching humans the skills needed to survive in their harsh environment, while Celtic art portrays crows speaking into the ears of men, as they were considered to have prophetic powers. Odin, the principal deity of the Norse pantheon, was long known as the 'raven god' and his two birds, Huginn (thought) and Muninn (memory), kept him informed of events, so ensuring his continual wisdom. Aesop, in his familiar fable 'The crow and the pitcher', also depicts the crow as a highly inventive species, one capable of solving challenging problems.

While it may have been easy in the past to dismiss these stories as the myths and fables that they are, recent field and laboratory observations have shown that the corvids' longstanding reputation for intelligence is well founded. Crows, for example, have developed a greater number of foraging innovations, including the use of tools, than any other bird group. Carrion Crows in Japan have learned to crack open particularly hard nuts by dropping them among the traffic. To avoid being run over, the birds wait beside pedestrian crossings and retrieve the shell's contents once the traffic lights have turned red. In Britain, some Rooks have discovered novel ways to feed from rubbish bins at motorway service stations. Single birds will pull at the bin liner bag with their bill, tuck the excess under their foot, and repeat this action until they reach the discarded food at the bottom of the bag. They are also known to work in pairs and will feed on the refuse as it emerges from the bin, or will spill it onto the ground so that their colony members can join in the feast.

New Caledonian Crows are renowned for their sophisticated tool-making skills: behaviour once believed to be the sole prerogative of primates. Such implements, ranging from hooked sticks to barbed leaves, are used to obtain their favourite food, the nutritious grubs of longhorn beetles.[10] The crows poke and irritate their prey so that the grub grasps the tool in its jaws, and is hauled out. Only a few larvae are required to satisfy the bird's daily energy requirements, highlighting the substantial rewards available to competent tool users.

In 2016, Christian Rutz, from the University of St Andrews, discovered a second, highly dextrous, corvid tool-maker: the Hawaiian Crow or Alalā (Plate 25).[11] Although the species became extinct in the wild in the early

2000s and only survives in captivity, several lines of evidence suggest that the behaviour is innate: juveniles develop functional tool use without being taught, while proficient tool use is a species-wide phenomenon. Interestingly, New Caledonian Crows and Hawaiian Crows are distant relatives, which imply that their skills evolved independently. But why would tool use evolve in some corvid species and not others? It may be significant that both tool-makers come from remote tropical islands that have no native woodpeckers. As a result, there would be little competition for their preferred prey, grubs embedded in tree trunks. Furthermore, the islands have few predators, and there would have been no need for either species to spend time scanning for hawks and snakes, a fact that allowed them to adopt the head-down position required for tool use. It is also possible that the species' facial structure is relevant, since both possess straighter, blunter and shorter bills than other corvids. Such features would have made them adept at handling tools, and over time the beaks would have evolved further, making both species ever more proficient foragers.

Other aspects of corvid behaviour provide us with clues to their intelligence. Young corvids have a prolonged developmental period, typically lasting several months, before becoming independent from their parents: an upbringing that provides them with greater opportunities to learn the skills needed for survival. Corvids often live in complex social groups and will band together to ward off intruders. Rooks form large colonies, where juveniles can learn from both family members and unrelated birds. Furthermore, adult Florida Scrub Jays, a cooperatively breeding species, will often help raise offspring that are not their own, while the nests of Mexican Jays may contain chicks from several sets of parents.

For corvid societies to function, group members require the ability to recognise one another, engage in long-term relationships, and monitor each other's social status. In other words, corvids must have evolved the specialised neural systems needed to process such types of information. Many corvids have prodigious memories. The Clark's Nutcracker, for example, prepares for winter by spending several weeks during the autumn amassing large food stores. Typically, these harvests consist of more than 30,000 pine seeds which are buried in up to 5,000 caches located over a 20-square-kilometre area. The birds then rely on memory to find the seeds that are needed to sustain them through the North American winter. To cope with such a strategy, Clark's Nutcrackers have evolved an excellent memory for spatial information, and it is likely that they can remember the location of fixed landmarks such as trees and rocks.

Like humans, corvids exhibit episodic memory, a form of recall that allows the remembrance of experiences: the what, when, where and who from our past. Though difficult to test in birds, there is evidence that some species possess this type of memory. The California Scrub Jay, like Clark's Nutcracker, also caches food. In laboratory experiments, adult scrub jays were allowed to cache both a stable food source (nuts) and a rapidly biodegradable food source (wax moth larvae). After both a four-hour and a five-day delay, the jays were permitted to recover whichever food they wanted. Because the birds prefer wax moth larvae, they recovered more larvae than nuts after the four-hour delay. However, after five days, when the larvae would have decayed, the jays favoured the nuts. In other words, scrub jays can remember what they have cached, as well as when and where – all hallmarks of an episodic-like memory. Furthermore, the birds seem to form integrated memories of past events, rather than encoding the information separately, and can use them to plan for the future.[12]

Corvids are also capable of knowing what they know or thinking about thinking, a trait that behavioural scientists call metacognition. Arii Watanabe and his fellow psychologists from the Behavioural and Clinical Neuroscience Institute at Cambridge University devised an ingenious experiment to prove that this was the case.[13] California Scrub Jays were coached to observe a researcher conceal food, a wax worm, in one of two adjacent compartments. Each unit contained four beakers containing sand. In one, known as the 'forced choice' box, three beakers were covered with lids, and the larva was always hidden in the randomly pre-selected open beaker. In the second compartment, all the containers were open, and the larva was buried randomly. Two screens, each with a spy hole, separated the jay from the compartments so that the bird could watch the food being hidden. Immediately afterwards, the jay was allowed to choose a cup. During training, the 'forced choice' and 'free choice' tests were conducted at different times. However, during the live test, both compartments had food concealed simultaneously, and the jay could enter only one compartment. As a result, the birds were forced to choose a spy hole to look through during the concealment of food in order to gain the maximum amount of information needed to make the correct choice. If the birds were smart, they would soon realise that the best strategy for the future would be to watch the researcher placing food in one of the four open beakers, knowing that the larva could only be put in the open cup in the other compartment. The psychologists reasoned that if the jays were capable of metacognition, they would spend more time observing the 'free choice' compartment. This is exactly what the birds did. It seems, therefore, that corvids, like humans, can collect and store information to solve future problems.

The evolution of cognition

Corvid brains lack a cerebral cortex, but as we have seen, their cognitive skills are on a par with primates. To understand how their brains work, German neuroscientists Lena Veit and Andreas Nieder wired up crows with electrodes to record their brain activity while they performed a series of abstract reasoning tests or brain teasers on a computer.[14] An image was flashed on the screen for a short time, followed immediately by two more images – one the same as before and one different. The experiment required the birds either to find a match with the initial image or to identify the different image. After a short practice, the crows were able to accomplish the challenge, even when the images were unfamiliar. As they performed the tests, the researchers discovered that most of the resultant brain activity was located in the nidopallium caudolaterale (NCL), an area analogous to the human prefrontal cortex (PFC) where higher thinking occurs and executive decisions are made (Figure 19.2). The NCL is situated towards the middle of the bird's brain, and the researchers noticed that different areas became active depending on whether the chosen images were supposed to be the same or different. Eventually, the researchers were able to predict the crow's choice, even before the bird had a chance to submit its selection.

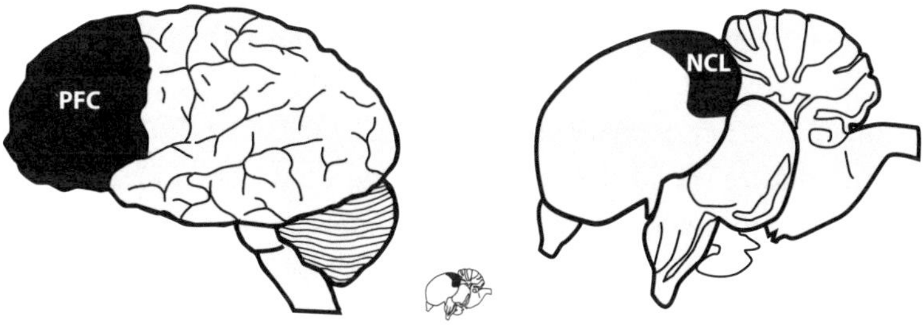

Figure 19.2 Side view of a human (left) and a pigeon brain (right). The pigeon brain in the lower middle part of the figure is to the same scale as the human brain. Birds lack a prefrontal cortex (PFC) and have instead the nidopallium caudolaterale (NCL), which performs similar 'higher' functions. Modified from Güntürkün (2005).[15]

Avian and mammalian brains share very few structures from our last common ancestor, which lived 300 million years ago during the Permian. Nevertheless, even though the anatomy is very different, the functional similarities between the NCL and PFC suggest that intelligence in primates and birds arose independently as the product of convergent or parallel evolution. At some point, the corvids' ancestors must have found themselves in an ecological niche where cognition enhanced their chance of survival, so their brains developed a cognitive ability to match that of primates. But, given that their brains are much smaller, how did evolution solve the problem?

The answer lies in the number of neuronal cells and the connections they make. In 2016, an international team, led by Pavel Němec, calculated the concentration of brain cells or neurones in a total of 28 different species of parrot and songbird.[16] The technique involved the separation of the various anatomical structures of the brain, followed by paraformaldehyde fixation, detergent separation, and counting the cellular components in the resultant suspensions. It turns out that the brains of songbirds and parrots contain very large numbers of neurones, at cellular densities considerably exceeding those of mammals, including primates. For example, a Goldcrest's body mass is around nine times smaller than that of a mouse, but its brain has at least twice as many neurones. But it is the corvids and parrots which harbour the greatest number of cells, at a total level comparable to those of small monkeys. Crucially, these 'extra' neurones are located in the forebrain, in an area known to be responsible for computational capacity, advanced behaviour and cognitive complexity. In other words, avian brains have the potential to provide a much higher 'cognitive power' per unit mass than do mammalian brains. One of the selective pressures for corvid intelligence may have been the deteriorating climate during the late Miocene to Pliocene, when innovative foraging strategies would have increased the chances of survival.

Flying animals also tend to have smaller genomes than their non-flying relatives. This is certainly the case for both birds and bats, and it is likely that the metabolic requirements for powered flight constrained their genome size. Recent studies reveal that animal genomes have been in a state of flux throughout evolutionary history, with the amount of DNA gained being counterbalanced by genetic deletions.[17] In the case of flying birds, however, the genetic loss appears to have outstripped any gains. The importance of this finding is that compact genomes are associated with smaller cell sizes. As a result, the greater surface-to-volume ratio of avian cells allows for faster transport of nutrients and signals across cell membranes. Since signals do not have to travel so far, birds can process information much faster than mammals. It is the combined effect of a greater number of neurones, and smaller neurones, that allowed the corvids to evolve their remarkable intelligence, while at the same time keeping their bodies light enough for flight.

Before we move on to discuss the Passerida, the largest global radiation of songbirds, we need to highlight another core corvoid clade: the enigmatic and flamboyantly decorated birds-of-paradise. For their story provides a fascinating insight into the combined effects of vicariance, sexual selection and diet.

CHAPTER 20

The Bird-of-Paradise's Story

SEXUAL SELECTION

Europeans became obsessed with the Paradisaeidae after skins of the Lesser and Greater Birds-of-Paradise were brought home from the Moluccas in 1522, by the surviving crew of Ferdinand Magellan's expedition.[1] Indeed, by the end of the 1540s, mounted birds-of-paradise were a 'common site in the cabinets of Europe and Turkey', according to Pierre Belon's *L'Histoire de la Nature des Oyseaux* (1555).[2] Naturalists and collectors of the day were beguiled by the bird's strange anatomy, for the legs had been removed by local traders who mistakenly believed that Westerners were only interested in the birds' colourful plumages. The bizarre conjecture that the birds lacked feet resulted in the notion that the birds spent their entire lives in perpetual flight, living off the ether's dew, only to grace *terra firma* on death. It was their supposed ethereal existence, between heaven and earth, that earned the birds their befitting name, 'birds-of-paradise'. Others, only slightly less gullible, knew they were being fooled and believed that the specimens had been cobbled together from different animal parts by unscrupulous dealers.[3] Today, of course, we know better.

Plate tectonics and vicariant speciation

The Paradisaeidae originated approximately 24 million years ago among the string of small islands and archipelagos that emerged in the seas to the north of Australia, in the area of Papua-Melanesia.[4] The 41 currently recognised species are core members of a large assemblage of crow-like passerines. Remarkably, despite being so utterly different, they have all descended from a common ancestor, one that probably resembled the drably plumaged Paradise-crow, a species restricted today to Halmahera and the islands of the North Moluccas. The fact that such a nondescript ancestor could have undergone such unparalleled diversity is the result of New Guinea's unique combination of tectonic activity and abundant food supplies.

New Guinea's landmass, the second-largest island in the world after Greenland, was never part of any continent but was created in recent geological times by the relentless forces of plate tectonics. As the northerly-moving Australian plate collided with the Pacific plate, its leading edge buckled, rising out of the sea to form the southern half of New Guinea, an area that includes most of the island's central mountains. To the north, a sequence of volcanic island arcs, formed by rising magma between the two subducting plates, was carried southwards as if on a geological conveyor belt. Eventually, these islands collided and docked ('accreted' in geological jargon) with the newly formed New Guinean landmass to form the northern part of the present-day island. Once fused, the 30 or so fragments of continental crust added to the rising central mountains, as well as forming the Huon Peninsula, and the Foja and Adelbert ranges in the north. This convergence process is ongoing, and several million years into the future the island of New Britain will have been added to New Guinea's mainland as the Australian plate bulldozes its way ever northwards.

It was against the backdrop of extreme geological turmoil, amid the jostling of volcanic island arcs and fast-growing mountains, that the birds-of-paradise first made their appearance. However, in contrast to their closest relatives, the monarch flycatchers, crows and drongos which dispersed to other continents and remote oceanic islands, the birds-of-paradise did not venture far and remained in the vicinity of New Guinea. The most distantly scattered birds were the forebears of the Paradise-crow, the oldest of the bird-of-paradise lineage, one that diverged 17 million years ago.[4] Today the Paradise-crow is found on the islands of the North Moluccas, some 300 kilometres to the west of New Guinea, although, interestingly, they did not fly there, but were carried. It is now known that Halmahera and the adjacent islands used to be part of mainland New Guinea. As the Pacific plate sheared past the Australian plate, fragments of the continental margin detached, creating islands that spilled out into the surrounding ocean,[5] carrying with them their cargo of ancestral birds-of-paradise.

Similar geological processes underpin the evolution of two highly ornate sickletails from the genus *Diphyllodes,* the Wilson's Bird-of-Paradise and the Magnificent Bird-of-Paradise (Plate 26). Currently, the two species are separated by the narrowest of barriers, the 3-kilometre-wide Sagewin Strait, although this was not always the case. Millions of years ago, the clash of tectonic forces prised a fragment of land, together with its population of early sickletails, from the western tip of New Guinea. As the newly formed island drifted away, the isolated population evolved into one of the most striking species on Earth, the Wilson's Bird-of-Paradise. The male presents a riotous palette: vivid yellow nape, blue skullcap, crimson back and coverts, iridescent emerald-green breast and shiny curlicue tail feathers. Not surprisingly, Wilson's Bird-of-Paradise is regarded by many birders as their most sought-after species. Indeed, the ornithologists Tim Laman and Edwin Scholes have described the species as 'a paragon of the endless forms for which the birds-of-paradise are renowned. Their extremes in size, shape, color and behavior seem to have been literally embodied in the males of this one species.'[6] Later, as a result of continuing geological forces, the island split and a 450-square-kilometre splinter, the Indonesian island of Batanta, headed back towards the mainland. Despite this splinter approaching to within 3 kilometres of the larger island of Salawati, the population of Wilson's Bird-of-Paradise has not been able to cross the strait. As a result, Batanta's birds remain isolated from their closest relatives on Salawati and the New Guinea mainland, the Magnificent Bird-of-Paradise. The Sagewin Strait must surely be one of the smallest geographical barriers in the avian world to have resulted in vicariant speciation.

According to the most recent phylogeny, the Paradise-crow's nearest kin is the bizarrely vocal Trumpet Manucode (genus *Phonygammus*), a species that evolved approximately 14 million years ago.[4] Four million years later, the closely related genus *Manucodia* diverged, forming an assemblage of medium-sized birds, with glossy black to purple-green plumages. Manucodes, in general, are widespread and inhabit the mainland's lowland forests, although two species, the Trumpet Manucode and Glossy-mantled Manucode, are found in the Aru Islands, located to the southwest. Both species may have flown there, but it is now thought more likely that they reached the islands during the Pleistocene ice age when the sea levels fell, and a land bridge formed connecting Aru to New Guinea. The adjacent Kai and Tanimbar Islands are surrounded by much deeper water and, since no land bridges ever formed, birds-of-paradise have been unable to gain a foothold on these otherwise ecologically similar islands. Another example of a land-bridge effect is seen in the Magnificent Riflebird, which inhabits both New Guinea and the rainforests of Cape York in Australia. When sea levels rose and the Torres Strait formed, 50 thousand years ago, Magnificent Riflebirds were left on either side of the divide. Though still considered a single species, the two populations are now evolving separately, since the birds are reluctant to fly across the open water.

The formation of New Guinea's fragmented and fast-growing cordilleras, which now stretch from east to west, not only divided the ancestral birds-of-paradise but also provided the opportunity for altitudinal speciation. The central mountains, for example, form a well-defined boundary between the low-lying habitats of the Lesser Bird-of-Paradise to the north, and those of the Greater Bird-of-Paradise to the south. In contrast, the sicklebills, which diverged around 12 million years ago, became restricted by altitude in the central mountains – the Brown Sicklebill living above 2,000 metres while the Black Sicklebill inhabits the forests below. Other birds-of-paradise, species that evolved on the volcanic islands after they had fused to the mainland, have highly restricted distributions since they are surrounded by vast lowland basins of unsuitable habitat. The rarely seen Bronze Parotia, for example, is confined to the remote Foja Mountains in West Papua, a fact that delayed its identification as a new species until 2005. The accreted mountains of the Huon Peninsula, lying to the northeast, possess a number of endemic species, including the Emperor Bird-of-Paradise (Plate 27). This species, adorned with unusual flank plumes and iridescent green head and throat, diverged 6 million years ago from the rest of its genus, all of which remain in New Guinea's lowland areas.

But geology alone cannot account for the bewildering diversity in appearance and behaviour that characterise the Paradisaeidae. Their extravagant plumages, complexity of colours, piercing calls and minutely choreographed displays all scream one thing: female choice, or 'sexual selection'.

Female choice, male handicap

It was Charles Darwin's realisation that nature does not select, individuals do, which led to the concept of sexual selection.[7] He argued that while natural selection was concerned with staying alive, sexual selection was about attracting mates, a preference unconstrained by the functional laws that govern evolutionary form. Although his classic example was the flamboyant tail of the peacock, Darwin's theory applies equally well to other species, including the birds-of-paradise. For, like the peahen, the female bird-of-paradise's choice of mate appears to be quite arbitrary, limited only by the individual's sensory and cognitive abilities. So, while the males' long tails, wiry end-feathers or elongated head-plumes may seem arbitrary or just unimportant to us, they are everything to the female. And it is because such preferences can operate outside the limitations imposed by Darwin's 'survival of the fittest' that the absurd ornamentations and behaviours of male birds-of-paradise have evolved. This view, however, was not widely accepted at first.

In the years following his seminal publication *The Descent of Man, and Selection in Relation to Sex* (1871), Darwin's ideas were deemed too controversial, and his theory fell into disrepute. Indeed, most biologists, including Alfred Russel Wallace, hated the apparent randomness of Darwinian sexual selection and believed that female choice of costly ornamentation was incompatible with natural selection. Instead, the Wallacean view emphasised that any ornamentation or dexterity in display must co-exist with superiority in utilitarian qualities. A notable exception in the early twentieth century, however, was the British geneticist and statistician Ronald Fisher, who used mathematical reasoning to resolve the paradox and showed for the first time how 'beauty for beauty's sake' might come about.[8] In essence, Fisher realised that there must be a genetic basis for both female preference and male ornamentation, and that if both coevolved then runaway selection for a particular trait could result. The increase in both the selected trait and the female preference for it would continue 'exponentially, or in geometric progression' until practical physical constraints limited any further exaggeration.

To better understand the concept of 'Fisherian runaway', let us consider the case of the ornamental head-wires of the King of Saxony Bird-of-Paradise (Plate 28). This iconic highland species is one of the most bizarrely adorned on Earth, with the male boasting a pair of half-metre-long ornamental head-plumes – head-wires, or flags – that are unlike any other known feather. Each one, emblazoned with 40–50 sky-blue pennants, projects from its head and can be waved seductively in front of prospective females.

Now imagine, if you will, the ancestral population of King of Saxony Birds-of-Paradise, somewhere in the central highlands of New Guinea, and suppose that one of the males has developed two small protruding head feathers, rooted behind each eye. Since this aberrant trait is the result of a random mutation, possibly one base change in a billion, it will be passed on to the next generation. If its possession is neutral and does not affect the bird's survival, it may, in time, become more widespread within the population. Now imagine that the females of the species, at some later point, develop a genetic preference for males with small protruding head feathers. The reasons can be arbitrary or even whimsical – maybe the ornamentation just looks attractive or 'sexy' to them. Since there will be a cohort of males with head feathers, the females are likely to choose the ones with the most prominent plumes. Crucially, and this is the nub of Fisher's theory, the resulting progeny will possess two sets of genes, one for longer head feathers and one for their preference. Of course, only the genes coding for head feathers will be expressed in males, while the preference genes will only manifest in females. Furthermore, since males with the longest head feathers will have the greatest success in attracting females, they will father more offspring, and the long plumes will become more common in subsequent generations. The important point is that female King of Saxony birds will tend to favour head-wires that are slightly longer than average, however long that average has become. But this raises the stakes and leads to a phenotypic 'arms race'. In the future, to stand out from the competition, a male would have to have even longer feathers, or some additional feature to gain the attention of a mate. The feathers, for example, could develop unilateral vanes or be adorned with odd looking projections or be wafted enticingly about the head.

I think you can see where this discussion is heading. If females evolved a preference for feathers with all three modifications, then over evolutionary time male King of Saxony Birds-of-Paradise would sport long, mobile and highly ornate head-wires, just as they do today. In the words of Richard Dawkins, the male's bizarre 'flags' can be viewed as the 'end-product of explosive, spiralling evolution by positive feedback.'[9] But female choice can be fickle. The preference for today's head-wires could just as easily change in the future for no other reason than it's what females have come to prefer.

However, Darwin's idea of 'aesthetic evolution' by mate choice remains, even today, an anathema to most scientists.[10] Instead, many theorists prefer alternative explanations: neo-Wallacean ideas in which female choice has a beneficial genetic effect for her progeny. It may be, for example, that the burdensome appendages are a male's means of advertising his health, one free of disease and infection. In effect, this means that females favour traits that are disadvantageous for males. Such an idea was championed by the charismatic and energetic Israeli biologist Amotz Zahavi, who proposed what has become known as the 'handicap principle'.[11] Essentially, Zahavi, who died in May 2017, regarded all sexual signals as having a cost, a downside for their owner. In the case of the male King of Saxony Bird-of-Paradise, the head-wires indicate not only underlying vigour but also that their possession is consuming considerable resources. In other words, the male birds are saying 'I have survived in spite of these enormous head-wires, and I am, for this reason, fitter and a better catch than all the other males.' As long as the advantages of 'healthy' genes outweigh the cost of the handicap, then the benefit to the offspring will be greater than for the progeny that result from random mating.

I can fully appreciate Zahavi's reasoning, after observing the remarkable display flight of Africa's Long-tailed Widowbird, a member of the weaver family (Ploceidae). During the breeding season, the male undergoes a remarkable Cinderella-like transformation, changing from an inconspicuous brown individual to an astounding bird with jet-black plumage and up to eight tail feathers of nearly half a metre in length (more than twice its body length). He also develops orange and white epaulettes (shoulders) and a bluish bill. Once the male's new outfit is complete, he displays during the breeding season by flying just above the bushveld's grasses, keeping his wings above his body, beating them slowly and erratically. At the same time, he trails his absurdly long tail feathers vertically, fluttering in the wind, in the form of a deep, long keel. This performance, which can be seen by females from a great distance, is clearly energy-sapping, and the male gives the impression of relief as he finally drops to the ground or clings to a prominent vantage point. The possession of such long feathers is obviously not beneficial for the male, as it makes flying more difficult while making it easier for predators such as falcons to pick him off. Even on the ground, he appears encumbered and drags his appendages through the grass while feeding. Indeed, just to produce the new complement of feathers every year involves a substantial investment of metabolic resources.

Nevertheless, despite these obvious disadvantages, there appears to be no limit to the female's desire for long tails. In a classic experiment by Malte Andersson, the tail length of several males was manipulated by cutting off tail feathers from some individuals and gluing them onto others. Amazingly, Andersson's results revealed that females preferred males with longer tails than those that occur in the natural setting.[12] These findings are often quoted in support of Zahavi's 'handicap principle', because only the fittest males will be able to produce the largest tails and survive long enough to pass on their 'good' genes. However, Zahavi's idea, as originally presented, was not just vague but flawed, and it took the input of Oxford mathematician Alan Grafen to rescue the theory by showing that such systems can be evolutionarily stable.[13] As a result, the aesthetic versus good-gene debate – the Darwinian versus neo-Wallacean polarisation – continues to this day.

The courtship displays of birds-of-paradise can be astonishingly complex, and for the male King of Saxony Bird-of-Paradise, head-wire waving is merely the opening act of a highly ritualised performance.[14] If a female shows interest, the male then reveals his mantle cape and breast shield feathers, after which he makes a descending invitation flight to a nearby springy vine, where an even more complex display takes place. Here, in the understorey, he bounces vigorously to vibrate the vine, which in turn bounces the female perched above. This unique bouncing show, with all feathers displayed, is followed by a dance consisting of upper-body rotation and side-to-side waggling of the head to cause the head-wires to swirl about. Each element of the male's elaborate and bizarre courtship has coevolved as the result of the arbitrary whims of female choice, a fact best understood in terms of the Fisherian runaway model, discussed above. But can such courtship displays tell us anything about the evolutionary relationships of the birds-of-paradise?

According to the American ornithologist Edwin Scholes, they can.

Scholes believes that the complex dance routines can be broken down into a series of smaller individual units, and that these can be combined to create a single choreographed piece. As a result, males can create novel performances by just reorganising and modifying the individual steps. Scholes also believes that it is their modularity that underscores courtship evolution, since by simply rearranging existing components, new dances can evolve quickly. If females prefer these new combinations, then these courtships will eventually become standard behaviour for that species. The dances' modularity may also help chart the evolution of the Paradisaeidae, since, once the relationship between the dance elements is known, then the phylogeny of the various species could be deduced. Indeed, this approach suggested that the King of Saxony Bird-of-Paradise is most closely related to the genus *Parotia*, a fact confirmed recently by molecular genetics.[4]

At first sight, manucodes appear to be the most normal of all the birds-of-paradise, given that both sexes look the same and that they live together in pairs with the males helping to raise the young. But sexual selection has created a hidden secret, one that cannot be seen, but which can only be heard. For male manucodes have evolved a greatly elongated windpipe that coils around the abdomen, just beneath the skin, before curling back to the lungs (Plate 29). Their resulting low-pitched calls penetrate the dense forest, helping to keep pairs in contact and to defend their territories from rivals. The most unusual vocalisations belong to the Curl-crested Manucode, a species restricted to the forests of several islands lying off the east coast of New Guinea. Their evocative, mellow, fluting calls are a characteristic sound of the islands' interiors, a vocalisation that has been honed by aeons of female selection

But why have the world's most extraordinary birds evolved in New Guinea, and why have they not dispersed to other nearby regions? In part, it relates to the island's lack of large predators. New Guinea was never connected to Asia, and its mammals have either had to fly across or make their way from Australia when sea levels were much lower. Over time, the island's dense forests have favoured smaller mammals – echidna, tree kangaroos and bats – most of which are insectivorous or herbivorous and pose no threat to birds. However, the basic explanation relates to the birds' diet of fruit. Unlike the manucodes, most of the 'showy' Paradisaeidae feed on complex fruits, berries and capsular fruits that are widespread, highly nutritious and available throughout the year in the lush rainforests. As a result, the birds do not have to spend much time and energy foraging and can devote more effort to matters relating to sex. In *The Bowerbird's Story* we saw how a frugivorous lifestyle is linked to elaborate bower-building and courtship rituals. A similar diet, in the same part of the world, has enabled the birds-of-paradise also to evolve a 'many-female', or polygynous, breeding system. For polygyny requires females to spend lengthy periods observing intricate courtship displays to ensure they pick the 'best' male. Females also have to raise their families by themselves, as the males spend most of their day tending display sites and courting other females. Even adolescent males practise their complex courtship routines to prepare themselves for the rigours of the female selection process to come. So, without New Guinea's abundance of all-year-round high-energy fruit, the evolution of the birds-of-paradise's promiscuous breeding systems and the males' absurd exaggerations, could never have happened. As David Attenborough aptly concluded, 'just as fruit plays a significant role in the biblical view of paradise, so it has created a paradise on earth for the birds.'[15]

However, it is because of their complex and unique relationship with the environment that the birds-of-paradise are restricted to New Guinea. Their strong attachments to display sites and the fact that males and females spend long spells apart make successful long-distance dispersal and the establishment of new breeding populations highly unlikely.

CHAPTER 21

The Starling's Story

STRUCTURAL COLOURS

The starling's story is one of structural colours – how they evolved and how they promoted a dramatic diversification of a group of African birds. But first we need to discuss the origins of the starling family (Sturnidae) and how they came to inhabit the African continent in the first place.

The Passerida

As we have already highlighted (see Figures 16.1 and 19.1), the core corvoids emerged from the proto-Papuan archipelago around 45 million years ago. But this was only the first of several early songbird radiations. The core corvoids were followed by two smaller dispersals – the 'transitory oscines' – and then by the Passerida, the largest of all songbird radiations. It was from the latter group that the starlings evolved, although they would require a further 20 million years to do so.

The 'transitory oscines' first gave rise to three small clades that remained within Australasia: the berrypeckers and longbills (Melanocharitidae), satinbirds (Cnemophilidae), and Australasian robins (Petroicidae). The berrypeckers and longbills consist of a variety of small songbirds with generally dull plumage that are restricted to the mountain forests of New Guinea. The females of two species, the Fan-tailed and Streaked Berrypeckers, appear unique among the oscines in that they exhibit a reversal of the usual pattern of sexual dimorphism, with females being larger and bulkier. The three species of satinbird are also restricted to the central montane areas of New Guinea. They all build dome-shaped nests, with the female tending their single chick without any contribution from the male.

The Australasian robins comprise 49 species that occupy a broad range of wooded habitats from subalpine to tropical rainforest. They are 'perch-and-pounce' insectivores that can cling sideways onto tree trunks and survey

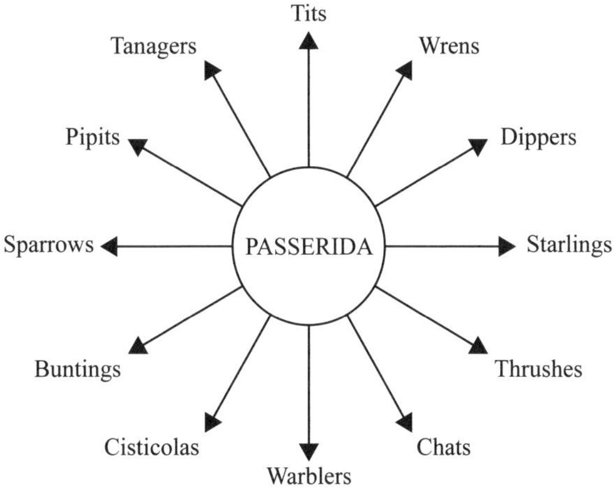

Figure 21.1 The Passerida, the largest songbird radiation, gave rise to over 3,500 species, from tits to tanagers.

the ground below without moving. The best known is the Black Robin, a species that occurs exclusively on the Chatham Islands, a small archipelago lying 700 kilometres east of New Zealand. By 1976, only two females and seven males remained, the lowest known population of any bird on Earth at the time. Despite only one of the females (nicknamed 'Old Blue' because of her coloured ring) being fertile, the population has slowly increased with intensive conservation efforts and now stands at around 250 individuals.

The second divergence spawned the picathartes and rockjumpers in Africa as well as the elusive Rail-babbler in Malaysia, Borneo and Sumatra.[1] Indeed, it was my interest in understanding the biogeography of this unusual group that first exposed me to the field of avian evolution and led to the research for this book. But it was the subsequent dispersal event, the Passerida, that was the most significant, as it spread rapidly across the world and gave rise to over 3,500 extant species.[2] They include not just the starlings, but also thrushes, warblers, white-eyes, finches, tanagers and many others (Figure 21.1)

The early history of the Passerida is conflicting and obscured by many gaps and uncertainties that result, in part, from the group's extremely rapid divergence. While some families must be closer relatives than others, the available data do not allow their precise evolutionary relationships to be resolved – a situation termed 'soft polytomy' by phylogeneticists.[3] Nevertheless, it is widely accepted that the Passidera emerged from Australasia, sometime during the Eocene,[4] and reached the Old World by one of two possible routes. Either they used the islands on the Sunda Shelf as stepping stones to Asia, or they crossed via the now submerged plateaus (Kerguelen, Crozet and Broken Ridge) to reach Africa via the Indian Ocean.[5] Whichever route they took, the common ancestor of starlings, oxpeckers (Buphagidae) and mockingbirds (Mimidae) appeared around 23 million years ago, amid the extensive forests of the northern hemisphere.[4] There then followed a prolonged period of global cooling, a climatic change that resulted in the population's fragmentation and subsequent radiation. Eventually, these early birds gave rise to the oxpeckers in Africa (22 million years ago), mockingbirds in North America (21 million years ago) and the starlings and mynas in Africa (16 million years ago).[6]

Surprisingly, members of the Sturnidae never dispersed to the Americas, and the continent's current superabundance of Common Starlings resulted from a bizarre introduction programme. In the 1890s, Eugene Schieffelin, an eccentric drug manufacturer from the Bronx, released 100 starlings in New York's Central Park. His idea was to bring to North America all the birds mentioned in Shakespeare's plays, and although bullfinches, skylarks and song thrushes failed to survive, the starlings adapted quickly. Since then the birds have spread to all parts of the United States, most of Canada, and parts of northern Mexico, with an estimated population today of around 200 million. The starlings' success relates to their exploitation of a large variety of habitats, nest sites and food sources, while their aggression and gregariousness make it difficult for native birds to compete. Sadly, the New World was not their only territorial conquest, for successful introductions also took place in the West Indies, New Zealand, Australia and South Africa at around the same time.

The present-day restriction of oxpeckers to Africa may not have always been the case, as environmental changes elsewhere could have considerably reduced their range. For oxpeckers have a highly specialised manner of feeding that involves the removal of ticks and other parasites from the skin of large grazing animals. As a result, the diverse megafaunal communities that once roamed across Eurasia and the Americas could have supported a widespread oxpecker-like ancestor. The rapid extinction of these herbivores at the end of the Pleistocene, however, would have led to a parallel extinction of any symbiotic birds. It has even been suggested that the association with large mammals is ancestral to the entire radiation, rather than merely an acquired behaviour confined to oxpeckers. Most of the starlings and their relatives, for example, regularly forage around the feet of large ungulates, and many will perch on animals while they feed. Although such associations are uncommon in the New World, several populations of Galápagos Mockingbird are known to exhibit oxpecker-like behaviour.[7] In addition to fruit and marine arthropods, they will also feed on skin parasites, drink blood and pick at the wounds on marine iguanas, nesting seabirds and even sea-lions.[8]

During the early Miocene, the African climate became significantly drier, a change that resulted in a marked reduction in the continent's forests. Since only a minority of extant African starlings are strictly arboreal, it is likely that such climatic changes and the subsequent emergence of grasslands kick-started the clade's speciation. The population then underwent two further radiations, one that colonised Madagascar and a much larger one that re-invaded the Palaearctic and the Orient around 14 million years ago.[9] Interestingly, the two species that are probably most familiar to western birders, the Common Starling and the Spotless Starling, belong to an isolated lineage at the base of the Eurasian radiation. It seems likely that both species evolved after their common ancestor became split following the formation of southern glacial refugia during the Pleistocene. Ongoing climate change, especially over the last 6 million years, encouraged further diversification of both the African and Eurasian starlings by creating extensive open habitats.[10]

Two additional traits, prying and flocking, deserve consideration, as they are germane to the evolution and speciation of the Eurasian clade. Prying or open-bill probing is unique to this lineage and is dependent on the evolution of powerful protractor muscles in the jaw. This anatomy enables the birds to insert their closed bills into a substrate and then open them forcefully to dislodge hidden prey. The thrust required is the exact opposite to that needed by other birds, which need the greatest force when closing the bill, for example when cracking seeds or holding prey. The anterior region of the skull also coevolved to enable the eyes to have a binocular view of the space between the parted beaks. Although both structural modifications are present throughout the clade, they are most prominent in the Common Starling and the White-cheeked Starling. Since both species lie some distance apart phylogenetically, the anatomy required for extreme prying must have evolved twice, which suggests a high evolutionary lability or plasticity for these traits.[11]

Most members of the Sturnidae move nomadically in flocks, especially when tracking their food sources, and roost in large numbers during the non-breeding season. These behavioural traits offer safety from predation, prevent heat loss at night, and facilitate the exchange of information. Importantly, the family's flocking behaviour contributed to their successful colonisation of new regions and may explain why many mockingbirds and starlings inhabit remote islands and archipelagos, often with small populations. For example, the Rarotonga Starling occurs on just one volcanic island in the Cook Archipelago, while the Atoll Starling is restricted to a few tiny Pacific islets.

The speciation of African starlings, however, involved more than climate and habitat change. It was also dependent on the evolution of iridescent feathers.

Structural colours

Until I visited Tsavo East National Park in Kenya, my encounters with starlings had been limited to sightings of the ubiquitous and mischievous Common Starling. Despite having a widespread distribution and possessing a distinctive wheezing and whistling call, Common Starlings are nevertheless a frequently overlooked species. They appear uniformly black at a distance, and it is only when observed at close quarters and in certain light conditions that their glossy feathers, with subtle violet and green iridescence, can be fully appreciated. My awareness of their ephemeral beauty, however, did little to prepare me for the extravagant palettes adopted by their African cousins (Plate 30). It was while waiting at Tsavo's entrance gate, with the famous dust-red elephants in attendance, that I first became beguiled by their remarkable plumages. The birds' descriptive names – Superb, Golden-breasted, Greater Blue-eared and Violet-backed – failed to convey the range of spectral hues on offer as I watched them jostle for drinking water in the midday sun. But why are there so many species of starling in Africa, and why do they sport feathers with such wildly different colour combinations? Before answering these questions, we should say a few words about the underlying mechanisms of colour production.

All animal colours are the result of two basic mechanisms: pigment colours and structural colours. Pigments produce their effect by selectively absorbing and reflecting specific wavelengths of visible light. If no light is reflected, we see black. If it's all reflected, we see white. Gradations in between give rise to many of the colours observed in the avian world, from the browns of thrushes to the reds of woodpeckers and redpolls. All avian pigments are derived from three classes of organic compound: melanins, porphyrins and carotenoids. For example, the sparrow's earthy shades and the black plumage of corvids are the effects of different melanin molecules (from the Greek *melanos*, meaning 'dark'), the same pigments that give human skin and hair its colour. The vivid greens and reds of turacos are produced by porphyrins, while the canary's yellow plumage is the product of dietary carotenoids. Crucially, each of these substances has important biological functions other than giving feathers their colour. Melanins act as structural components and their expression in feathers makes them stronger and more resistant to wear – which explains why many white species have black wing-tips, for extra strength. Porphyrins are essential components of the oxygen-carrying protein haemoglobin and several detoxifying enzymes in the liver, while carotenoids have antioxidant activity and are important stimulants of the immune system.

Not surprisingly, the expression of these compounds is highly conserved, and the chance of evolving different colour combinations is small because it would require the development of novel and complex metabolic pathways. But such innovations must have occurred in the past. For example, many lineages independently acquired the ability to alter yellow dietary carotenoids to form orange and red pigments without significantly affecting the individual's health. But blue pigments never evolved. This spectral omission suggests that blue may be extremely challenging or even impossible for birds to create, a concept referred to as the 'blue rose' hypothesis. (Horticulturists have long considered blue roses to be the unattainable 'Holy Grail' of the flowering world).[12] Furthermore, anthocyanins, the naturally occurring blue pigments, are degraded during digestion and so are unavailable for use by birds. The creation of blue feathers, therefore, required a different solution: structural colours.

Structural colours are produced by optical effects – interference, refraction or diffraction – when light interacts with regularly spaced nanometre-scale structures with varying refractive indices. If the reflected light is randomly scattered, white light is perceived, whereas if the wavelengths are ordered, vivid metallic iridescence can result. To see this effect, just twist a compact disc in daylight and view the bright spectrum of colours produced by the light scatter from the regularly spaced grooves. It is the latter's periodicity that allows the reflected rays to amplify each other and create the strong colours that readily change depending on the angle of view. Similarly, the bright blue iridescence of a male Morpho butterfly is produced by corrugated ridges present on the scales of its wings.

The structural colours of feathers, however, are generated by the nanoscale periodicity of their barbule constituents: melanosomes, keratin and air. Melanosomes are melanin-containing packages that self-assemble during the development of feathers to form stable, ordered structures. They are truly minute: the majority are only 200–600 nanometres in diameter, such that 200 could fit across a human hair (a nanometre is one-billionth of a metre).[13] Melanosomes lie embedded in a thin layer of beta-keratin just beneath the barbule's surface. Beta-keratins are filamentous proteins that readily cross-link to provide robust, lightweight bundles that give feathers, as well as beaks and claws, their strength. Many birds combine keratins with the melanin-containing melanosomes to increase the range of available colours, since the two proteins have different refractive indices and structural periodicities. Interestingly, keratins alone can produce blue structural colours. The male Eastern Bluebird, for example, generates its striking hues solely from air trapped within the barbules' beta-keratin channels. In fact, blue colour-producing nanostructures have evolved independently many times in over 20 bird families, including kingfishers, jays, manakins and honeycreepers.[14]

The evolution of further structural complexity has enabled the generation of even greater spectral ranges. For example, the male Lawes's Parotia, a member of the bird-of-paradise family, has evolved a unique cross-section structure of its barbules to create colours that change dramatically with feather orientation. The species' basic nanostructure produces a bright orange-yellow reflection, but since each barbule is V-shaped in cross-section, its sloping surfaces also act as reflectors of blue light. As a result, small movements of the feathers during courtship displays can cause the colour to switch from yellow-orange to blue-green suddenly, so increasing the chance of catching a female's eye.[15] Other species combine the effects of pigments and nanostructures. The characteristic green plumage of parrots, for example, is the result of laying a yellow pigment over a blue reflective layer of melanin and keratin.

Should readers remain sceptical that the blue colouration of birds is structural in nature, try the following simple experiment. Take any blue-coloured feather and shine a light on it from above, and you will observe bright blue. However, if you shine the light from underneath, the feather will appear dirty-brown owing to its melanin and keratin content acting as pigments.

But how is it that African starlings have generated such a diverse palette of colours compared to other avian families, when all feathers possess melanosomes and keratin? It was a mystery that caught the attention of Matthew Shawkey, a biologist working at the University of Akron, Ohio, and an authority on the optics and evolution of animal colours. Shawkey's approach was to determine the structure of the birds' melanosomes and relate the findings to the African starlings' phylogenetic tree.[16] The results indicated that the common ancestor of all starlings possessed relatively simple melanosomes, solid rod-like structures, that can still be found in some extant species. They include all 11 red-winged starlings (genus *Onychognathus*), a clade characterised by dark iridescent sheens. However, the researchers determined that 6 million years ago, as the early starlings began to form new species, different types of melanosome emerged, ones with shapes that interacted in novel ways with light. Some species developed flattened, oval-shaped melanosomes that allow for a more tightly packed arrangement. Others evolved air-containing, tube-like structures that offered a greater number of interfaces for the scattering of light. The most complex structures to emerge, however, were platelet-shaped, being both flattened and hollow: forms that can form colour-producing single layers, multilayers or alternating platelet–keratin layers.[17] Surprisingly, each starling species possesses only one of the four types of iridescence-producing melanosomes (Figure 21.2).

The American team concluded that the remarkable spectrum of colours possessed by African starlings results from the clade's unique possession of all

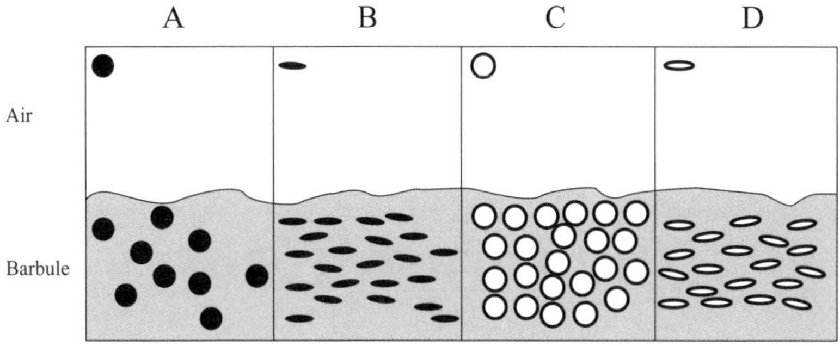

Figure 21.2 Morphology of the four types of melanosome responsible for the iridescence in African starlings, and which underpinned the clade's rapid speciation: (A) solid rods; (B) solid platelets; (C) hollow rods; (D) hollow platelets. Redrawn with permission from Maia *et al.* (2013).[17]

four types of melanosome. They also postulated that it would take only minor tweaks to the shape or spatial arrangement of any particular melanosome to produce a new range of metallic colours and, potentially, a new species. Such a conclusion makes sense, as the genetic changes required to produce such changes (e.g. alteration of the thickness of melanosome wall, the amount of air content, or the number of layers) are much more likely to evolve than those required to synthesise new pigments. Indeed, it is now known that the three complex melanosomal forms emerged on many occasions during the last 6 million years. And yet, although they can evolve from one to another, they have not been shown to revert to their ancestral state. This finding surprised Rafael Maia, one of the American investigators: 'I thought that maybe you'd have a lot of changing back and forth, but actually, once these complex structures evolve, they stick.'[18] It is likely that this unidirectional pathway of melanosome evolution, from simple to complex, may have contributed to the starlings' diversification, since by thwarting a return to the ancestral state it forced phenotypic change. Furthermore, not only did novel colours emerge that were unattainable with pigments, but the colours also became significantly brighter, by up to twofold. According to Shawkey, 'evolving these new melanosomes was like inventing the wheel for these birds – it allowed starlings to reach new colours at an incredibly fast rate.'[19] Indeed, it transpires that African starlings have evolved new colours 10–40 times faster than their cousins with simple melanosomes.

Sexual selection suddenly had another key ornamental innovation to play with, especially as plumage colour provides an 'honest' marker of a starling's fitness. Iridescent colours may be costly to produce and maintain, and only the healthiest birds can produce the most brilliant and vibrant hues. Since African starlings rely on colour for social communication and courtship, any sudden change could have acted as a barrier to genetic exchange and encouraged the rapid evolution of the dozens of species we see today.

CHAPTER 22

The Thrush's Story

SWEEPSTAKE DISPERSALS

To this day, the song of the Common Blackbird evokes memories of my halcyon days of childhood (Plate 31). Lying tucked up in bed with the windows open and the curtains fluttering gently in the evening breeze, I would fall asleep to the males' languid and fluty tones as each bird sang in defence of its home patch. Now and then, a loud and persistent metallic '*chook chook chook*' – most likely in response to next door's cat – would gain my full attention. Indeed, it seemed as though every garden in our neighbourhood possessed a resident 'Blackie', and their measured phrases dominated the spring's soundscape. Only later would I learn that these avian performances often reflect an ongoing struggle with newly arrived competitors from the continent. The English poet William Ernest Henley (1849–1903) encapsulated my youthful sentiment when he wrote:

> The nightingale has a lyre of gold,
> The lark's is a clarion call,
> And the blackbird plays but a boxwood flute,
> But I love him the best of all.

I am convinced that it was the blackbirds' beguiling refrains, which emanated from nearby rooftops, hedges and television aerials, that kindled my nascent passion for the natural world. And yet the birds' love of our suburban gardens is intriguing – for, as we will discuss, their drab colouration and low-frequency song appears to have evolved millions of years ago in the tropical rainforests of Central America.

Blackbirds are members of the thrush family (Turdidae), all of which are characterised by an elongated muscular structure, the turdine thumb, that protrudes from the syrinx.[1] This strange anatomical feature, as well as their cryptic spotted juvenile plumage, is shared by the Old World flycatchers (family Muscicapidae) and suggests a close evolutionary relationship. Most thrushes are proficient songsters, and some have vocalisations that rank among the most beautiful in the world. All 169 recognised species are plump, small to medium-sized passerines that belong to nearly 20 genera. The largest of these, *Turdus*, with 87 species, is one of the most widespread, with representatives on all continents except Australia and Antarctica (although nineteenth-century colonists did introduce Common Blackbirds to Australia and New Zealand).

Phylogenetic studies by Gary Voelker's team at the University of Memphis favour a Eurasian origin for the *Turdus* thrushes. It is most likely they emerged within the western Palaearctic, since the oldest extant taxa, the Mistle Thrush and the Song Thrush, are distributed throughout the area.[2] Such a conclusion is not altogether surprising, as the Turdidae are known to have evolved from the large Passerida radiation that dispersed to Asia from the proto-Papuan archipelago (see Figure 21.1). However, the thrushes' subsequent evolutionary history turns out to be unexpectedly complex.[3] Around 6.6 million years ago, an intercontinental dispersal occurred that led to the colonisation of Africa and the evolution of a limited number of species on islands in both the Atlantic and Indian Oceans, including São Tomé, Principe and the Comoros.

There then followed a remarkable biogeographical event. A population of the early African thrushes succeeded in making an east–west crossing of the Atlantic Ocean, against the prevailing winds and currents, to colonise the Caribbean basin.[3] So unlikely was this that scientists refer to it as a 'sweepstake dispersal', a term coined by the late American palaeontologist George Gaylord Simpson.[4] Although originally used to explain mammalian biogeography, the description is now applied to any chance floral or faunal dispersal across a major geographical barrier. For the ancestral thrushes, the crossing of the Atlantic Ocean around 6 million years ago seems to have been a matter of luck, a chance event that was never to be repeated. No other New World oscine reached the Americas via the transatlantic route – a fact that emphasises the extraordinary nature of this aspect of the early thrushes' story. In contrast, all other extant songbirds of Central America and the Caribbean evolved from ancestors that reached America from Asia, via the Beringian land bridge.

What factors facilitated the thrush's unique east–west crossing remains unclear. The fact that the Atlantic Ocean was narrower 6 million years ago, however, can be ignored. Indeed, although continental drift is constantly pushing America away from Europe and Africa, it is doing so only at the rate of growth of a human fingernail, approximately 3–5 centimetres a year. At the time of the *Turdus* dispersal, the journey would have been only 300–400 kilometres, or approximately 10 per cent, shorter. Voelker, however, believes that a freak storm may have been a key factor. Indeed, modern Atlantic

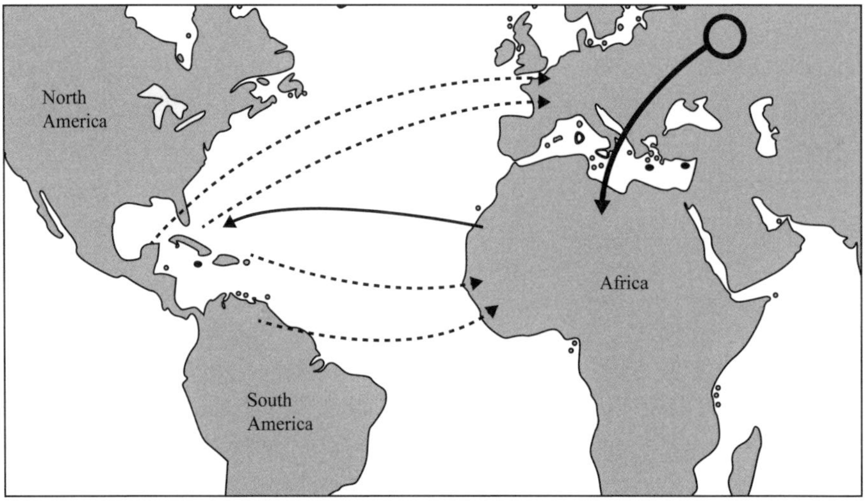

Figure 22.1 Five transatlantic sweepstake dispersals of the thrush family: (solid line) original east–west dispersal, (dotted lines) four reverse dispersals. The black circle indicates the origin of the ancestral thrushes, with the attached black arrow indicating the first intercontinental movement to Africa. Modified from Voelker *et al.* (2009).[3]

storm systems are often of such intensity that dust from the Sahara has been documented in the New World.

During the next million years, the early Caribbean thrush population spread out over all of America, from Alaska to Tierra del Fuego, to produce many species, including the American Robin in North America and the Austral Thrush in Argentina and Chile. Crucially, four separate populations then re-crossed the Atlantic Ocean during a narrow temporal window – between 4.7 and 5.7 million years ago – and returned to the Old World (Figure 22.1). It was from these four 'reverse sweepstake dispersals' that most of the extant species in Europe and Africa evolved. They include the Common Blackbird, a species widespread across Europe and the Middle East, and the Ring Ouzel, an unusual thrush for being an upland specialist. The migratory Fieldfare and Redwing of the western Palaearctic, as well as the non-migratory African Thrush, also have Caribbean ancestries. However, as highlighted previously, the Song Thrush and Mistle Thrush evolved from a population of early thrushes that remained in Eurasia.

Reverse dispersals

It has been argued that sweepstake dispersals rarely lead to successful colonisations, since they usually involve single, or just a few, individuals. In such a situation, the most likely outcome would be death before a self-sustaining population could be established.[5] Throughout the evolution of the genus *Turdus*, however, there has been a tendency to move in flocks, a fact that would make successful colonisation after a sweepstake dispersal more likely.[3] Recent evidence suggests that this can still happen, as the Fieldfare, which breeds in woodland and scrub in northern Europe and Asia, colonised Greenland after a flock reached the island early in the last century.[6]

In contrast to the initial east–west dispersal, the four return crossings could have been aided by the palaeoclimatic conditions that prevailed before the South and North American continents joined. Storms that formed in the eastern Pacific and Caribbean tended to move eastwards across the Atlantic, a factor that would have favoured the return dispersals to the Old World. This weather system, however, changed dramatically when the gap between the two continents closed.

Beneath the waterway that linked the Pacific and Atlantic Oceans – the Central American seaway – two tectonic plates collided, forcing the Pacific plate slowly beneath the Caribbean plate. The pressure and heat generated by this collision led to the formation of submarine volcanoes, some of which grew tall enough to break through the ocean surface and form islands approximately 15 million years ago. Over the next 10 million years, more and more volcanic islands formed, while the movement of the tectonic plates pushed up the sea floor, eventually forcing some areas above sea level. At the same time, the continents' coasts were eroded by strong ocean currents, and vast amounts of sediment – sand, soil and mud – were deposited between the newly emerged islands, filling in all the gaps. The resultant Panamanian isthmus linked the two continents and ended what George Simpson famously described as South America's 'splendid isolation'.[7] Indeed, the loss of the Central American seaway was one of the most important geological events to happen during the last 60 million years, since it not only played a major role in promoting the region's biodiversity but also greatly affected the Earth's climate.

The obvious effect of the seaway's closure on the region's biota was the increased migration of animals and plants between the two continents – what has been termed the Great American Interchange. In North America today, the many species of opossum, armadillo and porcupine can all be traced back to ancestors that crossed the land bridge from the south, while the ancestors of llamas and raccoons trekked in the opposite direction. As we have discussed, the direction of traffic for birds was primarily from the south to the north – an event that greatly transformed the tropical avifauna of the New World.[8] However, the gradual shoaling and ultimate closure of the Central American seaway also significantly altered the Earth's climate. The lack of water exchange between the Pacific and Atlantic Oceans around 4.7 million years ago resulted in altered salinities as evaporation in the tropical Atlantic and the Caribbean left their waters saltier and put freshwater vapour into the atmosphere.[9] The Trade Winds developed and transported the water vapour westwards across the isthmus, where it fell as rain in the Pacific. As a result, the Pacific became relatively fresher while salinity increased in the Atlantic. All these changes, together with the re-routing of currents to form the Gulf Stream, effectively eliminated the favourable storm patterns for reverse transatlantic dispersals. It is significant, therefore, that no subsequent intra-Caribbean or west–east Atlantic lineage diversification occurred after the formation of the Panamanian isthmus.

But how did Voelker's team deduce that most of the members of the genus *Turdus* evolved from Caribbean ancestors? And how can they be so sure that the initial, as well as the subsequent dispersals, were not the result of an interchange via the Beringian route? As unlikely as it may seem, the answers to these questions were deduced by analysing the family's genetic material. By comparing homologous sequences of DNA from 65 species of the genus *Turdus*, Voelker and his colleagues were able to construct a robust phylogeny. The resultant tree showed that four of the Old World clades are more closely related to extant Central American species than they are to each other. In other words, these four clades must have evolved independently from early Caribbean species, rather than descending directly from African or Eurasian ancestors. Furthermore, using the Beringian connection hypothesis to explain the relationships between Old and New World species would require a minimum of seven continental extinctions.[3] Given the long-term persistence of thrushes in central and eastern Eurasia, it seems improbable that no lineage was able to survive the varied North American habitats that existed during the last 6 million years. According to Voelker, therefore, the transatlantic route is a far more parsimonious explanation than the Beringian path. Indeed, the latter would require repeated extinctions across both Europe and North America, for which there is no current evidence.

The conclusion that palaeoclimatic changes played a significant role in the sweepstake dispersals of thrushes is dependent on the accuracy of the dates assigned to the various nodes or divergence points. So how were these derived? In practice, such dates are best obtained by calibrating against a reliable fossil. However, for the genus *Turdus* this approach is not possible, as the few existing specimens cannot be attributed to any extant species. An alternative method is to apply a molecular clock: one that relies on a fixed rate of DNA mutation, usually a 2 per cent divergence per million years for the commonly used avian genes. The universality of this method, however, has now been questioned.[10] Voelker's team, as a result, calibrated their time tree against a vicariant event, one that led to the speciation of two African thrushes. The tropical forest of central Africa attained its maximum eastern extension, reaching the coast of Kenya, between 5 and 3 million years ago. It then underwent a rapid retraction to the west that resulted in the separation of an ancestral population of thrushes. One group became restricted to the high-altitude mountains in Kenya, the Taita Thrush, while the other population was limited to the lowland forests of the rift system and became the Abyssinian Thrush. Since both species are sedentary, it is reasonable to assume that their speciation took place no later than 3 million years ago. Using this fixed date, Voelker then calibrated the rest of the thrushes' phylogenetic tree and translated the relative divergence times into absolute ones.

As a youngster, I was unaware that the Common Blackbird's low-frequency song provides a clue to its evolutionary origins. Foliage absorbs high-frequency sound waves, so the Blackbird's deep, baritone notes are much better at penetrating dense vegetation. This characteristic trait is now known to have been honed millions of years ago in the tropics to facilitate the defence of territories and the winning of mates. As I ponder this amazing fact, it seems appropriate that a Song Thrush should be singing from a nearby tree, while out of my study window I spot a Blackbird hopping about on our lawn. Although well aware that both species belong to the genus *Turdus*, I hadn't realised the remarkably different evolutionary pathways that have led each to my tended patch of suburbia.

CHAPTER 23

The Sparrow's Story

HYBRIDISATION AND SPECIATION

After viewing the exhibits of Rome's Galleria Borghese, we strolled along the wide shady lanes of its surrounding parkland, adorned with beautiful fountains and imposing statues. On one of the many lawns, beneath some overhanging trees, a male sparrow hopped about, chirping and behaving as all sparrows do. Except that this individual appeared to be subtly different from the sparrows in our garden: it possessed bright white cheeks and a chestnut-coloured head and nape, in contrast to the more familiar dusky-grey cheeks and grey crowns back home. For the bird was an Italian Sparrow, a species restricted to the Italian peninsula and a few nearby Mediterranean islands. This regional endemic is now known to have evolved around 7,000–8,000 years ago as a stable hybrid between the ubiquitous House Sparrow and the Spanish Sparrow (Plate 32). The existence of this otherwise unremarkable passerine has intrigued evolutionary biologists, and the study of its genome has provided valuable insights into the roles and mechanisms of hybridisation in speciation. As a result, the Italian Sparrow fully deserves its inclusion in the *dramatis personae* of *The Ascent of Birds*. However, before highlighting the hybrids' contribution to science, we should first discuss how the parents of the Italian Sparrow came to meet in Europe.

The Old World or 'true' sparrows (genus *Passer*) arose in Africa during the late Oligocene around 20 million years ago, a fact that explains why that continent has the highest number of extant species.[1] Furthermore, the Cape Sparrow or 'mossie', found in southern Africa south of Angola, is the oldest living member of the genus. According to the indefatigable English ornithologist, Denis Summers-Smith, all the Eurasian sparrows descended from a 'black-bibbed' African ancestor that spread out of the continent, via the Rift and Nile Valleys, around 1 million years ago.[2] This intercontinental radiation may have been driven by climate changes that favoured the development of grasslands in place of forests. The early Eurasian arrivals included a species very similar to the present-day House Sparrow, a fact suggested by the 400,000-year-old fossil jawbones unearthed in a cave near Bethlehem in Israel.[3] Moreover, all the early House Sparrow fossils (older than 10,000 years) stem from the Middle East and belong to a species known as *Passer predomesticus*. This 'predomestic' House Sparrow was most likely migratory and thrived in grassland habitats where it fed on wild seeds. The northern hemisphere then entered a prolonged ice age, a period characterised by southerly advances in glaciation and cold temperatures that divided the early sparrow population into two distinct groups. Sometime between 12,000 and 10,000 years ago, when the Earth rewarmed, the sparrow populations spread out again, but now as two separate species: the Spanish Sparrow in northwest Africa and southern Spain, and the House Sparrow in southeast Europe and the Middle East.

The next episode in the House Sparrow's story involves the creation of a novel ecological niche as the result of human activity. Around 10,000 years ago, the nomadic peoples of the Fertile Crescent, an area stretching from the Tigris to the Nile, radically changed their behaviour. Over centuries or even millennia, traditional hunter–gatherer lifestyles were left behind in favour of permanent settlements and reliable food supplies. The historian Yuval Harari envisages this most important event in human history to have evolved as follows.[4] When temperatures increased after the last ice age, the increased rainfall favoured the growth and spread of cereals such as wheat and barley. Neolithic peoples then began to eat more grain and, as a result, inadvertently encouraged their spread. Since wild grains are impossible to digest without processing, the harvest would have been carried back to temporary camps, with some inevitably falling by the wayside. Over time, more and more cereals grew along favourite human trails and near their encampments. The use of fire to clear forests and woodland would have further benefited cereals by increasing the availability of sunlight, water and nutrients. Then, in areas where wheat and barley became particularly abundant, and game was plentiful, the Neolithic peoples gave up their nomadic way of life and become sedentary agriculturalists living in permanent settlements.

Denis Summer-Smith believes that these fundamental changes in our ancestors' behaviour would not have gone unnoticed by the local sparrow population. As a result, the bird's diet altered to accommodate the more accessible cultivated grasses and other foods that were being grown. They also came to prefer holes and crevices in buildings and other human constructions as nesting places. The availability of all-year-round food and the relatively secure nesting sites removed the need to migrate south in winter and led to the birds becoming obligate commensals. Interestingly, the transition from a diet of wild seeds to one consisting mainly of cultivated grain had a significant effect on

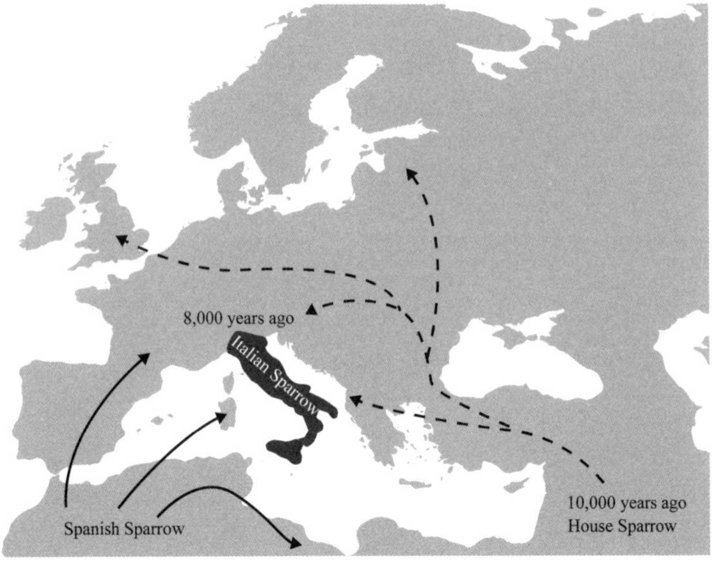

Figure 23.1 Geographical dispersal of House Sparrow (dotted lines) and Spanish Sparrow (solid lines). The Italian Sparrow evolved as a hybrid between the Spanish and House Sparrows around 8,000 years ago.

beak and skull morphology. To better cope with the increased size and hardness of cereals, the local sparrows evolved a more robust skull and a larger and more pointed beak.[5] At this point, the 'domestic' House Sparrow underwent a massive population and range expansion as the agricultural civilisations rapidly spread throughout the Palaearctic and Oriental regions.[6] Eventually, in the area around the Italian peninsula, they came into contact with the Spanish Sparrow that had begun moving northeastwards (Figure 23.1).

The Italian Sparrow's origin has long been debated. It used to be considered, together with the Spanish Sparrow, as a geographical colour variant of the House Sparrow. Others argued for it being a subspecies (*Passer domesticus italiae*), while some believed it to be a full species that had undergone a recent divergence from either the Spanish or the House Sparrow. It was the German ornithologist Wilhelm Meise who first suggested in 1936 that the Italian Sparrow might be an example of a stable hybrid.[7] Such an idea was unacceptable to many, since hybridisation was regarded as a process that led to sterile or subfertile offspring that would be actively selected against. One only has to think of the mule, a sterile hybrid resulting from the crossing of a mare and a donkey. Even if hybrids were fertile, they would most likely disappear, since they would be less able to compete with parental species, a process termed post-mating isolation. Should hybrids be successful, they would still probably fail in the long run, since they would most likely be absorbed back into the parental species in a phenomenon termed despeciation. Indeed, Darwin was so concerned that this scenario could negate speciation by natural selection that he devoted a whole chapter to the subject in *The Origin of Species*. Hybridisation, therefore, was seen by many biologists as a natural barrier that maintained each species' integrity and, as such, was of little evolutionary significance. Ronald Fisher echoed the prevailing sentiment when he wrote, 'the grossest blunder in sexual preference which we can conceive of an animal making, would be to mate with a species different from its own.'[8]

As so often happens in science, firmly held ideas can be found wanting. The countercharge was led by the botanists, who became convinced that hybridisation commonly underpins speciation in plants. In fact, it has subsequently been shown that 'somewhere between a third and a half of all the green things on this earth, and, at least, half of the world's flowering plants, arrived by the mixing of genes from separate species.'[9] For plants fertilised by the vagaries of wind and insects, the advantages of hybridisation are evident. Seeds must germinate where they land, or succumb. The greater the genetic variability, the more chance that some seeds will succeed and that the species will survive, particularly in times of environmental stress. In plants, both progenitor sets of chromosomes are frequently amalgamated to provide a hybrid with twice the chromosome complement: a condition known as allopolyploidy. Such a state may not only enhance the hybrid's robustness, by boosting its genetic variability, but also its reproductive isolation, since gametes, or sex cells, will have a mismatched number of chromosomes compared to either parental population. The genetic and evolutionary implications of such unions are complicated, and biologists are still struggling to understand them fully. Nevertheless, polyploidy can result in instantaneous or saltational (from Latin *saltus*, 'leap') speciation, a process that could have enabled the evolution of complex organisms and the catalysing of key evolutionary innovations. Fusion of genomes can be repeated many times, leading to some vascular plants having more than an eightfold increase in chromosomes. However, although 40–70 per cent of all plants are polyploids,[10] it is a highly unusual state of birds, being confined primarily to the domestic chicken (*Gallus domesticus*).

Hybrids with the same number of chromosomes, termed homoploid or recombinational hybrids, have been described in flowering plants and more recently in animals, including several species of fish, insect and flea.[11] The obvious question, therefore, is whether such crossings could have contributed to the evolution of birds. Early field and laboratory observations confirmed that avian hybrids do regularly occur in the wild. Ernst Mayr estimated, after examining thousands of museum skins, that perhaps one in 60,000 wild birds is a hybrid.[12] However, this evaluation was for old and well-established species, and it was argued that hybrids could be more common among younger lineages. In 1975, Wilhelm Meise supported this contention when he reported that 2 per cent of more recent species regularly hybridise and a further 3 per cent do so occasionally.[13] Peter and Rosemary Grant (see *The Tanager's Story*) undertook an extensive review of the literature and concluded that just over 9 per cent of all avian species have hybridised in the wild.[14] Among some bird families, the incidence is even higher. Astonishingly, three-quarters of British ducks (family Anatidae) and all four British gamebirds are known to hybridise with at least one other species.[15]

Of course, the occurrence of such hybrids does not equate to speciation, and many evolutionary biologists still regard such crossings as evolutionary cul-de-sacs. Sceptics emphasised that while plant fertilisation is a random process that favours cross-fertilisation, birds 'consciously' choose mates of the same species and hybridisation will be a rare event. In other words, wild birds have a very strong tendency for 'like-with-like' or assortative mating when living alongside closely related species. This argument, however, is only partially valid. Under certain circumstances, birds can become confused and less discriminatory when it comes to mate selection. In the first few hours of life, birds pass through a brief receptive period in which their future choice of mate is determined by the species they first encounter. This sexual preference is the result of psychological processes known as imprinting. As a result, early exposure may lead to nestlings imprinting on whatever species raised them. Indirect fostering can also have the same effect. A bird may learn an 'incorrect' song by hearing it from another species singing near its nest and later use it to attract a mate from the 'wrong' species. Birds can also pay a high cost regarding the time and effort spent in searching for a mate, especially if the species is rare. In these circumstances, it has been shown that such high demands may result in birds switching their behaviour from one of 'always mate with your own species' to 'accept the first male/female you encounter'. Indeed, the species does not have to be rare for this to happen. As the season progresses, the number of unmated females of a given species will become scarcer, and males may become less choosy as their chances of acquiring a mate lessen with time.[16]

As recognised by early scientists, naturally occurring hybrids are often infertile or suffer a reduction in fitness. Indeed, there are many examples from the avian world. The progeny from Western and Eastern Meadowlark matings are sterile, while hybrids from Pied and Collared Flycatchers have a significantly reduced fertility. In the Carrion and Hooded Crows' contact zone, hybrid females lay eggs with extremely thin shells, making them susceptible to breakage and therefore reducing survival. The progeny of Fischer's and Rosy-faced Lovebirds have problems of fitness as well as fertility, including the development of gout and bizarre behavioural traits that limit the success of nest-building.[17]

Intuitively, these findings make sense. A hybrid's novel combination of genes will not have been subjected to the relentless cull of natural selection and, as such, would be expected to be less well adapted. One particular group of interacting genes that may be important in determining a hybrid's fitness are those that control cellular respiration. They are located in both the mitochondria (the cell's 'powerhouse') and the nucleus, and any incompatibility in their coded proteins can lead to a bottleneck in the cell's energy production, leading to a build-up of dangerous free radicals – highly unstable and reactive toxins. Excessive amounts of these toxins will overrun the body's natural defence mechanisms, leading to irreversible cellular damage and death: a fact that could explain the high incidence of disease and infertility in hybrids.[18]

Introgressive hybridisation

Given these limitations, is it possible for cross-species breeding to have influenced avian evolution? Surprisingly, the answer appears to be yes. Biologists now realise that hybrids can act as 'go-betweens' that enable genes to jump from one species to another and so increase the genetic and phenotypic variation of one of the parental populations. This process, known as introgressive hybridisation, occurs mainly through male hybrids, as the female sex-chromosome contains a high concentration of infertility factors.[19] The swapping of genes between closely related species is likely to have been an important determinant in facilitating rapid evolution in the past, especially during times of sudden environmental stress. Intriguingly, the evolutionary outcome does not depend on the long-term survival of the hybrids. Most fertile first-generation hybrids will be too rare to breed with each other and will be far more likely to back-cross with one of the parental species, a choice often dictated by the song type of their fathers. Furthermore, the subsequent offspring will be fitter than any hybrid–hybrid progeny since they will contain a selection of characters that have already been tested by natural selection in the original species. As a result, the back-crossing of hybrids enables a varying number of genes from one species to enter the gene pool of a second and so introduce genetic novelty. If the new phenotype has an advantage in a changing environment, then those individuals will be more likely to survive and potentially undergo speciation. It is also possible that some species carry 'fossilised' genes transferred from another species that no longer exists, although this scenario is difficult to prove. It has even been suggested that our own species (*Homo sapiens*) has a complicated history of introgression, one that involved the acquisition of DNA from at least two extinct hominid lineages.[20]

Homoploid hybridisation

In contrast to introgressive hybridisation, which involves back-crossings, hybridisation can also result in speciation if the hybrids become reproductively isolated from their parent populations. This state, known as homoploid hybridisation, is believed to be rare in nature.[21] And yet we now know that this form of hybridisation underpins the origin of the Italian Sparrow, a conclusion that raises an interesting paradox. If gene flow has occurred between the House and Spanish Sparrows, how is it that the genetic material of the hybrid has remained isolated and distinct from its parents? Put another way, why has despeciation of the Italian Sparrow not occurred?

Professor Glenn-Peter Sætre, who heads the 'Sparrow Group' within the Department of Biology at the University of Oslo, has long been intrigued by the mechanisms of speciation. Ever since his team proved that the Italian Sparrow was a homoploid hybrid, Sætre has sought to explain its biogeography and the basis for its genetic stability.[22] His conclusions – a nexus of human activity, geographical isolation and genetic incompatibility – go a long way to explaining the species' emergence. However, the team's findings have a wider significance and suggest that hybridisation may be more common in nature than previously thought. Indeed, it is possible that many types of flora and fauna have a homoploid hybrid origin. So let us look at the Italian Sparrow and its genetics in a little more detail.

Agriculture developed in Italy somewhat earlier than in the rest of Europe, around 8,000 years ago in the 'boot-heel' region of the country's southeast.[23] The arrival of farmer-colonists from the Aegean would have favoured sparrow hybridisation since the population of front-wave co-spreading House Sparrows would have been small in number and mates would have been in short supply. As we have discussed, a thinly spread population can disrupt assortative mating, a fact supported by recent studies. House and Spanish Sparrows, for example, fail to interbreed when both occur in large numbers, but they do so readily in areas where one or both species are rare, as in parts of north Africa and the Cape Verde Islands. The interbreeding of the newcomers from the Middle East with the established Spanish Sparrows from the south led to a hybrid swarm that subsequently dispersed northwards throughout Italy, following the expansion of agriculture. Geographical isolation then played its part, a fact relevant today as the Italian Sparrow remains sheltered from gene flow by the massive geographical barriers provided by the Alps and the Mediterranean Sea. Indeed, even after the rest of Europe was colonised by House Sparrows, mountain and ocean obstacles have kept the Italian population isolated for long enough for it to have become a stabilised hybrid species.[24]

Geographical isolation, however, is not the whole story. In 2014, Sætre's team identified a genetic or intrinsic component to the hybrid's reproductive isolation: a barrier that results from the inheritance of a subset of its parents' 'incompatibility' genes. In other words, a pool of sparrow genes provides a reproductive barrier not only between the Spanish and House Sparrow but also between the Italian Sparrow and its parental species. This finding represents the first evidence in the animal kingdom that the inheritance of a selection of parental genes can contribute to the persistence of a hybrid species. The function of such genes remains to be determined, although they are known to be located on the sex chromosomes (Z-chromosome) and within the mitochondrial DNA.[25] Interestingly, the hybrid's mitochondrial DNA holds a further evolutionary clue, as it derives mainly from the House Sparrow. Since mitochondrial DNA is inherited from females (sperm, unlike ova, do not contribute mitochondrial DNA), this suggests that Italian Sparrows were mainly the offspring of male Spanish Sparrows and female House Sparrows.[26]

The number of avian species known to have evolved through homoploid hybridisation is small, but slowly growing. For example, recent genetic analysis of the yellow-rumped warbler complex (*Setophaga* spp.) suggests that at least one species has evolved as the result of hybridisation.[27] Sætre believes that 'should a sorting mechanism similar to that described for the Italian Sparrow prove to be more pervasive, the circumstances promoting homoploid hybridisation may be broader than currently suspected, and indeed, there may be many cryptic hybrid taxa separated at two boundaries by sorted inherited incompatibilities.'[28] Homoploid hybridisation could have been more frequent in the past, especially during the early phases of avian evolution when rapid continental dispersals would have resulted in small populations for most species.

Given the availability of cheaper and faster genome sequencing, we should not have long to wait before the role of hybridisation in avian evolution is clarified. In the interim, why not take a moment to ponder whenever you next see a hybrid in the wild – it might just be a new species in the making.

CHAPTER 24

The Zebra Finch's Story

EVOLUTION OF BIRDSONG

The Zebra Finch is commonly found over much of central Australia, as well as on the Lesser Sunda Islands to the northwest (Plate 33). This colourful and energetic species is not a true finch, but a member of the family Estrildidae, a clade that originated 16.5 million years ago during the Himalayan uplift.

The northward-moving Indian plate first made contact with the Asian plate around 50 million years ago, although its maximal effect did not occur until the Miocene. The ongoing result was the formation of the Himalayas and the Tibetan plateau, which in turn led to the establishment of the southern Asian monsoon system, the region's large rivers, and the Chinese deserts. It is likely that these dramatic environmental upheavals encouraged the ancestors of the Zebra Finch to radiate from their origins to reach Africa, south Asia, Australia and the South Pacific islands.[1] At first, the early estrildid finches were unable to colonise the southern continent as the area remained covered in rainforest and lacked suitable habitat. But Australia was drifting northwards, and as it did so, its climate changed. By 10 million years ago, the continent's Gondwanan rainforests had begun to recede, to be replaced by eucalyptus and acacia trees and large tracts of open grassland. These environmental upheavals provided the trigger for several waves of invasion by songbirds from southern China, among the first of which were the ancestors of the seed-eating Zebra Finch.[2]

Despite these fascinating evolutionary insights, it is the Zebra Finch's role as the avian equivalent of the laboratory rat that concerns us here. For several aspects of the species' biology – long-term pair-bonding, early breeding with 3–4 clutches per year, and its relative longevity – have made it the preferred subject for many biological studies. In particular, the Zebra Finch is the only tractable laboratory model suitable for the investigation of the genetics and evolution of birdsong. It is an ongoing story, but one that suggests that the brain circuits controlling complex traits like vocal learning have only a limited number of ways to evolve. At first sight, this may seem an arcane field of research, but it has turned out to be highly relevant to the understanding of human speech and its disorders. For this reason, the Zebra Finch's story is an important one, well beyond its central role in highlighting the evolution of birdsong.

Vocal skills and the evolution of the syrinx

The oscines' defining and most endearing feature is their ability to sing, an apomorphic characteristic prosaically encapsulated in their everyday collective name 'songbirds'. The beauty and sheer exuberance of their melodies, especially the evocative cadences of the Atlantic Canary, Eurasian Skylark and Common Nightingale, have inspired poets and musicians throughout history. The complexity of such performances can be stunning. As Mark Cocker highlights, a male Nightingale possesses 'as many as 250 different phrases compiled from a repertoire of 600 basic sound units'. Furthermore, 'in the passages of song, the phrases are drawn together and sequenced in a variety of ways so that each performance is a unique composition, never to be repeated.'[3]

In contrast to most other birds, oscines are absolute masters of vocal learning, or mimicry, an innate ability that can result in a marked embellishment of their fundamental rhythmic baseline. On hearing a vocal duel between Song Thrushes in spring, Cocker wrote:

> It is a fabulous noise that gains momentum as the season draws on, with a vocalist adding new motifs to his repertoire. A bird borrows elements from the others that it can hear, and you can imagine these scraps of melody being passed all round the country as one song thrush tosses the sound-torch to its song thrush neighbour.'[4]

Vocal learning reaches its apogee with the insatiable mimicry of the Marsh Warbler, a species that breeds in temperate Europe and winters in sub-Saharan Africa. The male's song is an ornithological collage, a musical jukebox, composed of various snippets from up to 200 avian sound-bites, learned en route from their breeding territories to their wintering grounds. For Marsh Warblers routinely incorporate phrases from common European species, including tits, raptors and warblers, coupled with flourishes borrowed from African species such as drongos, doves and bee-eaters.[5]

Vocalisation is likely to have evolved early after the oscine–suboscine split, as mimicry is a well-developed trait in several basal oscines. Indeed, the world's most celebrated vocal mimic is one such passerine, the Superb Lyrebird. This shy, ground-stalking species can imitate not just forest sounds, but also mechanical noises such as camera shutters, chainsaws and car alarms. Mimicry also plays a role in the courtship of another basal oscine, the Satin Bowerbird. Until recently, it has always been assumed that female bowerbirds selected mates based on their building ability – the symmetry and decoration of the bower (see *The Bowerbird's Story*). It now seems that architecture is not everything. Gerald Borgia's team at the University of Maryland has revealed that male Satin Bowerbirds ultimately make a more intimate appeal to females by approaching closely and 'whispering' an enticing mimicry of other Australian birds.[6]

Such vocal skills are now known to have evolved over millions of years as the result of sexual selection. But why should the female's choice of mate be based on a male's adeptness at singing and mimicry? Put another way, how can vocalisation be a characteristic that favours a species survival? The answer, it seems, is that vocal prowess mirrors genetic quality. A male oscine's vocal learning skills are a reflection of its juvenile health, freedom from disease and parasitic infection, at a time when learning abilities develop. Also, males with more complex songs have superior cognitive skills and a greater learning capacity.[7] The female, therefore, is selecting a mate with 'good' genes and, as a consequence, is enabling the male's singing proficiency to be passed on to the next generation. Rival males try and outdo each other, engaging in a form of vocal 'arms race' where the most varied performance, containing the best mimicry, gains the optimal territory and, with it, the chance to mate. As we saw in *The Bird-of-Paradise's Story* in relation to other aspects of male performance, female choice eventually leads to a runaway selection for vocalisation, as the genes underpinning the trait are the ones most likely to survive.

The song and mimicry skills of oscines result from the evolution of two linked structures: a complex syrinx and a dedicated neural control network, known as the song system. The syrinx in oscines is located at the base of the trachea, or windpipe, where it bifurcates into the two main bronchi that enter the lungs. It is a complex structure composed of extrinsic and intrinsic muscles that control the airflow, and membranes, or labia, that act as sound-generating structures. The mechanism of sound production is very similar to that used by ourselves, except that the oscine's syrinx is a bilateral structure. It has two sound sources instead of one, with each half possessing a pair of membranes. Both sides can vibrate independently and are, to some extent, separately controlled.[8] When a bird sings, the air from the lungs is forced through the syrinx, vibrating the membranes to cause a sound. Songbirds have up to four pairs of intrinsic muscles that can alter the organ's configuration, an anatomical arrangement that correlates with song complexity and vocal learning. Also, evolutionary modifications to the membranes' extracellular tissue (in effect, the amount of collagen and elastic fibres they contain) have facilitated the acoustic distinctiveness of oscines. It appears that the degree of asymmetry of the protein content in opposite membranes and the variation in their layering structure equates to the range of possible frequencies.[9]

Nobody knows how or why the early passerines split to give rise to the oscines, a suborder with a more complex syrinx. A likely scenario is that at some time during the early Palaeogene, a primitive passerine evolved the ability to produce a greater range of frequencies or combinations of sounds. Such an event would have resulted from recombinations or mutations in the founder's genetic code that ultimately affected the syrinx's anatomical structure and function. For example, the novel genotype could have resulted in a slight change in the position of the syrinx, or altered the arrangement or number of the syrinx's muscles, or modified the biochemical composition of the membranes. These phenotypic changes were then subjected to positive selection, probably by female choice, so that subsequent generations inherited the modified syrinx. Evolution is an ongoing process, and genetic changes are occurring continuously. Any further mutations that enhanced a male's singing ability would have become increasingly standard within the population. Eventually, after thousands or even millions of generations, a more sophisticated syrinx evolved, one with a capacity for diverse song and accurate mimicry. While something like this almost certainly occurred, it should be stressed that it remains an imaginary account, since the precise evolutionary pathways are unknown. What is certain, however, is that the oscine's syrinx evolved sometime during the Palaeogene.

The song system

The ability of oscines to vocalise is also dependent on a set of specialised brain cells, or nuclei, located in the forebrain, that are collectively known as the song system. The evolution of these discrete, interconnected nerve cells was a definitive event in the history of songbirds and probably occurred at the same time as the development of their complex voice box. In addition, the song system must have evolved soon after the suboscine–oscine split since, although it is present in all oscines, it is missing from suboscines.[10] As we will see, however, at least one suboscine, the Eastern Phoebe, possesses a vestigial homologue of the song system, a finding that could have evolutionary implications.[11]

Although the song system's structure is exceedingly complex, a simple outline will help readers appreciate its evolutionary significance. The controlling network consists of seven discrete structures that interconnect to form two major neural pathways. The first pathway, the cortical motor pathway, originates in the high vocal centre or HVC nucleus and controls the vocal and respiratory muscles indirectly, via a specialised cluster of nerves termed the robust nucleus of the arcopallium. This pathway ultimately connects to the syrinx and is essential for the production of learned vocalisations. The second pathway, known as the anterior forebrain pathway, acts in conjunction with area X and facilitates the acquisition and imitation of songs. Crucially, the anterior forebrain pathway also underpins the syntax and social context of learned vocalisations. As young male Zebra Finches acquire their vocal skills, the volume and number of neurones in the various song nuclei increase compared to the rest of the brain and the brains of non-learning females. As one might predict, damage to the pathways when the bird is young disrupts its ability to vocalise, such that it may never be able to acquire its distinctive song.

Vocal learning is an involved process and takes many weeks to master. It commences with a phase similar to the babbling of human infants, called 'subsong', that consists of variable, low-amplitude sounds. From this raw material, imitations of its parents emerge, copied mostly from the father. Indeed, if young oscines are removed from the nest and reared in isolation, their song development is curtailed, leaving the bird with nothing but its innate subsong. Through trial and error, the fledgling's imitations become recognisable and, once perfected, they become less and less variable. Nevertheless, readily identifiable dialects may occur in some species, acting as local cultural traditions. Vocal learning by sensory feedback ensures that by the time the bird is sexually mature, its song is robust enough to defend a territory and woo a mate. It is now known that the accuracy of a songbird's mimicry depends on the release of a cocktail of growth factors from the song system nuclei, the most important of which is brain-derived neurotrophic factor (BDNF). This protein is essential for nerve regeneration, and its administration to juvenile birds markedly improves their song-learning abilities.[12] Whether sexual selection by females is in any way governed by male levels of BDNF, however, remains to be seen. Like most birds, oscines possess an extensive repertoire of unlearned calls that are important in close-range communication. Recent experiments, incorporating wireless brain monitoring of Zebra Finches, have highlighted that it is the nerve cells controlling these innate calls that have evolved to give rise to the complex neural networks.[13]

The songbirds belong to one of only three related avian groups known to have acquired vocalisation through imitation rather than instinct, the others being parrots and hummingbirds. While parrots are synonymous with vocal mimicry, hummingbird vocalisation has only recently been demonstrated. The restriction of this behavioural trait to these three clades is surprising, since hummingbirds are not closely related to either parrots or songbirds. What is even more remarkable is that they have all evolved the same neural network: a song system consisting of seven interconnected forebrain nuclei. Psittacine brains, however, appear more complicated and contain a song system within a song system. While their 'core' song system is similar to that of songbirds and hummingbirds, their 'shell' song system is unique – and this may account for the advanced mimicry skills of parrots.[14]

So what do these findings tell us about the evolution of birdsong? The most likely explanation is one of convergent evolution, whereby avian vocalisation evolved on three separate occasions over the last 65 million years. According to the neurobiologist Erich Jarvis, this could have occurred if an adjacent, pre-existing system in the avian brain that controls motor learning skills, which also has seven active areas, duplicated to take control of vocalisation.[15] Indeed, he also believes that brain duplication could be a general mechanism to explain the evolution of other complex behavioural traits. If convergent evolution is indeed the explanation, it implies that songbirds, parrots and hummingbirds all evolved a core system independently, and that parrots went on to develop an extra shell system. An alternative hypothesis is that the song system evolved only twice, in hummingbirds and the common ancestor of parrots and passerines, and that it was subsequently lost in suboscines. Supporters of the latter scenario emphasise that the Eastern Phoebe has a rudimentary song system, while several bellbird species have been documented to develop distinct dialects.[16] Future experiments, however, will be necessary to resolve these two hypotheses.

Human parallels

One rationale for discussing the anatomy of the song system in some detail is that many of the structures are homologous to areas in the human brain known to be essential for the production and comprehension of the spoken word. Area X, for example, is homologous to the human basal ganglia, with similar connections, cell types and response to neurotransmitters, while the HVC nucleus is a homologue of Broca's area. This remarkable neural parallelism between birdsong and human expression has allowed scientists to use the Zebra Finch as a model system to understand the molecular control of speech and communication disorders. Already a number of genes, whose expression levels change when finches sing, have been linked to medical conditions. The best understood is the Forkhead box protein 2 gene (*FoxP2*), a 'master gene' that encodes a transcription factor that controls the expression of hundreds of other song-related genes. Mutations in its DNA binding site result in a severe speech disorder in humans, known as developmental verbal dyspraxia and characterised by incomprehensible talk with linguistic and grammatical deficits.[17] Not surprisingly, *FoxP2* has been popularly dubbed the 'language or grammar gene'. Other avian vocalisation genes, ones that encourage neuronal connections between the song centre and the nerves that control the syringeal muscles, have been linked to autism and dyslexia.[18] It may soon be possible to develop novel therapeutic strategies for these speech disorders, because scientists now have the skills to insert human disease-causing genes into avian genomes and explore their function in more detail.

Recently, Andreas Pfenning, at the Duke University Medical Center, together with other members of the International Avian Genome Consortium, investigated how many genes might be involved in song learning.[19] Using laser micro-dissected song nuclei, they compared gene expression levels or transcriptomes in samples taken from all three groups of vocalising birds, as well as from two non-vocalising species, a dove and a quail. Also, laser-captured samples were obtained from donated human brains as well as from our non-vocalising cousins, the Rhesus Macaque. What Pfenning's team discovered was that an identical set of more than 50 genes showed the same expression levels in the song system of vocal learning birds and the homologous speech areas in humans. Such changes did not occur in the brain tissue of birds that do not vocalise, or in the non-human primates that do not speak. In other words, if a gene's activity was increased in humans, it was also increased in songbirds, hummingbirds and parrots but not in any other avian order. What is surprising is that the same genes should be involved in bird and human vocalisation, since our common ancestor lived more than 300 million years ago. Such a finding supports Jarvis's belief that brain circuits for complex traits may have only a limited number of ways in which to evolve.

To conclude the Zebra Finch's story, let me quote Fernando Nottebohm, the Argentinian neurobiologist who discovered the avian song system:

> It may well be that our best understanding of how complex skills are acquired and how broken circuits can be fixed will come not from humans, or other primates, but from the way birds learn their song.[20]

CHAPTER 25

The White-eye's Story

SUPERTRAMPS AND GREAT SPECIATORS

You will not find a white-eye adorning the pages of the glossy picture-book *100 Birds to See in your Lifetime.*[1] Nor, I suspect, would you see an example included in any such publication in the future, even if its title were extended to include *1,000 Birds You Must See Before You Die.* Sadly, white-eyes are not regarded as charismatic or mesmerising enough to earn a place on most birders' ultimate wish-list. This fact can be explained, in part, by their remarkable physical similarity, despite being widely scattered throughout the Old World, in Asia, Africa, Australia and on many oceanic islands. They are mostly small gregarious warbler-like birds, olive-green above, yellowish-white below, and with a distinctive broad ring of white feathers around the eye – hence their English name. However, they do not all have a white eye-ring, and many island forms have lost the yellow lipochrome pigment and so manifest as various shades of brown and grey. White-eyes also lack sexual dimorphism and fail to exhibit seasonal variation, factors that add to their unobtrusiveness. And yet, the 130 species that constitute the family Zosteropidae deserve a higher profile and a greater appreciation. For white-eyes have provided biologists with unique insights into the mechanisms of passerine evolution, and their current distribution has acted as a catalyst for what has become known as the 'great speciator' hypothesis.

My fascination with the white-eye's story stems from a summer birding trip to the remote eastern islands of Papua New Guinea. After an idyllic sojourn on the D'Entrecasteaux Archipelago, we anchored for the night in the lee of a small uninhabited islet, lying a short distance to the south of the much larger Normanby Island. The following morning, our target species was the Louisiade White-eye (Plate 34), a bird rumoured to inhabit the sanctuary of the isle's densely wooded summit. Although rarely seen, this dainty species is typically found on the eponymously named archipelago lying to the east, and so we were attempting to locate it near the westernmost point of its range. Buffeted by squalls and with rollers breaking far below, we scrambled and hacked our way through the ridge's entangled vegetation without encountering a single individual. It was only after reaching the top of the isle that we encountered a small population, no more than 10–15 birds, flitting nervously among the swaying branches. I had seen white-eyes before in India and west Africa, but these continental sightings had left little impression, unlike the insular birds before me. I was both excited and baffled. Why was this isolated population living on an island no bigger than several football pitches while their nearest relatives resided out of sight to the east? More puzzling was why the birds had settled here and had never made the short crossing to nearby Normanby Island, where there exist plenty of suitable habitats? As I was to learn, the paradox of white-eyes' biogeography has been recognised for over half a century, a conundrum that has taxed minds far abler than mine.

Morphological studies

The study of white-eyes' speciation was initiated by the activities of the eccentric and fabulously wealthy collector Walter Rothschild. To satisfy his insatiable cravings for new and rare species, Rothschild hired legions of explorers and adventurers to collect specimens from the far corners of the world. One of the most successful was the intrepid Englishman Albert Stewart Meek, who became the first naturalist to study the birds of New Guinea and the adjacent Solomon Islands. Among his valuable shipments sent back to London towards the end of the nineteenth century were the skins of several of the region's white-eyes, each one carefully labelled with its island of origin. However, it was not until a study of these specimens was undertaken some years later by another of Lord Rothschild's appointments, the German ornithologist Ernst Hartert, that the unusual biogeography of white-eyes was first recognised. For Hartert realised that many of the birds were not only new to science but that several had very restricted distributions, limited to isolated volcanic islands. They included the white-eyes from Gizo, Kulambangra, Ranongga and Vella Lavella in the Solomon Islands as well as those from Tagula Island in the Louisiade Archipelago off New Guinea. It was Hartert's meticulous attention to detail that enabled another of his scientific achievements, the description of narrowly defined subspecies. He recognised that the morphological variations he identified in many species had a geographical basis, and he questioned whether such forms or subspecies could reflect the early phases of speciation. For example, could the four subspecies of Louisiade White-eye, confined to different islands, be evolving into full species?

Lord Rothschild's lifelong obsession led indirectly to the next episode of our story. In 1928, he recruited a 23-year-old German graduate, Ernst Mayr, with instructions to sail for New Guinea to collect further specimens for his private collection. Despite his youth, Mayr was thoroughly prepared for the task, having studied all the relevant skins in the major collections of Europe. After an arduous but successful year in New Guinea, the battle-hardened ornithologist immediately joined the Whitney South Sea Expedition, an American-financed undertaking based at the time in the Solomon Islands. Mayr, no doubt, was influenced by a hortatory letter from Hartert that emphasised 'there is no other place in the world more favourable for the study of speciation in birds than the Solomon Islands.'[2] During his time in the field, Mayr became intrigued by the unusual distribution of white-eyes, and particularly by the marked differences in behaviour and morphology that occurred between different islands, some separated by as little as 2 kilometres. What were the possible factors that could allow the white-eyes to evolve across geographies where other species showed little or no diversification?

Mayr approached the problem by first proposing a new definition for a species. In his magnum opus *Systematics and the Origin of Species* (1942), he had come to the conclusion that a species is not just a population of similarly looking individuals, but one that can only breed among themselves and with no others.[3] When populations within a species become isolated, by whatever means, they may start to differ from the original population through genetic drift and natural selection, and over time may evolve into new species (see *The Storm Petrel's Story*). But Mayr was confronted with a paradox. How could the white-eyes show such a high degree of speciation across so many oceanic islands, when their dispersal ability should aid gene flow and limit their differentiation? A similar problem was evident in many groups of birds, but especially in the white-eyes. Was there something special about white-eyes?

With the enigma unresolved, Mayr encouraged a young physiologist, the nascent polyglot and polymath Jared Diamond, to continue his preliminary field studies in northern Melanesia. Diamond set sail in 1964 on what would be the first of many expeditions to New Guinea and the surrounding islands, and his decision to do so resulted in a career change from medicine to evolutionary biology. Like his mentor, he became intrigued by the marked variation and speciation of the area's birds, not just the white-eyes, but also the whistlers, monarchs and myzomelas. However, he was particularly intrigued by the Louisiade White-eye since, unlike many other species on adjacent islands, it still possessed a strong ability to cross open water. Further observations led Diamond to regard such species – ones with a high dispersal and a broad tolerance to different habitats, but with a low ability to compete with other species – as 'supertramps'.[4] Typically, they are among the first to reoccupy small defaunated islands, although they become crowded out by the subsequent arrival of species with a greater competitive ability. Although the term supertramp encapsulates the high dispersive nature or vagility of the Louisiade White-eye, it is not an appropriate description for the other family members, species that have highly restrictive distributions. Nevertheless, it does provide a logical explanation for my puzzlement as to why the Louisiade White-eye has not been able to make the short crossing to Normanby Island. Their aptitude for colonising remote islets has come at a price, and that price is the inability to compete with specialists found on larger landmasses.

After more than a decade of collaborative work, Mayr and Diamond coined the term 'great speciator' to describe the marked geographic variation and speciation of genera like *Zosterops*. While they could not explain the scientific basis of the hypothesis, they questioned whether 'great speciators' possessed some intrinsic trait that underpinned their distribution patterns.[5]

Genetic studies solve the paradox

Sadly, despite living to over 100, Ernst Mayr died in 2005, four years before his ideas were supported by the molecular age of scientific enquiry. For it was in 2009 that Robert Moyle at the University of Kansas, together with his American collaborators including Diamond, reported the results of the most detailed study of white-eyes ever undertaken.[6]

Using nuclear and mitochondrial DNA analysis, the research team was able to determine both the timing and rate of white-eye speciation by reconstructing their molecular phylogenetic relationships. The results were a shock. All white-eyes, irrespective of geographical location, possessed a strikingly similar genetic profile, a finding that implies a remarkably recent evolution. Indeed, the data from the crown genus *Zosterops*, comprising over 80 species, indicates that an explosive diversification occurred during the Pleistocene and Holocene epochs. In other words, nearly all the world's white-eyes, from Africa to Oceania, evolved within the last 2 million years, yielding a per-lineage diversification rate of 2.24 taxa per million years. Amazingly, these results are the highest reported for any avian family on Earth. In effect, for every one of the many different geographical locations that was populated by the ancestral white-eyes, a new species evolved within 500,000 years. While several avian taxa are known to have evolved faster, for example the Cassia Crossbill (see *The Crossbill's Story*), the importance of the white-eyes is that their rapid speciation involved the whole genus and occurred across all the different areas that were colonised. The next fastest group are the *Setophaga* warblers, at 0.5 species per million years. Very few other vertebrates have greater speciation rates. Most notable among those that do are the cichlid fishes that live in Lake Malawi and Lake Victoria in the African Rift Valley (see *The Storm Petrel's Story*) – but there is a significant difference. Cichlid diversification occurred to meet the challenges of climate shifts and geological changes within a highly restricted geographical range. In contrast, the hemispheric spread and speed of radiation of white-eyes argue against such variables acting as drivers for their speciation. In the words of one team member, 'at this geographic scale, there is no one thing from the outside that could have made this happen; there is something special about those birds.'[7]

After ten years in the field and five more in the laboratory, the scientists concluded that the marked diversification of white-eyes is the result of an unusual combination of intrinsic 'life-history' traits. The longstanding paradox of the 'great speciator' appears to stem from the bird's extreme sociability, rapid morphological evolution, short generation times and loss of dispersal ability. Social birds – most white-eyes live in groups of 5–20 that roam about together, feeding as they go – will have an increased potential for successful colonisation. Indeed, as we have seen, it is the absence of this trait that underpins the highly restricted distribution of the birds-of-paradise. Furthermore, studies of recently founded white-eye populations have confirmed their tendency for rapid morphological and behavioural change. Remarkably, populations of Silvereyes that reached New Zealand in the 1850s are already morphologically distinguishable from their source populations, while the songs of many closely related species are highly divergent. The white-eyes on Gizo, for example, have a loud, high-pitched elaborate song with unpredictable calls, while just 12 kilometres away, the birds on Vella Lavella have a plodding and lower-pitched song.[8] White-eyes also have a relatively short generation time and may breed within six months of hatching, a feature that increases the chances of genetic variation (bowerbirds and birds-of-paradise are only sexually mature after 3–4 years).

But it is the rapid changes in white-eye's willingness to disperse that appears to be the most relevant trait. Dramatic increases enable populations to travel further and more widely than other species, while the sudden loss of vagility results in communities becoming marooned after they reach small islands. For example, despite Silvereyes reaching islands up to 1,500 kilometres from their source populations, many closely related species are unable to cross gaps as little as a few kilometres. The implication is that the important water-crossing ability of the founder population has been rapidly selected out to enable them to 'settle down' and evolve into the island species we see today. The endemic white-eyes of Gizo, Kulambangra and Ranongga in the Solomon Islands must have lost their ability to fly across open water soon after colonisation.

The realisation that insular animals and plants can lose their capacity to disperse is not new – indeed, it was first highlighted in 1849 by Sir Joseph

Hooker in relation to Galápagos weeds.[9] However, for most insular species the reduced vagility results from evolutionary changes in anatomical structure. Birds and insects on islands may evolve reduced or absent wings – paradigms being the flightless rails of Oceania and the Hymenoptera of Hawaii – while plants can develop heavier seeds that are less effective at long-distance dispersal.[10] Remarkably, reduced vagility can develop in less than a decade and within only a few generations, particularly when selection is against the high cost of dispersal. Extreme examples include dandelion populations that have recently colonised several islands off British Columbia and flowering plants that grow on 'virtual islands' around trees planted amid concrete pavements.[11]

But insular white-eyes have a typical avian wing structure with a normal power of flight. Their mechanical ability is quite sufficient for overwater dispersal, so an additional component, a behavioural one that is subject to natural selection, is required to explain why the birds do not disperse. It was Jared Diamond who first appreciated the counterintuitive fact that the white-eyes' reduced vagility must be due to 'behavioural' or 'psychological' flightlessness.[12] Despite still being able to travel significant distances overland, many white-eyes must have developed a fear of crossing open water only after reaching their new habitats. Similar behavioural changes are thought to account for the failure of the Seychelles Warbler to recolonise nearby islands. Sadly, this trait enhances the warbler's risk of extinction and exemplifies the inability of natural selection to see into the future.[13] In the case of white-eyes, their psychological flightlessness results from a balance shift between the evolutionary benefits and costs of dispersive behaviour. Increased vagility occurs when habitat becomes vacant, while reduced dispersal results when new habitats become unavailable or fully occupied. Robert Moyle and his colleagues highlight an analogous cost–benefit relationship for past human dispersive behaviours. Polynesians, for example, underwent an explosive expansion over the Pacific Ocean from Hawaii to New Zealand and Easter Island, an activity that was suddenly abandoned around 1,000 years ago. Since there were no more vacant Pacific islands to colonise, there was no longer any demographic payoff to offset the hazards of such voyages. Of course, the analogy only runs so far, because human colonisation involves mainly cultural rather than genetic modifications whereas changes to island *Zosterops* populations are entirely genetic.

After decades of speculation, the 'great speciator' paradox of white-eyes has finally been resolved. It is the outcome of a complex interplay of sociality, short generation times, marked plasticity of morphology and song, and rapid changes in dispersal behaviour.

Ernst Mayr was right all along: white-eyes are indeed very unusual birds.

CHAPTER 26

The Crossbill's Story

ADAPTIVE RADIATION AND COEVOLUTION

The evolution of complex traits is one of the deepest problems in Darwinism. How can novel adaptations arise from small and seemingly insignificant beginnings? How does an organ such as the vertebrate eye, which relies on the interaction of many complex functioning components, evolve by many small incremental changes?

Darwin acknowledged the problem in *The Origin of Species*:

> To suppose that the eye with all its inimitable contrivances for adjusting the focus to different distances, for admitting different amounts of light, and for the corrections of spherical and chromatic aberration, could have been formed by natural selection, seems, I freely confess, absurd in the highest degree.

Creationists often quote, or misquote, this sentence in their argument for 'intelligent design'. But they should read on, for Darwin's opening sentence is merely a rhetorical ploy used to engage his readership. In fact, he continues to explain exactly how the eye could have evolved by gradual degrees:

> If numerous gradations from a simple and imperfect eye to one complex and perfect can be shown to exist, each grade being useful to its possessor, as is certainly the case; if further, the eye ever varies and the variations be inherited, as is likewise certainly the case; and if such variations should be useful to any animal under changing conditions of life, then the difficulty of believing that a perfect and complex eye could be formed by natural selection, though insuperable by our imagination, should not be considered as subversive of the theory.

In other words, Darwin realised that if you lack an eye, you are blind, but the possession of an inferior eye might make all the difference, as the ability to detect a predator's movement, however slight that ability, might make the difference between life and death.[1] The increased chance of survival makes it more likely that the genes for a poorly performing eye are inherited by future generations, giving scope for additional modifications and improvements.

In his book *Darwin on Trial*, published in 1991, the evolution-basher Phillip E. Johnson wrote: 'The prevailing assumption in evolutionary science seems to be that speculative possibilities, without experimental confirmation, are all that is really necessary.'[2] It is ironic, therefore, that as the lawyer's support for creationism went to press, Craig Benkman and Anna Lindholm provided the perfect scientific response. The two American biologists demonstrated for the first time, just as Darwin had hypothesised, that complex and improbable traits can have potential selective benefits that increase as their expression increases. However, their evidence related not to eyes, but the mandibles of crossbills.[3]

Twisted beaks

Crossbills belong to the finch family (Fringillidae), and their genus name *Loxia* (derived from the Greek *loxos*, meaning 'cross-wise') refers to their remarkable beak structure. Their lower mandible twists either to the right or left as it passes over the upper mandible. According to legend, crossbills deformed their beaks while trying to extract the nails from Jesus' hands and feet, while the red feathers were the result of staining with Christ's blood. They are the most specialised of all the finches, feeding almost entirely on seeds obtained from the cones of pine, spruce and larch. An essential characteristic of their food source is that the cones are closed or partially closed, which minimises the risk of competition from other species. Furthermore, because conifer seeds are available all year round and are subject to random crop failures, crossbills have evolved a complicated, nomadic lifestyle. They are even found nesting in the depths of winter, when incubating females can survive coverings of snow. The adult birds secure the seeds by inserting their beaks between the scales of the cones and forcing the lower mandible sideways, in the direction of its twist, to create enough force to push the scales apart. Once the seed is exposed, it is scooped out by the bird's protrusible tongue, held in a groove in the upper mandible – one that perfectly mirrors the seed size – and broken by pressure from a ridge on the lower mandible.

The genus *Loxia* evolved amid the extensive conifer forests of the northern hemisphere around 9 million years ago.[4] Their common ancestor

Figure 26.1 Beaks of three crossbill species and their primary food sources. *Top*, a Parrot Crossbill with a pine cone; *middle*, Common or Red Crossbill and a spruce cone; *bottom*, Two-barred Crossbill and a larch cone. Reproduced from *Finches* by Ian Newton, Collins, 1972.[7]

probably relied on conifer seeds that, like today, underwent irregular cycles of 'boom and bust'. During a time of particularly severe seed shortage, an ancestral cohort was forced to leave the forests and seek food elsewhere. They developed much smaller beaks and fed on the seeds of birch and alder, and eventually gave rise to the redpoll complex, although only after the Pleistocene glaciations. Genetic analysis suggests that the three groups of redpoll, the Arctic, Common and Lesser Redpolls, probably reflect a single gene pool, but with geographical polymorphisms due to differences in climate.[5] Residents in warmer southern areas, the Lesser Redpoll, tend to be smaller than those further north, having an increased surface-to-volume ratio (Bergmann's rule), while populations in cooler and drier habitats, the Arctic Redpoll, tend to be less heavily pigmented (Gloger's rule).[6] Alternatively, it may be that the three populations are too recent to have allowed enough genetic drift to create the differences required for full speciation.

But it was the population left behind in the boreal forests that provides our main storyline. For these ancestral birds became larger and evolved earlier than the redpoll complex, around 3–4 million years ago, to produce at least four species of crossbill. Crucially, each species evolved bill shapes that matched their food source. The Parrot Crossbill, for example, developed the largest and most heavily built beak, which gave them access to the toughest of pine cones, even closed ones. The Hispaniolan Crossbill and the Red or Common Crossbill, with their medium-sized bills, target pine and spruce cones respectively, while the Two-barred Crossbill, a specialist of the small, soft cones of larch, has the daintiest beak of all (Figure 26.1). Crossbills, therefore, can be regarded as the Galápagos finches of northern America and Europe, with different taxa evolving slightly different bill shapes according to their diet.

To determine whether small morphological differences can have adaptive significance, Benkman and Lindholm studied the bills of seven captive Red Crossbills. They were all of the same ecotype, a population mostly confined to the Western Hemlock (*Tsuga heterophylla*) forests that stretch along North America's west coast from northern California to Alaska. In their experiment, the scientists straightened the beaks of four individuals by trimming the crossed part of their mandibles with nail clippers. The other three birds were left as controls. This manipulation, akin to cutting our fingernails and toenails, was not distressing to the birds, as their keratinous bills lack nerve endings. Those individuals with straightened bills appeared to be just as efficient at extracting seeds from dry, open cones as the three controls, but they could no longer cope with closed cones. Day by day, as their crossed bills slowly regrew, the time taken to extract seeds from closed cones improved. After a month, when their beaks had completely regrown, the birds' extracting skills had returned to normal.[3] In other words, there had been a progressive improvement in performance as the expression of the trait increased, just as Darwin had speculated for the evolution of the eye.

Of course, the clipping experiments are not an exact representation of the evolutionary process, since only a single element of the species' feeding apparatus was modified. Crossbills have coevolved numerous morphological, as well as behavioural, adaptations to enable their bills to function optimally. These include a horny, grooved palate that allows the efficient handling of seeds, a robust and asymmetric jaw musculature, a cartilage-containing tongue, a flexible jaw hinge that allows lateral abduction, and a honed instinct for hunting cones. Furthermore, the laboratory studies do not explain how the array of adaptations came together during evolution, nor do they tell us whether morphological evolution is driven by behaviour or the other way round.[8] Nevertheless, this simple but ingenious experiment does reveal how the crossbill's unique bill structure could have arisen by progressive selection over generations, with each new generation performing slightly better with closed cones than their predecessors.

But how did the reproductive isolation of different populations of crossbill occur, to allow subsequent speciation? Why didn't interbreeding muddle the gene pool and prevent new taxa from forming? Benkman postulates that the refinement of their bills went hand in hand with the emergence of various call types, and that this led to assortative mating and subsequent speciation. His conclusion was reached after spending many hours in the field observing the behaviour and feeding habits of thousands of crossbills.

Benkman noted that whenever a flock arrives in a tree, the birds usually remain quiet. Sometimes, however, one or two individuals begin to call, as if enquiring how their flock mates are doing – if you like, a sort of secret code. If the rest of the flock remains quiet and continues foraging, then the birds cease calling and continue to feed on other cones. Benkman interprets this behaviour as indicating that the early callers had simply picked a poor cone that did not reflect the overall quality of the tree. In contrast, if other crossbills joined in, creating a crescendo of calls, then the flock would fly off and seek an alternative food source. The evolution of flock cooperation enables crossbills to find seed-rich trees with the minimum of energy expenditure – an important factor in the survival game, as 'time is food is life'. Benkman believes that it was the evolution of distinct call types that allowed crossbills to flock with similar birds, since it is much easier for individuals to recognise group members by call than by subtle differences in beak structure. As crossbills flock all year round and choose mates within flocks, assortative flocking leads indirectly to assortative mating, a factor that may account for their rapid adaptive radiation.[9]

Flocking also facilitates the species' characteristic 'irruptions' – periodic mass movements when crossbills search for new feeding grounds, often thousands of kilometres away, whenever seeds are in short supply. Some invasions may lead to sporadic breeding and even colonisation. For example, one recent irruption of the Parrot Crossbill, from the Russian and Scandinavian forests, resulted in a small colony being established in Abernethy Forest in the Scottish Highlands. DNA studies have revealed that the Parrot Crossbill is both genetically and morphologically very similar to the Caledonian Forest's other two species – the Common Crossbill and the Scottish Crossbill. The later has only recently been accepted as a full species, a decision based on its distinctive 'excitement' call, a metallic 'jip' that requires a sonogram for confirmation.

Idaho's new endemic bird

Craig Benkman's 35-year obsession with crossbills has enabled him to confirm the presence of many geographical races or ecotypes within the Red Crossbill complex. Indeed, there may be as many as nine in North America and one in Central America, with each population possessing a subtle but significantly different call type that is reflected in its genetic makeup.[10] One of these, the South Hills Crossbill, has proved to be highly informative from an evolutionary point of view and has recently (2017) been promoted to the rank of full species.[11] It is now formally known as the Cassia Crossbill, after the Idaho county that encompasses the two mountain ranges in which the species occurs – South Hills and the Albion Mountains. Working with various researchers over many years, Benkman has shown that this small population of crossbills has become genetically isolated, despite considerable yearly immigration of other crossbills from surrounding areas. The South Hills population, numbering only 4,000 birds, differs from other crossbills in that they are resident and do not migrate. Pollen fossil records show that their food source, the Rocky Mountain Lodgepole Pine (*Pinus contorta*), became more common at higher altitude around 6,000 years ago when the climate became warmer. Crucially, squirrels, a major competitor of crossbills, never moved into the mountain ranges – hence the new species' scientific name, *Loxia sinesciuris* ('the crossbill without squirrels').

Over time, the ancestral South Hills birds became dependent on the seeds of the dominant pine since, like most Red Crossbills today, they preferred smaller, shorter cones with thinner scales at the tip. However, the birds' insatiable demand for seeds led the Lodgepole Pines to evolve in the opposite direction and develop larger, wider cones, with heavier distal scales. In effect, the crossbills were exerting a selection pressure on the trees to increase their seed defences. To keep pace, the birds then evolved a more robust beak that, in turn, led to an altered contact call that was no longer recognisable by other crossbills. Simply put, natural selection favoured pines with thicker scales to keep the crossbills out, which in turn favoured crossbills with larger bills to get in (Plate 35). Today, the South Hills birds can easily out-compete their less well-adapted and nomadic relatives with different calls. Consequently, the chances of interbreeding are limited, and even if offspring were produced, they would have intermediate-sized bills and die of starvation before having the opportunity to pass on their genes.

The evolution of a reproductively isolated resident population of crossbills that exhibits genome divergence, despite other crossbills freely flying through their territory, is an example of sympatric speciation (see *The*

Storm Petrel's Story).[12] This unusual state has resulted from an evolutionary arms race between crossbill and pine, or predator and prey.[13] It is also the fastest known speciation event in the avian world, occurring in less than 6,000 years. According to Benkman, 'this indicates that the level of reproductive isolation characteristic of recognised species can evolve rapidly even in the continued face of potential gene flow.' His long-term study of crossbills has shown that coevolutionary processes can be potent forces for rapidly generating biodiversity, although how frequently they contribute to avian speciation remains to be determined.[14]

The Cassia Crossbill is not the only *Loxia* species to have evolved as a result of predator–prey interactions. The endemic Hispaniolan Crossbill is the product of 6,000 years of coevolution with the island's Hispaniolan Pine (*Pinus occidentalis*). Compared to Cuba, where there are no crossbills, the pines on Hispaniola have evolved scales that are over 50 per cent thicker. During the same time, the island's crossbills have developed beaks that are 25 per cent deeper than those of their mainland relatives, the Two-barred Crossbill (known in North America as the White-winged Crossbill). It is likely that the beak morphology of the Hispaniolan birds reflects a longer crossbill–pine interaction, one that has been less disrupted by Pleistocene events because of the area's lower latitudes.[15] However, there is one crucial difference between the speciation of Hispaniola's crossbills and that of the South Hills population. The Caribbean birds have evolved in geographical isolation, rather than sympatrically as in the case of the Idaho birds.

Sadly, the leading character in our story, the Cassia Crossbill, will probably become extinct over the next 100 years as climate change and loss of suitable habitat exert their deleterious effects.[16] Lodgepole Pine cones typically remain closed for years, but the recent hot summers are causing them to open and shed their seeds prematurely. The species' demise could be even quicker, should a catastrophic forest fire or a major beetle infestation occur.

CHAPTER 27

The Tanager's Story

THE FINAL FLOURISH

As we have seen, the dispersal of songbirds from their proto-Papuan birthplace reached all continents except Antarctica: a radiation that came to dominate the Earth's avifauna. In the process, populations became fragmented and reproductively isolated, so spawning new species, lineages and families. Some groups underwent explosive multidirectional long-distance dispersals while others used islands as stepping stones for a more gradual expansion. Along the way, mountainous areas provided hotspots of diversity, especially those of central and southwestern China, a reflection of their complex topography and the resultant panoply of ecological niches.[1]

After crossing the Beringian land bridge into North America during the Miocene, the songbird clade rapidly diverged and radiated across the New World. The largest of these branches was that of the nine-primaried oscines (superfamily Emberizoidea): a lineage that contains nearly 8 per cent of all extant bird species. Their name refers to the presence of just nine primary feathers on each wing compared to the ancestral state of ten (in reality, they all have a tenth primary, but it is markedly reduced and largely hidden).[2] The Emberizoidea include the New World blackbirds and orioles (Icteridae), cardinals (Cardinalidae), Hawaiian honeycreepers (still formally classified within the Fringillidae), New World sparrows (Passerellidae), tanagers (Thraupidae) and New World warblers (Parulidae), as well as a ragbag of less notable groups. One offshoot even backtracked across Beringia and recolonised the Old World around 12 million years ago to give rise to the buntings (Emberizidae).[3] Even though other groups have undertaken reverse dispersals (see *The Thrush's Story*), the buntings are the only passerines to have done so through Beringia. This unique dispersal probably accounts for the relative paucity of buntings in western Europe compared to Asia and the Americas.

At first, the Panama seaway restricted the songbirds' southerly dispersal, but as the gap between the two American continents started to close around

13 million years ago, not just birds but many species of land and freshwater fauna began to cross from one side to the other. Although slow at first, the bi-directional flow speeded up, especially after the land bridge was completed around 3 million years ago. The 'great American biotic interchange' was the end of the 'splendid isolation' for the South American suboscines.[4] From now on, they faced competition from their newly arrived cousins from the north.

Despite being able to fly, the forest understorey-specialising suboscines were slow to disperse northwards and only did so after the Panamanian isthmus had fully formed. Furthermore, except for the tyrant flycatchers, very few suboscines made it further north than the border between Mexico and the United States. In contrast, the songbird generalists reached South America millions of years before the closure of the seaway, with taxa from several families, including finches, larks, pipits, sparrows and wrens, eventually reaching Tierra del Fuego. Remarkably, a single species, the South Georgia Pipit, even breeds within the biogeographical zone of Antarctica – the only passerine to do so.

Overall, the north–south interchange transformed the tropical avifauna of the New World and reshaped the patterns of biodiversity across many taxonomic families. But how is it that the invading songbirds were able to out-compete the suboscines and replace them in the forest canopy and open habitats?

It is commonly assumed that endemic lineages are better adapted to their home environment than new arrivals, but this is not always the case, as evidenced by many studies of island faunas. One explanation, according to the ornithologist and ecologist Robert Ricklefs, relates to their morphology: songbirds have longer legs and toes than suboscines and can engage in more active movement through vegetation. Their more generalised behaviour and diets, including greater intratropical migratory tendencies, would have been advantageous during the deteriorating climates of the Miocene and Pliocene.[5] Most suboscines feed only on large insects, often using sit-and-search tactics, while oscines make greater use of fruit, consistent with their dominance in forest canopies. It is only the wrens that resemble the suboscines, and this is the one group of songbirds that has been successful in the forest interiors of South America. Furthermore, the ability of oscines to respond more rapidly to resource fluctuations, as well as their shorter incubation times and overall faster pace of life, would have helped. Indeed, it has been argued that their higher metabolic rate acted as an 'intrinsically superior trait' and that this was the primary factor that underpinned their global dominance.[6]

The first songbirds to cross the Panama seaway gave rise to the tanagers, a family that accounts for around 4 per cent of all avian species and nearly 12 per cent of all Neotropical birds. After reaching northern South America, the ancestral population underwent rapid diversification, leading to a bewildering array of plumage colours, morphologies, behaviours and foraging strategies that include thick-billed seed-eaters, thin-billed nectar-feeders, aerial insect-foragers, foliage-gleaners, bark-probers and frugivores. Before the advent of molecular genetics, such extreme diversity resulted in taxonomic confusion, with convergent evolution concealing many relationships. It now turns out that some taxa that look like tanagers, such as the Hepatic Tanager and Scarlet Tanager, as well as the ant tanagers, are not tanagers at all, but cardinals. In contrast, the Coal-crested Finch and Plushcap, as well as the honeycreepers, seedeaters, conebills, saltators, Darwin's finches and flowerpiercers, have all been reclassified recently as tanagers, despite not looking at all like them.[7]

The Plushcap is named for its unusual dense, golden-yellow forecrown of stiff feathers that are less susceptible to wear and more resistant to moisture than typical feathers. Such a morphological feature may be an adaptation for its specialised feeding mode, in which it probes dense whorls of bamboo for small insects and plant material.[8] In contrast, the Coal-crested Finch is a fire-following specialist of Brazil's grasslands, with no close living relatives. Not only are its plumage colours and pattern unlike any other family member, but both male and female possess a crest, a relatively rare feature among the tanagers.

Most speciation events within the Thraupidae coincided with the accelerated uplift of the Andes which occurred between the Miocene and early Pliocene (12 to 4.5 million years ago), when habitat change, climate cycles and tectonic activity created many new opportunities for isolation. More recently, repeated invasions of Central America, encouraged by the formation of the Panamanian isthmus, led to further diversification. Sea-level changes may also have been a factor. For example, the Lesser Antillean Saltator, a taxon endemic to the Caribbean islands, and the Green-winged Saltator, a species restricted to the Brazilian Shield, diverged from a common ancestor half a million years ago. The Lesser Antilles were never connected to South America, except for the island of Trinidad, and it is likely that the lowered sea levels during the Pleistocene facilitated access from a mainland ancestor. Furthermore, the lack of genetic differences between four island populations of saltator suggests that this was a single, rapid colonisation event.[9]

Seedeaters

The seedeaters are now classified as tanagers, despite being relatively small birds with stubby, conical beaks adapted for feeding on seeds and grain (most tanagers are omnivores with a diet consisting mainly of fruit, nectar and insects). Eleven members of the *Sporophila* genus, colloquially known as 'southern capuchinos' due to the head colouration of the males, have provided fascinating insights into the mechanisms of speciation. These finch-like birds inhabit the grasslands of northeastern Argentina, Uruguay, Bolivia and Southern Brazil. Leonardo Campagna, an Argentinian evolutionary biologist working at Cornell University, became intrigued by this sympatric group of birds after noticing that despite their marked variations in plumage, they could not be separated by standard DNA analysis, despite using thousands of genome-wide polymorphisms. However, species-specific genetic variations must exist, given their marked phenotypic differences. This observation led the researchers to speculate that substitutions in one or more critical genes must have occurred so fast that the random, or stochastic, changes in neutral markers that they had used for lineage assignment had not kept pace.[10] But which genes could be involved, and how and why did they mutate so quickly?

A universal characteristic of seedeaters is their sexual dimorphism, or dichromatism, with males exhibiting a variety of colourful plumages that they use to attract females and to defend territories. While all 11 capuchino seedeaters have a cinnamon-based plumage, the presence or absence of white, black and grey patches characterises the different species. The females of these 11 species, in contrast, are indistinguishable morphologically, and cannot even be identified when in the hand. Of course, other birds can recognise their own kind, probably by detecting specific ultraviolet colour plumage patterns.[11] The unusual features of capuchinos – marked genetic homogeneity, distinct phenotypes and overlapping geographical distributions – led Campagna to hypothesise that the differences in male plumage must be the result of strong selection pressure at a few key genetic loci.

To find out if this idea was correct, the research group used high-throughput sequencing to compare the genomes from nine capuchino species. The results were conclusive, and offered a unique insight into the genetics of speciation. Ninety-nine per cent of the genome differences occurred in the same small areas of DNA: regions on different chromosomes that are involved in synthesising the melanin-related proteins that give feathers their colour. Most frequently, these changes did not affect the genes themselves, but rather the adjacent non-coding regulatory sequences. The important point is that although the regulatory regions vary from species to species, they control the expression of the same set of genes. Campagna believes that it is the altered expression of these melanin-related genes that underpins the phenotypic variation seen in the capuchinos.[12] Also, because speciation occurred so rapidly, it must have resulted from female choice, which in turn caused the males to fine-tune their gene expression very quickly.

While Campagna's team were collating their data, other groups reported differences in the pigmentation genes from several unrelated pairs of incipient avian species. For example, the different coloured subspecies of the Chestnut-bellied Monarch found on the Solomon Islands result from single amino acid substitutions in two melanin-related genes. The races' contrasting belly plumages, chestnut versus blue-black, may be sufficient to cause pre-mating isolation and future speciation.[13] Similar genetic differences have been reported for the Carrion and Hooded Crows, as well as for the Blue-winged/Golden-winged Warbler complex.[14] The importance of all these studies is that genetic alterations in one simple molecular pathway can result in phenotypic changes that have the potential to create new species.

While sexual selection accounts for the capuchinos' high overall rate of speciation, it does not explain why it should have been greatest to the south of the Amazon. One explanation is that the region was repeatedly and extensively flooded by sea water during the late Pleistocene, and that these events fragmented the early population, leading to isolated groups with little scope for genetic exchange. Also, Jon Fjeldså and Carsten Rahbek believe that the combination of open-habitat refugia and specific niche requirements could have kept small local populations apart in patches of suitable grassland.[15] The temperate climate during the last glacial maximum, some 27,000 years ago, resulted in savanna-like habitats throughout South America, while fluctuations in the south polar air currents, as a result of Quaternary glacial changes, led to cycles of reciprocal expansion and contraction of open habitats and rainforests.

As in most taxa, seedeater speciation does not seem to be the result of a single mechanism, but rather of a combination of factors, including sexual selection, marine ingressions and open-habitat refugia.

Darwin's finches

Darwin's finches are the most celebrated of all the tanagers and, ever since their discovery, have been a model of speciation and adaptive radiation. Indeed, the species living on the Galápagos Islands (another inhabits Cocos Island, 600 kilometres to the north) are 'among the most thoroughly studied wild animals in existence'.[16] Unlike other young adaptive radiations, such as the Hawaiian honeycreepers, Darwin's finches are intact, and no species has been lost from human activity. This fact makes the clade especially valuable for understanding the ecological and genetic basis of biodiversity. Despite different body sizes, beak shapes, songs and feeding behaviours, all have evolved from a common grassquit-like ancestor that dispersed from South America approximately 1.5 million years ago. Soon after their arrival, the ancestral population split into two: one branch produced the warbler-finches, the Cocos Finch and the Vegetarian Finch, while the second branch gave rise to a group of ground and tree finches. The result was that species evolved to exploit every available food source: seeds, buds, fruit, insects, pollen, nectar, and even blood (Plate 36).

The two warbler-finches, for example, look and behave like mainland warblers, with very thin and pointed beaks which they use to probe leaves of the Palo Santo trees (*Bursera graveolens*) to catch small insects and their larvae. Both species (recently split on account of different song, habitat and range) look so unlike finches that Darwin mistook the birds for true warblers.

The Sharp-beaked Ground Finch, however, has a slightly larger and more cone-shaped beak that it uses to collect both insects and small seeds, while the Genovesa Cactus Finch has a relatively robust and elongated beak for penetrating the firm covers of cactus fruits. The largest of all beaks, a massive, extremely deep and broad bullfinch-shaped structure, is possessed by the Large Ground Finch; it uses its beak to crush the large and hard seeds that are not accessible to other species. On two small islands, the Vampire Ground Finch uses its sharp arrow-shaped beak to peck at the wings and tails of boobies to drink their blood. Strangely, the seabirds do not seem to mind, and it is thought that the behaviour evolved from the pecking strategy that the finches used to remove parasites from the boobies' feathers. This oddly behaving finch also smashes boobies' eggs against the rocks to obtain the yolk, and drinks the blood of its own dead. In contrast, the Vegetarian Finch uses its broad and stout beak to strip back the bark of small branches to get at the underlying nutritious pulp, while the Woodpecker Finch uses a cactus spine or leaf stem as a tool to extract its prey.

Darwin's finches exhibit considerable morphological overlap, so that many congeners are difficult to tell apart, not just in the field but even in museum collections. And if that were not enough, recent research has revealed the existence of three additional species based solely on genetic differences. In 2015, Leif Andersson, a geneticist at Uppsala University in Sweden, and his colleagues were the first to sequence the complete genomes of all 14 recognised species.[17] When the genomic data were used to construct a phylogenetic tree, the researchers found that what had traditionally been accepted as two species, based on beak morphology, were in fact five species. The Sharp-beaked Ground Finch occupied three entirely different branches of the phylogenetic tree, while the Large Ground Finch sat on two separate divisions. In other words, genomics has shown that there is much more to Darwin's finches than meets the eye, with no less than 17 identifiable taxa.

One explanation for the number of species is their exceptionally fast radiation, even by tanager standards, one strongly influenced by geography and climate change. All the islands remain isolated, not just from the mainland but also from each other, so that few competitors or predators made the journey, especially during the early years of colonisation. Also, each island varies in height and habitat and supports a unique combination of animals and plants. New islands were formed by volcanic activity, while climatic oscillations caused by the El Niño phenomenon led to changes in the islands' flora and associated insect life, all of which created a range of different ecological niches. Furthermore, speciation has been facilitated by crossbreeding and the mixing of genes throughout the finches' evolutionary history. While hybrids of most avian species are sterile, the hybrid chicks of Darwin's finches are fertile and can mate with both parental species. The resulting offspring will associate with one or other of the parents through song or appearance despite possessing genes from both species. The importance of gene flow or introgression (see *The Sparrow's Story*) is that novel combinations of genes can be rapidly created, promoting the emergence of new taxa.

It was Peter and Rosemary Grant, a British husband and wife team now working at Princeton University, who first demonstrated just how fast evolution can operate. The two biologists have devoted their working lives to the study of the Galápagos finches, spending six months every year for over a quarter of a century, capturing and tagging birds, documenting beak and wing measurements, and taking blood samples for genetic analysis. In 1977, following a severe drought, the Grants witnessed a marked reduction in the population of Medium Ground Finches on the island of Daphne Major, due to a shortage of food. At the beginning of the year, the island had over 1,000 individuals, but 12 months later the population had fallen to less than 200. When the survivors were studied, they were found to be slightly bigger than the previous population average, with marginally larger beaks. It appeared that these individuals, unlike their smaller kin, were better able to access the island's only remaining food source – the big, spiky seeds of the plant *Tribulus*. Most survivors were male, as females are slightly smaller, and when the rains returned the following year and breeding recommenced, there were five males for every female. In the fierce sexual competition that followed, only the largest males succeeded in breeding, so the genes for larger bodies and beaks were more likely to be inherited. Remarkably, it seems that for the Medium Ground Finch the difference between life and death in times of drought depended on as little as half a millimetre in beak length.

Five years later, the Grants observed climatic changes that would again change the makeup of Daphne Major's resident birds. This time a prolonged rainy season due to one of the strongest El Niño events ever transformed the island's vegetation into a community of smaller seed-producing plants that favoured individuals with smaller beaks. The El Niño rains had altered the food supply, and only the smaller-beaked birds won the battle for survival – the opposite of what had happened a few years earlier.

For Peter and Rosemary Grant, the Medium Ground Finch had shown that natural selection can produce morphological changes exceedingly quickly. Indeed, they calculated that it would only take 23 bouts of severe drought to convert the phenotype of the Medium Ground Finch into that of its parrot-billed cousin, the Large Ground Finch.[16] Recently, developmental biologists have added a new twist to the Grants' meticulous field observations. Arkhat Abzhanov, a postdoctoral fellow at Harvard Medical School, studied embryos from six species of finch and determined when and where different growth-factor genes were expressed in the developing beak.[18] Three species had large bills for cracking seeds while three species had slender, pointed beaks for extracting nectar. The only protein to distinguish between the two cohorts was bone morphogenic protein 4 (Bmp4), a signalling protein typically associated with the development of the skull and other bones. The two bird groups differed in both the amount and the timing of Bmp4 expression. Those birds with the largest beaks made more protein and at an earlier stage than smaller-beaked individuals, with each species having its unique pattern of expression. Furthermore, artificially increasing the amount of protein, by local injection of a virus which contained the *Bmp4* gene, resulted in the beak becoming larger. These preliminary experiments suggest that Bmp4 is important in shaping birds' beaks, although what makes the growth factor more active in birds with big beaks remains unknown.

Beak development, like most biological traits, is an intricate process and controlled by more than one gene. In 2015, *ALX1*, a gene known to be involved in human craniofacial development, was also linked to beak shape in Darwin's finches, specifically to how blunt or sharp the bill becomes.[17] Researchers noted two variants or haplotypes of the gene in the birds: an ancestral form associated with pointed beaks, which evolved early, before the group's radiation, and a derived form that leads to blunt beaks. Individuals with both haplotypes have intermediate bill shapes.

Several months later, the same research team identified a gene linked to beak size rather than shape.[19] The gene, known as *HMGA2*, had previously been associated with variations in height, facial structure and tooth eruption in humans. The role of *HMGA2* in beak structure was confirmed after studying survivors from a severe drought that struck the Galápagos Islands in 2004 to 2005. At that time, many of the Medium Ground Finches with larger-than-average beaks starved to death as they were out-competed by a larger species that had recently colonised their island and were better able to eat large seeds. After the drought, the surviving Medium Ground Finches had smaller beaks than those that succumbed, since they were more suited to taking the smaller seeds that their competitors tended to avoid. Even today, the Medium Ground Finch has a smaller beak size than before the drought. By analysing DNA from birds that lived at the time, it was noted that the large-beak *HMGA2* variant was more common in the Medium Ground Finches that starved to death, and less common in survivors. In other words, changes at a single locus had facilitated a rapid diversification in beak shape.

Genetic mutations may not be the only molecular factors to explain how species evolve. Epigenetics, which is the regulation of gene function without a change in the genetic code, may also be important. The term 'epigenetics' was coined by developmental biologist Conrad Waddington of the University of Edinburgh after he noticed that developing fruit flies exposed to chemicals or temperature changes could be induced to produce different wing structures (*epi-* means 'above', 'on' or 'beside' in Greek). Although such anatomical changes were induced in a single generation, they were passed on to all subsequent lineages. It seems that Jean-Baptiste Lamarck's once-ridiculed idea that the environment can directly alter traits that can then be inherited was right after all. Indeed, biologists now understand how such processes may occur. One of the mechanisms involves acquired DNA methylation, in which chemical components called methyl groups (derived from methane) bind to DNA and regulate gene expression. When methyl groups are added to a particular gene, the gene is switched off or silenced, and no protein is produced. Environmental factors, including temperature and stress, can alter the extent of methylation, and these changes may be permanently programmed and inherited over many generations: a process known as epigenetic transgenerational inheritance. Crucially, such genetic modifications can account for the rapid emergence of new traits, events that are difficult to explain solely with classical genetics and neo-Darwinian theory.

Could epigenetics be involved in the evolution of Darwin's finches? To find out, Michael Skinner, working at Washington State University, undertook a detailed study of the patterns of methylation across the genomes of five finch taxa.[20] Not only did each species have a different methylation pattern, but the changes also increased with the phylogenetic distance between each species. There was also minimal overlap of the individual epigenetic sites among the species. Skinner and his colleagues then carried out a more focused study, looking at the methylation profiles of genes known to be involved in the development of beak shape, immune-system responses and plumage colour. Once again, the epigenetic patterns differed for all the gene groups, while the gene sequences were nearly identical.

For almost 200 years, Darwin's finches have provided us with a unique window into the mechanisms of evolution. But the more they have revealed, the more complex the story becomes. It is now apparent that a full understanding of the molecular control of speciation will require the integration of results from both genetic and epigenetic studies.

Flowerpiercers

The long and at times labyrinthine journey that has taken us from the ancient monotypic ostrich to the most recent and highly polytypic family of tanagers is now nearing completion. After around 100 million years of evolution, and approximately 220 families and over 10,600 species later, we have reached the terminal bifurcations, the outermost twigs and offshoots of our schematic 'bird tree'. The tanager's story, however, is not yet complete and promises one more surprise. Eight-and-a-half million years ago, a unique feeding strategy evolved in a group of ancestral Thraupidae living along the northern slopes of the Andean rainforest. It seems that this group abandoned fruit as their primary food source and switched instead to nectar, and in so doing gave rise to the 18 species we know today as flowerpiercers.

The events that led to their dietary change are lost in the mists of time, but it is not hard to imagine a likely scenario. A bird in search of fruit, possibly with a slightly longer bill than usual, may have accidentally pierced the tubular corolla of a nectar-containing flower and so benefited from its sugar-rich food supply. Subsequent generations became ever more dependent on nectar and, as a result, evolved increasingly efficient means of obtaining it. Their tongues developed a ventral groove, with the distal part splitting into a terminally frilled fork, while their bills grew longer and increasingly hooked (Plate 37). These traits allowed the birds to stabilise the flowers while penetrating the nectar-containing bases with their lower jaws and extracting the contents with their tongues. It is this unique combination of bill and tongue that has increased the flowerpiercers' access to a wide range of nectar sources which remain inaccessible to legitimate, non-robbing, visitors. Biologists regard flowerpiercers as 'nectar-robbers', or 'nectar stealers', since, unlike hummingbirds, they are not pollinators and provide no benefit to the plant.

The notion that the development of novel morphological features is accompanied by trade-offs is central to current evolutionary theory, although it has been surprisingly difficult to prove in the case of higher organisms. What this means, in reality, is that it is not possible for a 'superspecies' to evolve, one that is optimised to do everything, with a high degree of efficiency in all possible situations. But can this thesis be shown to apply to birds? In 2003, Jorge Schondube and Carlos Martinez de Rio, two American biologists, addressed this question by devising a simple experiment using captive Cinnamon-bellied Flowerpiercers.[21] They trimmed their bill lengths, with ordinary nail clippers, to mimic the ancestral state – a procedure that did not hurt the birds, as keratinous beaks lack nerve endings and will grow back within a month. They found, as they had conjectured, that birds with clipped bills ingested fruit far more efficiently, but at the expense of a reduced ability to rob flowers. Conversely, birds with intact bills were good flower robbers, but poor frugivores.

The phylogenetic relationships of the flowerpiercers have long puzzled ornithologists, and it is only after the availability of mitochondrial DNA sequence data that their position within the tanager family has been proven. Furthermore, the seminal study on their phylogenetics, conducted by Kevin Burns and William Mauck at the San Diego State University, has also shed light on the complex mechanisms that underpin their diversification, processes that we have met and discussed previously: Andean uplift, habitat change, climatic cycles and tectonic activity.[22] The main mountain-building processes in South America occurred between 7 and 4 million years ago, while further and more rapid Andean uplifts took place as recently as 2 million years ago. Intriguingly, these dates correspond well with the molecular timing of many flowerpiercer speciation events. Also, several early dispersals occurred from out of the Andean founder population: one to Central America, giving rise to two species, the Cinnamon-bellied Flowerpiercer and the Slaty Flowerpiercer, and at least one to the tepuis, the isolated table-top mountains scattered throughout the rainforests of Venezuela.

Glossy images of the best-known tepuis adorn tourist brochures, and uniquely convey the region's beguiling beauty and remoteness from our everyday experience. The largest, Auyán-Tepui, boasts the world's tallest waterfall, Angel Falls, which plummets nearly a kilometre to the forest floor, while the highest, the mystical 2,810-metre Mount Roraima, is the most accessible and lies at the triple border point of Venezuela, Guyana and Brazil. There are well over 100 tepuis, all relics of a vast sandstone plateau, stretching from the Andes to the Atlantic coast, that became eroded during the Cretaceous and possibly as late as the Quaternary. It now seems that the ancestors of two species, the Greater Flowerpiercer and the Scaled Flowerpiercer, reached the tepuis before much of this erosion had taken place when the habitat was more or less continuous with that of the Andes. Subsequent geological processes marooned these avian wanderers amid a sea of open savanna, and with no further opportunities for gene exchange, they evolved into separate species. Such scientific conclusions, based on the latest molecular techniques, had already been envisioned by the Scottish author Sir Arthur Conan Doyle. His popular novel *The Lost World* (1912) presciently depicts the tepuis' flora and fauna as remnants of ancient life forms, albeit dinosaurs, prehistoric reptiles and ape-men.[23]

As I stare at my computer screen and guide the cursor for the final time over the spiral-like graphics of the avian tree of life at www.onezoom.org, the terminal offshoot, bearing the very last leaf, comes into focus. Its blank green ovate form, once enlarged, informs me that 128,300 years ago the Grey-bellied Flowerpiercer diverged from the rest of its genus: a rather small and unobtrusive bird to occupy such a noteworthy place in our story. An endemic of Bolivia and northern Argentina, it is fairly common above 2,500 metres, being found in montane woodlands, scrub and gardens, where it feeds on fuchsia and other flowers.

I'll vouch that most visitors, strolling along the paths of La Paz's botanical gardens, will be unaware that the distinctive song issuing from a nearby epiphyte is that of a male Grey-bellied Flowerpiercer, a species that belongs to evolution's most recent avian lineage. Nor, I suspect, will they be cognisant of his impressive pedigree, a bird whose ancestors survived the Chicxulub impact, reached Australasia following the break-up of Gondwana, dispersed from the Old to the New World via Beringia, to become the latest line in the myriad of oscines that today populate the Neotropics. His rapid, jumbled warbling is delivered to attract a mate and so ensure that his genetic makeup, honed by millions of years of natural selection, is passed to the next generation and that the process of evolution continues – an unbroken chain of neornithine genetic causality that stretches from tinamou to tanager.

Now, that is what I call an ascent.

Postscript

The Sixth Extinction

Just when modern science is providing us with the tools to explore avian evolution and speciation in unprecedented detail, the future of birds is looking bleak. More than 160 species are known to have become extinct, or are very likely to have done so, during the last 500 years. More than 10 times this number may have gone from the Pacific islands since humans first arrived in their flimsy canoes over 3,000 years ago. Hawaii's loss alone accounts for around 30 per cent of all known avian extinctions, and the future looks no better. On the island of Kaua'i, for example, global warming has allowed avian malaria to expand into the birds' high-elevation ranges which were once too cold for mosquitoes to survive. The story of Guam is just as depressing, as the island has lost 60 per cent of its native taxa over the last 30 years following the accidental introduction of the Brown Tree Snake (*Boiga irregularis*). And in the Indian Ocean, human-induced extinctions have accounted for nearly 30 species on the Mascarene Islands (Mauritius, Réunion and Rodrigues), including rails, waterfowl and landbirds – the most famous examples being the flightless, pigeon-like Dodo of Mauritius and Rodrigues Solitaire, both mercilessly hunted by European sailors for sport and food.

Extinctions are not restricted to rare taxa living in exotic faraway places. The migratory Passenger Pigeon was once the commonest bird in North America, if not the world, with an estimated 136 million breeding adults in Wisconsin alone. Remarkably, the pigeon was hunted to oblivion, from billions to none, persecuted by the delusion that no amount of exploitation could endanger a species so abundant.[1] The rate of avian extinctions, sadly, shows no sign of abating, and may even be increasing, especially on continental landmasses. As we discussed in *The Manakin's Story*, the recent recognition of the mini-interfluves and fine-scale endemism of the Amazonian rainforest suggests that many cryptic endemics may have already been lost.

According to BirdLife International, one in eight species, over 1,300 taxa, are under the threat of extinction. Indeed, 19 species have been lost since I first started birding as a boy, including three this millennium. Spix's Macaw disappeared from the wild towards the end of 2000, the remaining two non-captive Hawaiian Crows died in June 2002, and the last Poo-uli (Black-faced Honeycreeper), also from Hawaii, died in captivity not long after. Many more species are presumed to have gone the way of the Dodo, including the Pink-headed Duck, Pohnpei Starling and Slender-billed Curlew. However, since there is still a slim chance that a few individuals might remain, the International Union for Conservation of Nature (IUCN) currently lists all three as 'critically endangered'.

Amid the doom and gloom, there have been the occasional flickers of light. Several species long thought to be extinct have miraculously 'reappeared'. The Cebu Flowerpecker was rediscovered in the Philippines after nearly a century of absence, although its future hangs in the balance due to catastrophic habitat loss. The New Zealand Storm Petrel (see *The Storm Petrel's Story*), recently relocated off Auckland, is faring better, while Jerdon's Babbler was rediscovered in Myanmar three years ago after a group of scientists recognised its call.

During the last two decades, a dedicated band of field biologists and conservationists have worked tirelessly to prevent the loss of several iconic species. Those given a second chance include the Abbott's Booby, California Condor, Chatham Island Robin, Hawaiian Goose, Kakapo and Whooping Crane.[2] Such heroic conservation efforts, however, may result in future populations harbouring genetic weaknesses. A small founder population, less than 20 individuals in some cases, inevitably results in reduced genetic variation, what biologists term a 'genetic bottleneck'. The robustness of any saved species, therefore, could be impaired, since its ability to survive future selection stresses, such as climate change or shifts in resource availability, may be reduced.

Some optimists, while conceding the disturbing trends, argue that it might be possible to create or resurrect species using cutting-edge genetic engineering. Indeed, an international team of scientists, coordinated by the American organisation Revive & Restore, have made plans to 'genetically rescue' the Great Auk, a species last seen alive in 1844, off the coast of Iceland. The ambitious project aims to sequence the Great Auk's complete genome, using DNA extracted from museum specimens. Once the genes that characterise the Great Auk have been identified and recreated in the laboratory, they will be edited into the cells of the auk's nearest living relative, the Razorbill. Since the Great Auk was much bigger than any extant alcid, any viable 'test-tube' embryos will need to be implanted into larger birds, possibly a species of goose. The project organisers have even identified a suitable release site, the Farne Islands, lying off the northeast coast of England. Other 'resurrection' projects overseen by Revive & Restore include the Carolina Parakeet, once the most colourful bird in the United States, as well as the Heath Hen, Passenger Pigeon and Woolly Mammoth. However, there are many hurdles to overcome before Great Auks are seen diving off our shores again. The field of avian reproductive biology, for example, is still in its infancy, and the ethics of applying reproductive technologies to extinct species should not be underestimated. Furthermore, even if genetic engineering works, the resulting Razorbill–Great Auk hybrid will need to be taught how to be a Great Auk! Of course, critics of 'de-extinction' argue – correctly in my view – that scientists can only restore a fraction of lost species, and that such an approach fails to address the environmental issues that led to the extinction of the majority of those taxa in the first place.

Although this postscript relates to birds, similar depressing accounts would not be out of place in publications covering other animal groups, including amphibians, mammals and insects, or indeed plants. The extinctions happening right now are, for the first time in the Earth's history, the responsibility of just one species, *Homo sapiens*. It is a process arcanely dubbed the 'Anthropocene defaunation'. There is no bolide, or massive volcanic eruption, to blame, only a litany of human impacts that include the acidification of oceans, habitat loss, climate change, deforestation, hunting, pollution, wind turbines, mining and drilling, and the introduction of alien species and diseases. Such events are happening so fast that most birds are unable to change their ranges or life cycles to adapt and so survive. Importantly, many vulnerable bird families contain a significant proportion of avian diversity and possess genotypic and phenotypic characteristics not present elsewhere. An extinction event may not only reduce the number of species, but it can also erode their evolutionary heritage, the history that an individual species represents, which in turn may lead to the homogenisation of the world's avifauna.[3]

One may think that life appears tenuous at times, but the history of the Earth would suggest otherwise. Despite five mass extinction events over the last 450 million years, life has not only survived but rebounded, evolving many new taxa on each occasion. Life, it seems, is impossible to wipe out completely. This realisation will be of little comfort to our grandchildren. For according to some, we are now facing another mass extinction, since the combined effects of 'human activities are leading to an unparalleled rate of species' loss. To paraphrase Elizabeth Kolbert, author of *The Sixth Extinction*, the average background rate of species loss has been very slow for millions of years, and speciation and extinction have more or less equalled out. But that's clearly not what is happening now. Any naturalist out in the field has watched something go extinct. Even children can name things that have gone extinct.[4]

A recent study by Jurriaan De Vos of Brown University, Rhode Island, has estimated that the current extinction rate is approximately 100 extinctions per million species per year, which is 1,000 times greater than at any stage during the last 60 million years.[5] Given the estimate of 8.7 million species on Earth,[6] it is a sobering thought that while you were reading this book, at least three species may have disappeared for ever. Indeed, some biologists believe that this grim wave of extinctions may yet involve humans in its relentless path.

Let us hope that the current awareness of biodiversity-related issues, and growing concern at the increasing pace of species loss, can galvanise humanity to address this most pressing of issues. It is our moral and ethical duty to do so.

But time is not on our side.

Glossary

Adaptive radiation – the rapid evolution from a common ancestor of several species that occupy different ecological niches.

Alleles – alternative forms of a gene at a particular locus.

Allen's rule states that body appendages and shape vary by climatic temperature by either minimising exposed surface areas to reduce heat loss or maximising exposed areas to increase heat loss.

Allochrony – an ecological term used to describe the situation were two sympatric species are not active simultaneously.

Allopatric – occurring in separate, non-overlapping geographical areas.

Allopolyploidy – having two or more complete sets of chromosomes derived from different species.

Anatidae – a bird family that includes ducks, geese and swans.

Anseriformes – an order that includes screamers, the Magpie Goose and Anatidae.

Anisodactylous – three toes turned forward, one backward: the pattern that characterises the order Passeriformes.

Apex predator – a predator residing at the top of a food chain, on which no other creatures prey.

Apomorph – any derived character occurring at a branching point and carried through one descending group in a phyletic lineage.

Apoptosis – a process of programmed cell death that occurs in multicellular organisms.

Archaea – a domain or kingdom of single-celled organisms.

Assortative mating – a non-random mating pattern in which individuals with similar genotypes and/or phenotypes mate with one another more frequently than would be expected under a random mating pattern.

Bacteriolytic – pertaining to the killing of bacteria by the destruction of their cell structure.

Benthic – relating to the bottom of the ocean.

Bergmann's rule states that within a broadly distributed taxonomic clade, populations and species of larger size are found in colder environments, and populations and species of smaller size are found in warmer regions.

Beringia – a series of landforms that once existed periodically between northeastern Asia and northwestern North America.

Biogeography – the study of the distribution of species and ecosystems in geographic space and through geologic time.

Bolide – an extremely bright meteor, especially one that explodes in the atmosphere.

Bootstrapping – in statistics, any test or metric that relies on random sampling with replacement to allow an assignment of accuracy for any given phylogram (defined in terms of bias, variance, confidence limits or similar measurement).

Caprimulgiformes – an order of birds that includes potoos, owlet-nightjars, frogmouths, nightjars and the Oilbird.

Cavernicolous – inhabiting caves, or cave-like places.

Cenozoic – the latest era of geologic time that includes the Palaeogene, Neogene and Quaternary periods, characterised by the formation of modern continents, glaciation, and the diversification of mammals, birds and plants.

Cere – a fleshy, membranous covering of the base of the upper mandible, especially of a bird of prey or a parrot, that protects the bird's nostrils, or nares, enabling them to breathe.

Cetaceans – a clade of mammals that includes whales, dolphins and porpoises.

Character displacement – the phenomenon where differences among similar species whose distributions overlap are accentuated in regions where the species co-occur but are minimised or lost where the species do not overlap; results from evolutionary change driven by competition for a limited resource (e.g. food).

Clade – a group consisting of an ancestor and all its descendants, a single 'branch' on the tree of life.

Cladistics – a system of classification based on the phylogenetic relationships and evolutionary history of groups of organisms.

Cladogenesis – an evolutionary splitting event where a parent species splits into two distinct species, forming a clade.

Coevolution – the process by which two or more interacting species evolve together, each changing as a result of changes in the other(s).

Colonisation – dispersal by a species to an area not previously inhabited by that species, with subsequent establishment of a viable population.

Commensalism – a form of symbiosis between two organisms of different species in which one benefits from the association while the other is largely unaffected.

Congener – a species that belongs to the same taxonomic genus as another species.

Convergent evolution – the development of similar structures or functions, despite evolutionary ancestors being unrelated.

Corvid – a member of the family Corvidae, a group of birds that includes crows, jays, ravens, magpies, jackdaws and rooks.

Cretaceous – a geological period extending from 145 to 66 million years ago.

Crocodilia – an order of mostly large, predatory, semi-aquatic reptiles that includes alligators, caimans and true crocodiles. They are the closest living relatives of birds.

Crown group – a group of living species and their ancestors back to the most recent common ancestor.

Cryptic species – two or more distinct but morphologically identical species that were previously classified as a single species.

Cursorial – an organism that has limbs adapted specifically for running.

Dentine – the main supporting structure of the tooth, the second-hardest tissue in the body after enamel.

Dichotomy – in phylogenetics, used to describe the dividing of an evolutionary tree into two branches.

Dispersal – the spread of species to a new area, a complex process that involves emigration and establishment.

Ecotype – a genetically distinct geographical variety, population or race within a species, which is adapted to specific environmental conditions.

El Niño – occurrence of unusually warm surface waters in the Pacific Ocean, and associated heavy rains.

Emberizoidea – a superfamily of over 800 passerines; formerly known as the New World nine-primaried oscines.

Eocene – the second epoch of the Palaeogene, occurring from 56 to 33.9 million years ago.

Epigenetics – external modifications to DNA that, although not altering its sequence, modifies the expression of genes.

Extra-pair copulation – a promiscuous mating behaviour in monogamous species.

Folivorous – feeding on leaves.

Foraminifera – single-celled organisms with shells for protection that live mostly on the sea floor.

Founder effect – the loss of genetic variation that occurs when a new population is established by a very small number of individuals from a larger population.

Free radical – an atom or molecule with an unpaired electron that makes it unstable and highly reactive; oxygen free radicals produced from mismatched respiratory proteins can lead to ageing, disease and cell death.

Frugivorous – feeding on fruit.

Galliformes – an order of heavy-bodied birds that includes turkeys, grouse, chickens, partridges, pheasants and junglefowl. Lesser-known members include megapodes, chachalacas, guans and curassows.

Galloanserae – a group that includes the Galliformes and the Anseriformes.

Gamete – a reproductive cell having the haploid number (half) of chromosomes that fuses with a gamete of the opposite sex to produce a zygote.

Genetic bottleneck – reduction in number of alleles associated with a short-term reduction in population size.

Genetic drift – the statistical drift over time of gene frequencies in a population, due to random sampling effects in the formation of successive generations.

Genome – the complete set of chromosomes of an individual.

Genotype – the genetic makeup, as distinct from the physical appearance, of an organism. See also *phenotype*.

Gloger's rule states that within a population or species, more heavily pigmented forms are found in warmer and more humid environments.

Gondwana – a southern supercontinent that started to break up 180 million years ago to produce the following landmasses we recognise today: Africa, Madagascar, Australia, New Zealand, South America and India.

Granivorous – relying on a diet of seeds.

Haplotype – a set of DNA variations, or polymorphisms, that tend to be inherited together.

Holarctic – a biogeographical region that includes the northern areas of the world and is divided into Nearctic and Palaearctic regions.

Holocene – a geological epoch that extends from 11,700 years ago to the present.

Holotype – a single physical example of an organism, used when the species was formally described.

Homologous – showing a degree of similarity, for example in position, structure, function or characteristics, that may indicate a common origin.

Hybrid swarm – a population of hybrids that have survived beyond the initial hybrid generation.

Hybridisation – the process of an animal or plant breeding with an individual of another species or variety.

Imprinting – a rapid learning that occurs during a brief receptive phase, typically soon after birth or hatching, and establishes a long-lasting behavioural response to a specific individual or object.

Infraorder – a taxonomic group that is below a suborder.

Insectivorous – species that feed on insects, worms and other invertebrates.

Introgression – transfer of alleles from one species to another as a result of hybridisation and back-crossing (also known as introgressive hybridisation).

Laurasia – the northernmost of two supercontinents that formed part of Pangaea.

Lithosphere – the crust and uppermost mantle, constituting the hard and rigid outer layer of the Earth.

Magic trait – an acquired characteristic that drives speciation because it also affects mating signals.

Mantle – a layer between the Earth's crust and the outer core.

Mesozoic – the geological era extending from 252 to 66 million years ago – the age of the reptiles.

Methylation – an epigenetic process by which methyl groups (CH_3^-) are added to DNA molecules, leading to altered rates of gene transcription.
Miocene – the first geological epoch of the Neogene period, extending from 23 to 5.3 million years ago.
Molecular clock – the concept that the difference (number of nucleotide substitutions) separating two homologous DNA sequences is proportional to the amount of time since they last shared an ancestor.
Molecular phylogenetics – a branch of phylogenetics that determines differences in DNA sequences to determine an organism's evolutionary relationship.
Monophyly – a group of species is monophyletic if they all share the same common ancestor, and share it with no other species.
Monotypic – having only one type, such as a family consisting of only one species.
Mya – million years ago.
Necrosis – a form of cell injury which results in the premature death of cells by autolysis.
Nectarivorous – feeding on nectar.
Neoaves – a clade that consists of all modern birds except Palaeognathae and Galloanserae.
Neogene – the second period in the Cenozoic era, extending from 23 to 2.5 million years ago.
Neognathae – a superorder of Neornithes that includes all modern birds, except the Palaeognathae. The name refers to the reduction of the median bones of the palate ('recent jaws').
Neornithes – all modern birds, comprising two superorders: the Palaeognathae (tinamous and flightless ratites), and the Neognathae (all other birds).
Neoteny – the retention of juvenile characteristics into adulthood. See also *paedomorphism*.
Neotropics – the biogeographical region from southern Mexico and the West Indies to southern South America.
Niche – the part of the environment that is used by organisms.
Node – a junction point in a phylogenetic tree.
Nucleotides – the building blocks of DNA.
Obligate – refers to an organism capable of functioning or surviving only in a particular condition or by adopting a particular behaviour.
Oligocene – the third and last epoch of the Palaeogene period, lasting from 33.9 to 23 million years ago.
Omnivorous – feeding on both animal and plant foods.
Operculum (Latin, meaning 'little lid') (plural *opercula*) – a skin flap that protects the ears of certain species of owl.
Order – a category in Linnaean classification above suborder and below class.
Ornithophily – the pollination of flowering plants by birds.

Orogeny – building of continental mountains by plate-tectonic processes that squeeze the lithosphere.

Oscines – a large suborder of passerine birds, the songbirds, that have numerous syringeal muscles, conferring musical ability.

Paedomorphism – the retention of juvenile characteristics into adulthood. See also *neoteny*.

Palaeocene – the earliest geological epoch in the Palaeogene, which lasted from 66 to 56 million years ago.

Palaeogene – the first period in the Cenozoic era, extending from 66 to 23 million years ago.

Palaeognathae – a superorder of Neornithes comprising birds with a primitive reptilian type of palate that includes ostriches, rheas, kiwis, emus, cassowaries and tinamous.

Paraphyly – a group of organisms that includes an ancestor but not all of its descendants.

Parvorder – a specific taxonomic category above superfamily, and below infraorder.

Passerines (*Passeriformes*) – a group of more than 6,000 species of small to medium-sized birds which have widely varied plumage and shape. They all have three toes pointing forwards, and one directed backwards which assists with perching, and are sometimes known as 'perching birds'.

Permian – a geological period that spans 46.7 million years from the end of the Carboniferous period, 298.9 million years ago, to the beginning of the Triassic, 252.2 million years ago.

Phenotype – the observable physical or biochemical characteristics of an organism. See also *genotype*.

Phylogenetics – the study of evolutionary relationships, especially through molecular sequencing data.

Pinnipeds – a group of marine mammals that are fin- or flipper-footed, including seals, sea-lions and walruses.

Piscivorous – a carnivorous diet that consists largely of fish, although it may include molluscs and crustaceans.

Pleiotropism – when one gene influences two or more seemingly unrelated phenotypic traits.

Pleistocene – a geological epoch, the first of the Quaternary period, extending from 2.5 million years ago to 11,700 years ago.

Plesiomorphy – an ancestral or primitive character.

Pliocene – the most recent epoch of the Neogene period, extending from 5.3 to 2.5 million years ago.

Ploidy – a multiple of the number of chromosomes in a cell.

Polyandry – a mating system in which one female mates with several males in a breeding season.

Polygynandry – a mating system in which both females and males have multiple mating partners during a breeding season.

Polygyny – a mating system in which a male mates with several females during a breeding season.
Polyphyly – a group of organisms that are classified into the same group but which came from different ancestors.
Polytomy – a term in phylogenetics used to describe the dividing of an evolutionary tree into three or more parts that cannot be fully resolved to dichotomies.
Polytypic – having or involving many types, such as a family with many species.
Population – a group of interbreeding individuals and their offspring.
Post-mating isolation – an incompatibility of two genomes that stops the normal development in a hybrid.
Psittacine – relating to or denoting birds of the parrot family.
Quaternary – the most recent geological period, extending from 2.5 million years ago to the present day.
Refugium (plural *refugia*) – an area that has escaped ecological changes occurring elsewhere and which enables species to survive after extinction in surrounding habitats.
Relictual – a species that inhabits a much smaller geographical area than it did in the past.
Relictual taxon – a sole surviving representative of a formerly diverse group.
Reverse colonisation – expansion of a species range from an area of low species richness to one of high species richness, specifically from island to continent.
Rictal bristle – one of a group of stiff, hair-like feathers that project from the base of a bird's bill.
Riparian – relating to, or situated on the banks of a river.
Scansorial – relating to the ability or propensity to climb.
Sexual dimorphism – a phenotypic difference between males and females of the same species.
Sink – for a species, an area in which extinction exceeds colonisation.
Sister species – taxa that are each other's closest evolutionary relatives.
Soft polytomy – the inability to reduce three or more branching points of a phylogram to a dichotomy due to insufficient information.
Spatial memory – ability to recall the location of objects in space.
Speciation – the evolutionary process by which new biological species arise.
Subduction – a process in which one tectonic plate moves under another tectonic plate, sinking into the Earth's mantle, as the plates converge.
Suborder – a taxonomic group that is below order and above infraorder.
Suboscines – a suborder of passerine birds, comprising the supposedly more primitive members of the order, with less well-developed vocal organs than the oscines.
Superorder – a taxonomic category ranking below a class or its subdivisions and above order.
Superspecies – a group of largely allopatric species which are descended from a

common evolutionary ancestor and are closely related but too distinct to be regarded as subspecies of one species.

Symbiotic – involving interaction between two different organisms living in close physical association.

Sympatric speciation – a process by which new species evolve from a single ancestral species while inhabiting the same geographical region.

Taxonomy – the science of naming, describing and classifying all organisms, whether plants, animals or microorganisms.

Teleost – ray-finned fishes that make up 96% of all extant species of fish.

Transcriptome – a set of all RNA molecules transcribed in one cell or a population of cells.

Ungulate – any of a large groups of mammals with hooves (a hoof being an enlarged toenail).

Vagility – the ability or tendency to move from place to place or disperse.

Vibrissae – long modified feathers located at the side of the beak that function like whiskers.

Vicariance – the separation of a group of organisms by a geographical barrier, such as a mountain or a body of water, resulting in differentiation of the original group into new species.

Zygote – the cell formed by the union of two gametes: a fertilised ovum.

Notes

Prologue: Evolution of an Idea

1. Thompson, H.S. and Fotso, R. 1995. Rockfowl: the genus *Picathartes*. *Bulletin of the African Bird Club*, **2** (1), 25–28.
2. Serle, W. 1952. The lower guinea bare-head crow (*Picathartes oreas*, Reichenow). *Nigerian Field*, **17**, 131–132.
3. Thompson, H.S.S. 2007. Family Picathartidae (*Picathartes*). In J. del Hoyo, A. Elliott and D.A. Christie (eds), *Handbook of the Birds of the World*, Vol. 12. Barcelona: Lynx Edicions, pp. 60–69.
4. Haffer, J. 1969. Speciation in Amazonian forest birds. *Science*, **165**, 131–137.
5. Jønsson, K.A., Fjeldså, J., Ericson, P.G.P. *et al.* 2007. Systematic placement of an enigmatic southeast Asian taxon *Eupetes macrocerus* and implications for the biogeography of a main songbird radiation, the Passerida. Biology Letters, 3, 323–326.
6. Low, T. 2014. *Where Song Began: Australia's Birds and How They Changed the World*. Melbourne: Penguin Random House.
7. See *Science*, 2014, volume **346**, for all eight papers.
8. See http://b10k.genomics.cn for further information.
9. Gill, F. and Donsker, D. (eds). 2017. *IOC World Bird List* (v 7.3). doi: 10.14344/IOC.ML.7.3.
10. Dawkins, R. 2004. *The Ancestor's Tale: a Pilgrimage to the Dawn of Life*. London: Weidenfeld & Nicholson.
11. See, for example, Pickrell, J. 2014. *Flying Dinosaurs: How Fearsome Reptiles Became Birds*. Sydney: NewSouth Publishing; Chiappe, L.M. 2007. *Glorified Dinosaurs: the Origin and Early Evolution of Birds*. Chichester: John Wiley & Sons.
12. Jetz, W., Thomas, G.H., Joy, J.B. *et al.* 2012. The global diversity of birds in space and time. *Nature*, **491**, 444–448.
13. Rosindell, J. and Harmon, L.J. (2012). OneZoom: a fractal explorer for the tree of life. *PLoS Biology*, **10** (10), e1001406. doi: 10.1371/journal.pbio.1001406.

Chapter 1: The Tinamou's Story

1. Desmond, A. 1994. *Huxley: the Devil's Disciple*. London: Michael Joseph, p. 359.
2. Huxley, T.H. 1870. Further evidence of the affinity between the dinosaurian reptiles and birds. *Proceedings of the Royal Geographical Society*, **26**, 12–31.
3. Dinosaurs may have originally evolved feathers as the result of 'aesthetic selection' by females. The resultant key innovation, a planar vane, enabled the display of complex pigment patterns (stripes, spots, dots, and spangles) for use as sexual displays. Only later did birds adapt these vaned feathers to create the aerodynamic forces required for flight. According to Richard Plum in *The Evolution of Beauty* (Doubleday, 2017) 'feathers did not evolve for flight; rather, flight evolved from feathers.' However, why wings (and

hence flight) evolved remains unclear. Mainstream hypotheses include the 'cursorial model', which states that wings evolved to aid running and leaping, while the 'tree-down model' envisages wings to have assisted in gliding from high vantage points.

4. Desmond, A. 1994. *Huxley: the Devil's Disciple.* London: Michael Joseph, p. 360.
5. Cocker, M. and Tipling, D. 2013. *Birds and People.* London: Jonathan Cape, p. 27.
6. Huxley, T.H. 1867. On the classification of birds; and on the taxonomic value of the modifications of certain of the cranial bones observable in that class. *Proceedings of the Zoological Society of London*, (1867), 415–472.
7. de Beer, G. 1956. The evolution of ratites. *Bulletin of the British Museum (Natural History), Zoology*, **4**, 57–70.
8. Bhullar, B-A.S., Marugán-Lobón, J., Racimo, F. *et al.* 2012. Birds have paedomorphic dinosaur skulls. *Nature*, **487**, 223–226.
9. Originally termed the K–T boundary, as the geological boundary lies between the Cretaceous and Tertiary (C had already been appropriated for Cambrian). However, since the Tertiary is no longer recognised by the International Commission on Stratigraphy, the K–T boundary is now referred to as the Cretaceous–Palaeogene boundary or K–Pg.
10. Modern birds are living dinosaurs with juvenile skulls. http://askwhy.co.uk/dinosauroids/?p=15987 (accessed December 2017).
11. Somel, M., Franz, H., Yan, Z. *et al.* 2009. Transcriptional neoteny in the human brain. *Proceedings of the National Academy of Sciences of the United States of America*, **106**, 5743–5748.
12. Brin, D. Neoteny and two-way sexual selection in human evolution. www.davidbrin.com/nonfiction/neoteny1.html (accessed December 2017).
13. Bhullar, B-A.S., Morris, Z.S., Sefton, E.M. *et al.* 2015. A molecular mechanism for the origin of a key evolutionary innovation, the bird beak and palate, revealed by an integrative approach to major transitions in vertebrate history. *Evolution*, **69**, 1665–1677.
14. Hess, H. 1962. History of ocean basins. In A.E.J. Engel, H.L. James and B.F. Leonard (eds), *Petrologic Studies: A Volume in Honor of A.F. Buddington.* New York: Geological Society of America, pp. 599–620.
15. Cracraft, J. 1974. Phylogeny and evolution of the ratite birds. *Ibis*, **116**, 494–521.
16. Dawkins, R. 2004. *The Ancestor's Tale: a Pilgrimage to the Dawn of Life.* London: Weidenfeld & Nicolson, pp. 235–245.
17. Harshman, J., Braun, E.L., Braun, M.J. *et al.* 2008. Phylogenomic evidence for multiple losses of flight in ratite birds. *Proceedings of the National Academy of Sciences of the United States of America*, **105**, 13462–13467.
18. Baker, A.J., Haddrath, O., McPherson, J.D. *et al.* 2014. Genomic support for a moa–tinamou clade and adaptive morphological convergence in flightless ratites. *Molecular Biology and Evolution*, **31**, 1686–1696.
19. Mitchell, K.J., Llamas, B., Soubrier, J. *et al.* 2014. Ancient DNA reveals elephant birds and kiwis are sister taxa and clarifies ratite bird evolution. *Science*, **344**, 898–900.
20. Christidis, L. and Boles, W. 2008. *Systematics and Taxonomy of Australian Birds.* Melbourne: CSIRO Publishing.
21. Choi, C.Q. 2014. Closest living relative of ancient elephant bird is tiny. www.livescience.com/45824-evolution-of-flightless-birds.html (accessed December 2017).
22. Huxley, T.H. 1901. *The Scientific Memoirs of Thomas Henry Huxley.* London: Macmillan. Vol. 3, p. 580.

Chapter 2: The Vegavis's Story

1. Nordenskiold, O. 1905. *Antarctica, or Two Years Amongst the Ice of the South Pole.* London: Hurst & Blackett.

2. Jablonski, D. and Chaloner, W.G. 1994. Extinction in the fossil record (and discussion). *Philosophical Transactions of the Royal Society B*, **344**, 11–17.
3. Fossils from the final frontier. www.abc.net.au/science/features/antarcticfossils (accessed December 2017).
4. Clarke, J.A., Tambussi, C.P., Noriega, J.I. *et al.* 2005. Definitive fossil evidence for the extant avian radiation in the Cretaceous. *Nature*, **433**, 305–308.
5. Clarke, J.A., Chatterjee, S., Li, Z. *et al.* 2016. Fossil evidence of the avian vocal organ from the Mesozoic. *Nature*, **538**, 502–505.
6. Renne, P.R., Deino, A.L., Hilgen, F.J. *et al.* 2013. Time scales of critical events around the Cretaceous–Paleogene boundary. *Science*, **339**, 684–687.
7. Anon. 2005. Cretaceous duck ruffles feathers. *BBC News*, 20 January 2005. http://news.bbc.co.uk/2/hi/science/nature/4187287.stm (accessed December 2017).
8. Ksepka, D. and Clarke, J. 2015. Phylogenetically vetted and stratigraphically constrained fossil calibrations within Aves. *Palaeontologia Electronica*, 18.1.3FC.
9. Jarvis, E.D., Mirarab, S., Aberer, A.J. *et al.* 2014. Whole-genome analyses resolve early branches in the tree of life of modern birds. *Science*, **346**, 1320–1331.
10. Alvarez, L.W., Alvarez, W., Asaro, F *et al.* 1980. Extraterrestrial cause for the Cretaceous–Tertiary extinction. *Science*, **208**, 1095–1108.
11. Renne, P.R., Sprain, C.J., Richards, M.A. *et al.* 2015. State shift in Deccan volcanism at the Cretaceous–Paleogene boundary, possibly induced by impact. *Science*, **350**, 76–78.
12. Gary, S. 2015. Asteroid impact may have boosted volcanic eruptions in double-whammy dinosaur extinction disaster. *ABC News*, 1 October 2015. www.abc.net.au/news/science/2015-10-02/asteroid-impact-and-deccan-volcano-eruption-mass-extinction/6783018 (accessed December 2017).
13. Tobin, T.S., Ward, P. D., Steig, E.J. *et al.* 2012. Extinction patterns, δ^{18} O trends, and magnetostratigraphy from a southern high-latitude Cretaceous–Paleogene section: links with Deccan volcanism. *Palaeography, Palaeoclimatology, Palaeoecology*, **350**, 180–188.
14 Randall, L. 2015. *Dark Matter and the Dinosaurs: the Astounding Interconnectedness of the Universe.* London: Bodley Head.
15. Schulte, P., Alegret, L., Arenillas, I. *et al.* 2010. The Chicxulub asteroid impact and mass extinctions at the Cretaceous–Paleogene boundary. *Science*, **327**, 1214–1218.
16. Wignall, P.B. 2001. Large igneous provinces and mass extinctions. *Earth-Science Reviews*, **53**, 1–33.
17. Robertson, D.S., Lewis, W.M., Sheehan, P. M. *et al.* 2013. K–Pg extinction: reevaluation of the heat-fire hypothesis. *Journal of Geophysical Research: Biogeosciences*, **118**, 329–336.
18. Lovegrove, B.G., Lobban, K.D. and Levesque, D.L. 2014. Mammal survival at the Cretaceous–Palaeogene boundary: metabolic homeostasis in prolonged tropical hibernation in tenrecs. *Proceedings of the Royal Society B*, **281**, 20141304. doi: 10.1098/rspb.2014.1304.
19. Babbitt, V.L. Differential survival across the K–T boundary: why the non-avian dinosaur eggs didn't hatch, and the reptile and bird eggs did. www.theeggsdidnthatch.blogspot.com (accessed December 2017).
20. Erickson, G.M., Zelenitsky, D.K., Kay, D.I. *et al.* 2017. Dinosaur incubation periods directly determined from growth-line counts in embryonic teeth show reptilian-grade development. *Proceedings of the National Academy of Sciences of the United States of America*, **114**, 540–545.
21. Hogenboom, M. 2014. Egg shape 'helped birds survive' asteroid impact. www.bbc.co.uk/news/science-environment-29895683 (accessed December 2017).
22. Benson, R.B.J., Campione, N.E., Carrano, M.T. 2014. Rates of dinosaur body mass evolution indicate 170 million years of sustained ecological innovation on the avian stem lineage. *PLoS Biology*, **12** (6), e1001896. doi: 10.1371/journal.pbio.1001853.
23. Levenson, R.M., Krupinski, E.A., Navarro, V.M. *et al.* 2015. Pigeons (*Columba livia*) as trainable observers of pathology and radiology breast cancer images. *PLoS ONE*, **10** (11), e0141357. doi: 10.1371/journal.pone.0141357.

24. Milner, A.C. and Walsh, S.A. 2009. Avian brain evolution: new data from Palaeogene birds (Lower Eocene) from England. *Zoological Journal of the Linnean Society*, **115**, 198–219.
25. Natural History Museum, London. Birds survived mass extinction that wiped out dinosaurs because of their larger brains. www.sciencedaily.com/releases/2009/01/090127165505.htm (accessed December 2017).
26. Zelenitsky, D.K., Therrien, F., Ridgely, R.C. *et al.* 2011. Evolution of olfaction in non-avian dinosaurs and birds. *Proceedings of the Royal Society B*, **278**, 3625–3634. doi: 10.1098/rspb.2011.0238.

Chapter 3: The Waterfowl's Story

1. Jarvis, E.D., Mirarab, S, Aberer, A.J. *et al.* 2014. Whole-genome analyses resolve early branches in the tree of life of modern birds. *Science*, **346**, 1320–1331.
2. Huang, Z., Yang, C., Ke, D. 2016. DNA barcoding and phylogenetic relationships in Anatidae. *Mitochondrial DNA*, **27**, 1042–1044.
3. Gonzalez, J., Düttmann, H. and Wink, M. 2009. Phylogenetic relationships based on two mitochondrial genes and hybridization patterns in Anatidae. *Journal of Zoology*, **279**, 310–318.
4. Johnson, K.P. and Sorenson, M.D. 1999. Phylogeny and biogeography of dabbling ducks (genus: *Anas*): a comparison of molecular and morphological evidence. *The Auk*, **116**, 792–805.
5. Lavretsky, P., Hernández-Baños, B.E. and Peters, J.L. 2014. Rapid radiation and hybridization contribute to weak differentiation and hinder phylogenetic inferences in the New World mallard complex (*Anas* spp.). *The Auk*, **131**, 524–538.
6. Sorenson, M.D., Cooper, A., Paxinos, E.E. *et al.* 1999. Relationships of the extinct moa-nalos, flightless Hawaiian waterfowl, based on ancient DNA. *Proceedings of the Royal Society B*, **266**, 2187–2193.
7. Lavretsky, P., Engilis, A. Eadie, J.M. *et al.* 2015. Genetic admixture supports an ancient hybrid origin of the endangered Hawaiian duck. *Journal of Evolutionary Biology*, **28**, 1005–1015.
8. Worthy, T.H. 2009. Description and phylogenetic relationships of two new genera and four new species of Oligo-Miocene waterfowl (Aves: Anatidae) from Australia. *Zoological Journal of the Linnean Society*, **156**, 411–454.
9. Livezey, B.C. 1995. Phylogeny and evolutionary ecology of modern seaducks (Anatidae: Mergini). *The Condor*, **97**, 233–255.
10. Fulton, T.L., Letts, B. and Shapiro, B. 2012. Multiple losses of flight and recent speciation in steamer ducks. *Proceedings of the Royal Society B*, **279**, 2339–2346. doi: 10.1098/rspb.2011.2599.
11. Ottenburghs, S.J., Megens, H-J., Kraus, R.H. *et al.* 2016. A tree of geese: a phylogenomic perspective on the evolutionary history of true geese. *Molecular Phylogenetics and Evolution*, **101**, 303–313.
12. Shafer, A.B., Cullingham, C.I., Côté, S.D. *et al.* 2010. Of glaciers and refugia: a decade of study sheds new light on the phylogeography of northwestern North America. *Molecular Ecology*, **19**, 4589–4621.
13. Ogilvie, M.A. 1978. *Wild Geese*. Berkhamsted: T. & A.D. Poyser.
14. Paxinos, E.E., James, H.F., Olson, S.L. *et al.* 2002. mtDNA from fossils reveals a radiation of Hawaiian geese recently derived from the Canada Goose (*Branta canadensis*). *Proceedings of the National Academy of Sciences of the United States of America*, **99**, 1399–1404.
15. Hawkes, L.A., Balachandran, S., Batbayar, N. *et al.* 2013. The paradox of extreme high-altitude migration in bar-headed geese *Anser indicus*. *Proceedings of the Royal Society B*, **280**, 20122114. doi: 10.1098/rspb.2012.2114.

16. Hawkes, L.A., Balachandran, S., Batbayar, N. *et al.* 2011. The trans-Himalayan flight of bar-headed geese (*Anser indicus*). *Proceedings of the National Academy of Sciences of the United States of America*, **108**, 9516–9519.
17. Scott, G.R., Hawkes, L.A., Frappell, P. B. *et al.* 2015. How bar-headed geese fly over the Himalayas. *Physiology (Bethesda)*, **30**, 107–115.
18. Wang, C., Dai, J., Zhao, X. *et al.* 2014. Outward-growth of the Tibetan plateau during the Cenozoic: a review. *Tectonophysics*, **621**, 1–43.
19. Liechti, F. and Schaller, E. 1999. The use of low-level jets by migrating birds. *Naturwissenschaften*, **86**, 549–551.
20. Graham Scott, personal communication.
21. Kevin McCracken, personal communication.
22. Projecto-Garcia, J., Natarajan, C., Moriyama, H. *et al.* 2013. Repeated elevational transitions in hemoglobin function during the evolution of Andean hummingbirds. *Proceedings of the National Academy of Sciences of the United States of America*, **110**, 20669–20674.
23. Natarajan, C., Hoffmann, F.G., Weber, R.E. *et al.* 2016. Predictable convergence in hemoglobin function has unpredictable molecular underpinnings. *Science*, **354**, 336–339.
24. Gould, S.J. 1990. *Wonderful Life: the Burgess Shale and the Nature of History*. London: Hutchinson Radius, p. 51.
25. Montgomerie, R. and Briskie, J.V. 2007. Anatomy and evolution of copulatory structures. In B.G.M. Jamieson (ed.), *Reproductive Biology and Phylogeny of Birds. Part A: Phylogeny, Morphology, Hormones and Fertilization*. Enfield, NH: Science Publishers, pp. 115–148.
26. Herrera, A.M., Shuster, S.G., Perriton, C.L. *et al.* 2013. Developmental basis of phallus reduction during bird evolution. *Current Biology*, **23**, 1065–1074.
27. Brennan, P.L.R. and Prum, R.O. 2011. The erection mechanism of the ratite penis. *Journal of Zoology*, **286**, 140–144.
28. McCracken, K.G. 2000. The 20-cm spiny penis of the Argentine lake duck (*Oxyura vittata*). *The Auk*, **117**, 820–825; McCracken, K.G., Wilson, R.E., McCracken, P. J. *et al.* 2001. Sexual selection: are ducks impressed by drakes' display? *Nature*, **413**, 128.
29. Coker, C.R., McKinney, F., Hays, H. *et al.* 2002. Intromittent organ morphology and testis size in relation to mating system in waterfowl. *The Auk*, **119**, 403–413.
30. Brennan, P.L.R., Prum, R.O., McCracken, K.G. *et al.* 2007. Coevolution of male and female genital morphology in waterfowl. *PLoS ONE*, **2** (5), e418, doi: 10.1371/journal.pone.00000418.
31. Birkhead, T., Wimpenny, J. and Montgomerie, B. 2014. *Ten Thousand Birds: Ornithology Since Darwin*. Princeton, NJ: Princeton University Press, p. 203.
32. Brennan, P.L.R., Clark, C.J. and Prum, R.O. 2010. Explosive eversion and functional morphology of the duck penis supports sexual conflict in waterfowl genitalia. *Proceedings of the Royal Society B*, **277**, 1309–1314. doi: 10.1098/rspb.2009.2139.
33. Briskie, J.V. and Montgomerie, R. 1997. Sexual selection and the intromittent organ of birds. *Journal of Avian Biology*, **28**, 73–86.
34. Prum, R.O. 2017. *The Evolution of Beauty: How Darwin's Forgotten Theory of Mate Choice Shapes the Animal World – and Us*. New York: Doubleday.
35. Naish, D. 2012. *Birds*. In M.K. Brett-Surman, T.R. Holtz and J.O. Farlow (eds), *The Complete Dinosaur*. Bloomington, IN: Indiana University Press, pp. 379–424.
36. Brennan, P.L.R., Birkhead, T.R., Zyskowski, K. *et al.* 2008. Independent evolutionary reduction of the phallus in basal birds. *Journal of Avian Biology*, **39**, 487–492.

Chapter 4: The Hoatzin's Story

1. Beebe, W. 1918. *Jungle Peace*. New York: Henry Holt & Company, pp. 123–140.
2. Jarvis, E.D., Mirarab, S, Aberer, A.J. *et al.* 2014. Whole-genome analyses resolve early branches in the tree of life of modern birds. *Science*, **346**, 1320–1331.

3. Gill, F. and Donsker, D. (eds). 2017. *IOC World Bird List* (v 7.3). doi: 10.14344/IOC.ML.7.3.
4. Grajal, A. 1995. Structure and function of the digestive tract of the hoatzin (*Opisthocomus hoazin*): a folivorous bird with foregut fermentation. *The Auk*, **112**, 20–28.
5. Godoy-Vitorino, F., Ley, R.E., Gao, Z. *et al.* 2008. Bacterial community in the crop of the hoatzin, a Neotropical folivorous flying bird. *Applied and Environmental Microbiology*, **74**, 5905–5912.
6. Wagner, A. 2014. *Arrival of the Fittest: Solving Evolution's Greatest Puzzle*. London: Oneworld, p. 4.
7. Ohno, S. 1970. *Evolution by Gene Duplication*. Berlin: Springer-Verlag.
8. Fleming, A. 1922. On a remarkable bacteriolytic element found in tissues and secretions. *Proceedings of the Royal Society B*, **93**, 306–317.
9. Kornegay, J.R., Schilling, J.W. and Wilson, A.C. 1994. Molecular adaptation of a leaf-eating bird: stomach lysozyme of the hoatzin. *Molecular Biology and Evolution*, **11**, 921–928.
10. Kornegay, J.R. 1996. Molecular genetics and evolution of stomach and nonstomach lysozymes in the hoatzin. *Journal of Molecular Evolution*, **42**, 676–684.
11. Mayr, G., Alvarenga, H. and Mourer-Chauviré, C. 2011. Out of Africa: fossils shed light on the origin of the hoatzin, an iconic Neotropical bird. *Naturwissenschaften*, **98**, 961–966.
12. Mayr, G. 2014. A hoatzin fossil from the middle Miocene of Kenya documents the past occurrence of modern-type Opisthocomiformes in Africa. *The Auk*, **131**, 55–60.
13. Mayr, G. and De Pietri, V.L. 2014. Earliest and first northern hemispheric hoatzin fossils substantiate Old World origin of a 'neotropic endemic'. *Naturwissenschaften*, **101**, 143–148.
14. de Oliveira, F.B., Molina, E.C. and Marroig, G. 2009. Paleogeography of the South Atlantic: a route for primates and rodents into the New World? In P.A. Garber *et al.* (eds), *South American Primates. Developments in Primatology: Progress and Prospects*. New York: Springer, pp. 55–68.
15. Houle, A. 1998. Floating islands: a mode of long-distance dispersal for small and medium-sized terrestrial vertebrates. *Diversity and Distributions*, **4**, 201–216.
16. Censky, E.J., Hodge, K. and Dudley, J. 1998. Over-water dispersal of lizards due to hurricanes. *Nature*, **395**, 556.
17. Longrich, N.R., Vinther, J., Pyron, R.A. *et al.* 2015. Biogeography of worm lizards (Amphisbaenia) driven by end-Cretaceous mass extinction. *Proceedings of the Royal Society B*, **282**, 20143034. doi: 10.1098/rspb.2014.3034.
18. See Alan de Queiroz's book *The Monkey's Voyage* for further discussions of dispersalism.
19. Poux, C., Chevret, P., Huchon, D. *et al.* 2006. Arrival and diversification of caviomorph rodents and platyrrhine primates in South America. *Systematic Biology*, **55**, 228–244.
20. Stehlin, H.G. 1909. Remarques sur les faunules de mammifères des couches éocènes et oligocènes du bassin de Paris. *Bulletin de la Société Géologique de France*, **9**, 488–520.

Chapter 5: The Penguins' Story

1. Slack, K.E., Jones, C.M., Ando, T. *et al.* 2006. Early penguin fossils, plus mitochondrial genomes, calibrate avian evolution. *Molecular Biology and Evolution*, **23**, 1144–1155.
2. Didymus, J.T. Giant prehistoric penguin stood over four feet, weighed 130 lbs. www.digitaljournal.com/article/320382 (accessed December 2017).
3. Fretwell, P. T., LaRue, M.A., Morin, P. *et al.* 2012. An emperor penguin population estimate: the first global, synoptic survey of a species from space. *PLoS ONE*, **7** (4), e33751. doi: 10.1371/journal.pone.0033751.
4. Ksepka, D.T. and Thomas, D.B. 2012. Multiple Cenozoic invasions of Africa by penguins (Aves, Sphenisciformes). *Proceedings of the Royal Society B*, **279**, 1027–1032.
5. Williams, C.L., Hagelin, J.C. and Kooyman, G.L. 2015. Hidden keys to survival: the

type, density, pattern and functional role of emperor penguin body feathers. *Proceedings of the Royal Society B*, **282**, 20152033. doi: 10.1098/rspb.2015.2033.
6. Clarke, J.A., Ksepka, D.T., Salas-Gismondi, R. *et al.* 2010. Fossil evidence for evolution of the shape and color of penguin feathers. *Science*, **330**, 954–957.
7. Li, C., Zhang, Y., Li, J. *et al.* 2014. Two Antarctic penguin genomes reveal insights into their evolutionary history and molecular changes related to the Antarctic environment. *GigaScience*, **3**, 27. doi: 10.1186/2047-217X-3-27.
8. Zhao, H., Li, J. and Zhang, J. 2015. Molecular evidence for the loss of three basic tastes in penguins. *Current Biology*, **25** (4), R141–R142. doi: 10.1016/j.cub.2015.01.026.
9. Ksepka, D.T. 2010. 5 things you never knew about penguins! https://blogs.scientificamerican.com/guest-blog/5-things-you-never-knew-about-penguins (accessed December 2017).
10. Joel Asaph Allen (1838–1921) was an American zoologist and ornithologist who became the first president of the American Ornithologists' Union and chief editor of their journal *The Auk*.
11. Symonds, M.R. and Tattersall, G.J. 2010. Geographical variation in bill size across bird species provides evidence for Allen's rule. *American Naturalist*, **176**, 188–197. Similarly, the length of exposed, non-feathered leg elements correlates with thermoregulatory costs in gulls and terns. See Nudds, R.L. and Oswald, S.A. 2007. An interspecific test of Allen's rule: evolutionary implications for endothermic species. *Evolution*, **61**, 2839–2848.
12. Fraas, A.P. 1989. *Heat Exchanger Design*, 2nd edition. New York: John Wiley & Sons, p. 2.
13. Thomas, D.B., Ksepka, D.T. and Fordyce, R.E. 2011. Penguin heat-retention structures evolved in Greenhouse Earth. *Biological Letters*, **7**, 461–464.
14. Fordyce, R.E. and Ksepka, D.T. 2012. The strangest bird: recent fossil discoveries reveal the surprising evolutionary history of penguins. *Scientific American*, **307**, 56–61.
15. Hospitaleche, C.A. 2014. New giant penguin bones from Antarctica: systematic and paleobiological significance. *Comptes Rendus Palevol*, **13**, 555–560.
16. Zitterbart, D.P, Wienecke, B., Butler, J.P. *et al.* 2011. Coordinated movements prevent jamming in an emperor penguin huddle. *PLoS ONE*, **6** (6), e20260. doi: 10.1371/journal.pone.0020260.
17. Smith, N.A. and Clarke, J.A. 2015. Systematics and evolution of the Pan-Alcidae (Aves, Charadriiformes). *Journal of Avian Biology*, **46**, 125–140.
18. Elliott, K.H., Ricklefs, R.E., Gaston, A.J. *et al.* 2013. High flight costs, but low dive costs, in auks support the biomechanical hypothesis for flightlessness in penguins. *Proceedings of the National Academy of Sciences of the United States of America*, **110**, 9380–9384.
19. Cherry-Garrard, A. 1922. *The Worst Journey in the World.* London: Constable & Company.

Chapter 6: The Storm Petrel's Story

1. OneZoom Tree of Life Explorer. www.onezoom.org (accessed December 2017); Rosindell, J. and Harmon, L.J. (2012). OneZoom: a fractal explorer for the tree of life. *PLoS Biology*, **10** (10), e1001406. doi: 10.1371/journal.pbio.1001406.
2. Flood, B. 2003. The New Zealand Storm-Petrel is not extinct. *Birding World*, **16**, 479–483.
3. Mason, C. 2013. Critically endangered NZ storm petrel found breeding. *New Zealand Herald*, February 25, 2013.
4. Fitzpatrick, B.M., Fordyce, J.A. and Gavrilets, S. 2008. What, if anything, is sympatric speciation? *Journal of Evolutionary Biology*, **21**, 1452–1459.
5. Monteiro, L.R., Granadeiro, J.P. and Furness, R.W. 1998. Relationship between mercury levels and diet in Azores seabirds. *Marine Ecology Progress Series*, **166**, 259–265.
6. Bolton, M., Smith, A.L., Gómez-Díaz, E. *et al.*, 2008. Monteiro's storm-petrel *Oceanodroma monteiroi*: a new species from the Azores. *Ibis*, **150**, 717–727.

7. Friesen, V.L., Smith, A.L., Gómez-Díaz, E. *et al.* 2007. Sympatric speciation by allochrony in a seabird. *Proceedings of the National Academy of Sciences of the United States of America*, **104**, 18589–18594.
8. Gavrilets, S. 2004. *Fitness Landscapes and the Origin of Species*. Princeton, NJ: Princeton University Press, pp. 368–387.
9. Mayr, E. 1963. *Animal Species and Evolution*. Cambridge, MA: Belknap Press, p. 451.
10. Coyne, J.A. and Price, T.D. 2000. Little evidence for sympatric speciation in island birds. *Evolution*, **54**, 2166–2171.
11. Phillimore, A.B., Orme, C.D., Thomas, G.H., *et al.* 2008. Sympatric speciation in birds is rare: insights from range data and simulations. *American Naturalist*, **171**, 646–657.
12. Price, T. 2008. *Speciation in Birds*. Greenwood Village, CO: Roberts & Company, p. 19.
13. Shaw, P.W., Turner, G.F., Idid, M.R. *et al.* 2000. Genetic population structure indicates sympatric speciation of lake Malawi pelagic cichlids. *Proceedings of the Royal Society B*, **267**, 2277–2280.
14. Genner, M.J., Knight, M.E., Haesler, M.P. *et al.* 2010. Establishment and expansion of Lake Malawi rock fish populations after dramatic late Pleistocene lake level rise. *Molecular Ecology*, **19**, 170–182.
15. Jiggins, C.D. and Bridle, J.R. 2004. Speciation in the apple maggot fly: a blend of vintages? *Trends in Ecology and Evolution*, **19**, 111–114.

Chapter 7: The Albatross's Story

1. Penhallurick, J. and Wink, M. 2004. Analysis of the taxonomy and nomenclature of the Procellariiformes based on complete nucleotide sequences of the mitochondrial cytochrome c gene. *Emu*, **104**, 125–147.
2. Mayr, E. 1942. *Systematics and the Origin of Species from the Viewpoint of a Zoologist*. New York: Columbia University Press.
3. Cracraft, J. 1983. Species concepts and speciation analysis. In R.F. Johnson (ed.), *Current Ornithology*. Plenum Press, pp. 159–187.
4. Agapow, P. M., Bininda-Emonds, O.R., Crandall, K.A. *et al.* 2004. The impact of species concepts on biodiversity studies. *Quarterly Review of Biology*, **79**, 161–179.
5. Penhallurick, J. 2012. The number of albatross (Diomedeidae) species. *Open Ornithology Journal*, **5**, 32–41.
6. Tudge, C. 2008. *Consider the Birds: Who They Are and What They Do*. London: Allen Lane, p. 108.
7. Brooke, M. 2004. *Albatrosses and Petrels Across the World*. Oxford: Oxford University Press.
8. Olson, S.L. and Rasmussen, P.C. 2001. Miocene and Pliocene birds from the Lee Creek Mine, North Carolina. In C.E. Ray and D.J. Bohaska (eds), *Geology and Paleontology of the Lee Creek Mine, North Carolina, III*. Washington, DC: Smithsonian Institute Press, pp. 233–369.
9. Olson, S.L. and Hearty, P.J. 2003. Probable extirpation of a breeding colony of short-tailed albatross (*Phoebastria albatrus*) on Bermuda by Pleistocene sea-level rise. *Proceedings of the National Academy of Sciences of the United States of America*, **100**, 12825–12829.
10. Reyes, A.V., Carlsen, A.E. Beard, B.L. *et al.* 2014. South Greenland ice-sheet collapse during Marine Isotope Stage II. *Nature*, **510**, 525–528.
11. Weimerskirch, H., Louzao, M., de Grissac, S. *et al.* 2012. Changes in wind pattern alter albatross distribution and life-history traits. *Science*, **335**, 211–214.
12. Mayr, G. and Smith, T. 2012. A fossil albatross from the Early Oligocene of the North Sea basin. *The Auk*, **129**, 87–95.
13. Meyers, R.A. and Stakebake, E.F. 2005. Anatomy and histochemistry of spread-wing posture in birds. 3. Immunohistochemistry of flight muscles and the 'shoulder lock' in

albatrosses. *Journal of Morphology*, **263**, 12–29.
14. Zolfagharifard, E. 2013. Mystery of how the wandering albatross travels 10,000 miles in a single journey without flapping its wings is solved. www.dailymail.co.uk/sciencetech/article-2509211/mystery-wandering-albatross (accessed December 2017).
15. Nevitt, G.A., Losekoot, M. and Weimerskirch, H. 2008. Evidence for olfactory search in wandering albatross, *Diomedea exulans*. *Proceedings of the National Academy of Sciences of the United States of America*, **105**, 4576–4581.
16. van Buskirk, R.W. and Nevitt, G.A. 2008. The influence of developmental environment on the evolution of olfactory foraging behaviour in procellariiform seabirds. *Journal of Experimental Biology*, **21**, 67–78.

Chapter 8: The Godwit's Story

1. Landys-Cianelli, M.M., Piersma, T. and Jukema, J. 2003. Strategic size change of internal organs and muscle tissue in the bar-tailed godwit during fat storage in a spring stopover site. *Functional Ecology*, **17**, 151–159.
2. Rattenborg, N.C., Voirin, B., Cruz, S.M. *et al.* 2016. Evidence that birds sleep in mid-flight. *Nature Communications*, **7**, 12468. doi: 10.1038/ncomms12468.
3. Amos, J. 2013. Savvy bar-tailed godwit equipped for climate challenge. *BBC News*. www.bbc.com/news/science-environment-25274483 (accessed December 2017).
4. Baker, A.J., Pereira, S.L. and Paton, T.A. 2007. Phylogenetic relationships and divergence times of Charadriiformes genera: multigene evidence for the Cretaceous origin of at least 14 clades of shorebirds. *Biology Letters*, **3**, 205–209.
5. Livezey, B.C. 2010. Phylogenetics of modern shorebirds (Charadriiformes) based on phenotypic evidence: analysis and discussion. *Zoological Journal of the Linnean Society*, **160**, 567–618.
6. De Pietri, V.L., Scofield, R.P., Tennyson, A.J.D. *et al.* 2016. Wading a lost southern connection: Miocene fossils from New Zealand reveal a new lineage of shorebirds (Charadriiformes) linking Gondwanan avifaunas. *Journal of Systematic Palaeontology*, **14**, 603–616.
7. Joseph, L., Lessa, E.P. and Christidis, L. 1999. Phylogeny and biogeography in the evolution of migration: shorebirds of the *Charadrius* complex. *Journal of Biogeography*, **26**, 329–342.
8. Kraaijeveld, K. and Nieboer, E.N. 2000. Late Quaternary paleogeography and evolution of Arctic breeding waders. *Ardea*, **88**, 193–205.
9. Ferns, P.N. and Siman, H.Y. 1994. Utility of the curved bill of the curlew *Numenius arquata* as a foraging tool. *Bird Study*, **41**, 102–109.
10. Mueller, J.C., Pulido, F. and Kempenaers, B. 2011. Identification of a gene associated with migratory behaviour. *Proceedings of the Royal Society B*, **278**, 2848–2856, doi: 10.1098/ rspb.2010.2567. For further discussion see Liedvogel, M., Åkesson, S. and Bensch, S. 2011. The genetics of migration on the move. *Trends in Ecology and Evolution*, **26**, 561–569. doi: 10.1016/j.tree.2011.07.009.
11. Winger, B.M., Barker, F.K. and Ree, R.H. 2014. Temperate origins of long-distance seasonal migration in New World songbirds. *Proceedings of the National Academy of Sciences of the United States of America*, **33**, 12115–12120.
12. Fricke, H.C., Hencecroth, J. and Hoerner, M.E. 2011. Lowland–upland migration of sauropod dinosaurs during the late Jurassic epoch. *Nature*, **480**, 513–515.
13. Hedenström, A. 2010. Extreme endurance migration: what is the limit to non-stop flight? *PLoS Biology*, **8** (5), e1000362. doi: 10:1371/journal.pbio.1000362.
14. Wiltschko, W. and Wiltschko, R. 1972. Magnetic compass of European Robins. *Science*, **176**, 62–64.
15. Wiltschko, W. and Wiltschko, R. 1996. Magnetic orientation in birds. *Journal of Experimental Biology*, **199**, 29–38.

16. Al-Khalili, J. and McFadden, J. 2014. *Life on the Edge: the Coming Age of Quantum Biology*. London: Bantam Press, p. 4.
17. Pollack, L. 2012. That nest of wires we call the imagination: a history of some key scientists behind the bird compass sense. www.ks.uiuc.edu/History/magnetoreception/PollackMagSensors.pdf (accessed December 2017).
18. Schulten, K., Swenburg, C.E. and Weller, A. 1978. A biomagnetic sensory mechanism based on magnetic field modulation coherent electron spin motion. *Zeitschrift für Physikalische Chemie Neue Folge*, **III**, 1–5.
19. Wiltschko, W., Möller, A., Gesson, M. *et al.* 2004. Light-dependent magnetoreception in birds: analysis of the behaviour under red light after pre-exposure to red light. *Journal of Experimental Biology*, **207**, 1193–1202.
20. Ritz, T., Adem, S. and Schulten, K. 2000. A model for photoreceptor-based magnetoreception in birds. *Biophysical Journal*, **78**, 707–718. For a review of the radical-pair mechanism, see also Hore, P.J. and Mouritsen, H. 2016. The radical-pair mechanism of magnetoreception. *Annual Reviews of Biophysics*, **45**, 299–344.
21. Hore, P.J. 2011. The quantum robin. *Navigation News*, November/December, pp. 15–17.
22. Ritz, T., Thalau, P., Phillips, J.B. *et al.* 2004. Resonance effects indicate a radical-pair mechanism for avian magnetic compass. *Nature*, **429**, 177–180.
23. Gegear, R.J., Foley, L.E., Casselman, A. *et al.* 2010. Animal cryptochromes mediate magnetoreception by an unconventional photochemical mechanism. *Nature*, **463**, 804–807.
24. Wiltschko, R. and Wiltschko, W. 2013. The magnetite-based receptors in the beaks of birds and their role in avian navigation. *Journal of Comparative Physiology*, **199**, 89–98.
25. Mei, Q. and Dvornyk, V. 2015. Evolutionary history of the photolyase/cryptochrome superfamily in eukaryotes. *PLoS ONE*, **10** (9), e0135940. doi: 10.1371/journal.pone.0135940.
26. Wiltschko, R. and Wiltschko, W. 2012. Magnetoreception. In C. López-Larrea (ed.), *Sensing in Nature*. New York: Springer, pp. 126–144.
27. Woodley, K. 2012. *Shorebirds of New Zealand: Sharing the Margins*. Auckland: Penguin.

Chapter 9: The Buzzard's Story

1. Jarvis, E.D., Mirarab, S., Aberer, A.J. *et al.* 2014. Whole-genome analyses resolve early branches in the tree of life of modern birds. *Science*, **346**, 1320–1331.
2. Prum, R.O., Berv, J.S., Dornburg, A. *et al.* 2015. A comprehensive phylogeny of birds (Aves) using targeted next-generation DNA sequencing. *Nature*, **526**, 569–573.
3. Johnson, J.A., Brown, J.W., Fuchs, J. *et al.* 2016. Multi-locus phylogenetic inference among New World vultures (Aves: Cathartidae). *Molecular Phylogenetics and Evolution*, **105**, 193–199.
4. Lerner, H.R.L. and Mindell, D.P. 2005. Phylogeny of eagles, Old World vultures, and other Accipitidae based on nuclear and mitochondrial DNA. *Molecular Phylogenetics and Evolution*, **37**, 327–346.
5. Mayr, G. 2009. *Paleogene Fossil Birds*. Berlin: Springer-Verlag.
6. Monti, F., Duriez, O., Arnal, V. *et al.* 2015. Being cosmopolitan: evolutionary history and phylogeography of a specialized raptor, the osprey *Pandion haliaetus*. *BMC Evolutionary Biology*, **15**, 255. doi: 10.1186/s12862-015-0535-6.
7. Negro, J.J., Pertoldi, C., Randi, E. *et al.* 2006. Convergent evolution of *Elanus* kites and the owls. *Journal of Raptor Research*, **40**, 222–225.
8. Helbig, A.J., Kocum, A., Seibold, I. *et al.* 2005. A multi-gene phylogeny of aquiline eagles (Aves: Accipitriformes) reveals extensive paraphyly at the genus level. *Molecular Phylogenetics and Evolution*, **35**, 147–164.
9. Haast, J. 1871. Notes on *Harpagornis moorei*, an extinct gigantic bird of prey, containing description of femur, ungual phalanges, and rib. *Transactions and Proceedings of the New Zealand Institute*, **4**, 193–196. Haast named the eagle *Harpagornis moorei* after George

Henry Moore (1812–1905), the owner of Glenmark Estate, where the bird's bones had been found. The genus name is derived from the Greek *harpax*, meaning 'grappling hook', and *ornis*, meaning 'bird'.

10. Bunce, M., Szulkin, M., Lerner, H. *et al.* 2005. Ancient DNA provides new insights into the evolutionary history of New Zealand's extinct giant eagle. *PLoS Biology*, **3** (1), e9. doi: 10.1371/journal.pbio.0030009.
11. Scofield, R.P. and Ashwell, K.W.S. 2009. Rapid somatic expansion causes the brain to lag behind: the case of the brain and behavior of New Zealand's Haast's Eagle (*Harpagornis moorei*). *Journal of Vertebrate Paleontology*, **29**, 637–649.
12. Tudge, C. 2008. *Consider the Birds: Who They Are and What They Do*. London: Allen Lane, p. 112.
13. Riesing, M.J., Kruckenhauser, L., Gamauf, A. *et al.* 2003. Molecular phylogeny of the genus *Buteo* (Aves: Accipitridae) based on mitochondrial marker sequences. *Molecular Phylogenetics and Evolution*, **27**, 328–342.
14. Venditti, C., Meade, A. and Pagel, M. 2010. Phylogenies reveal new interpretation of speciation and the Red Queen. *Nature*, **463**, 349–352.
15. Hedges, S.B., Marin, J., Suleski, M. *et al.* 2015. Tree of life reveals clock-like speciation and diversification. *Molecular Biological and Evolution*, **32**, 835–845.

Chapter 10: The Owl's Story

1. Jarvis, E.D., Mirarab, S., Aberer, A.J. *et al.* 2014. Whole-genome analyses resolve early branches in the tree of life of modern birds. *Science*, **346**, 1320–1331.
2. Houde, P. and Olson, S.L. 1992. A radiation of coly-like birds from the early Eocene of North America (Aves: Sandcoleiformes, new order). *Natural History of Los Angeles County, Science Series*, **36**, 137–160.
3. Prum, R.O., Berv, J.S., Dornburg, A. *et al.* 2015. A comprehensive phylogeny of birds (Aves) using targeted next-generation DNA sequencing. *Nature*, **526**, 569–573.
4. Burton, J.A. (ed.). 1992. *Owls of the World: Their Evolution, Structure and Ecology*, 3rd edition. London: Eurobooks.
5. Aliabadiar, M., Alaei-Kakhki, N., Mirshamsi, O. *et al.* 2016. Phylogeny, biogeography, and diversification of barn owls (Aves: Strigiformes). *Biological Journal of the Linnean Society*, **119**, 904–918.
6. Suárez, W. and Olson, S.L. 2015. Systematics and distribution of the giant fossil barn owls of the West Indies (Aves: Strigiformes: Tytonidae). *Zootaxa*, **4020**, 533–553.
7. Wetmore, A. and Swales, B.H. 1931. *The Birds of Haiti and the Dominican Republic*. Bulletin 155. Washington, DC: Smithsonian Institute, p. 239.
8. Wink, M., Heidrich, P., Sauer-Gürth, H. *et al.* 2008. Molecular phylogeny and systematics of owls (Strigiformes). In C. König and F. Weick (eds), *Owls of the World*. London: Helm, pp. 42–61.
9. Dantas, S.M., Weckstein, J.D., Bates, J.M. *et al.* 2016. Molecular systematics of the New World screech-owls (*Megascops*: Aves, Strigidae): biogeographic and taxonomic implications. *Molecular Phylogenetics and Evolution*, **94**, 626–634.
10. Sangster, G., King, B.F., Verbelen, P. *et al.* 2013. A new owl species of the genus *Otus* (Aves: Strigidae) from Lombok, Indonesia. *PLoS ONE*, **8** (2), e53712. doi: 10.1371/journal.pone.0053712.
11. de Kok-Mercado, F., Habib, M., Phelps, T. *et al.* 2013. Adaptations of the owl's cervical arteries in relation to extreme neck rotation. *Science*, **339**, 514.
12. Payne, R.S. 1971. Acoustic location of prey by barn owls (*Tyto alba*). *Journal of Experimental Biology*, **54**, 535–573.
13. Norberg, R.A. 1977. Occurrence and independent evolution of bilateral ear asymmetry in owls and implications on owl taxonomy. *Philosophical Transactions of the Royal Society B*, **280**, 375–408.

14. Newton, I., Kavanagh, R., Olsen, J. *et al.* (eds). 2002. *Ecology and Conservation of Owls*. Melbourne: CSIRO Publishing, p. 338.
15. Barn Owl Centre. Hearing capabilities. www.barnowl.co.uk/upload/docs/593/hearing_capabilities.pdf (accessed December 2017).
16. Knudsen, E.I. and Konishi, M. 1978. A neural map of the auditory space in the owl. *Science*, **200**, 795–797.
17. Wylie, D.R., Gutiérrez-Ibáñez, C. and Iwaniuk, A.N. 2015. Integrating brain, behaviour, and phylogeny to understand the evolution of sensory systems in birds. *Frontiers in Neuroscience*, **9**, 281. doi: 10.3389/fnins.2015.00281.
18. Gutiérrez-Ibáñez, C., Iwaniuk, A.N., Wylie, D.R. 2011. Relative size of auditory pathways in symmetrically and asymmetrically eared owls. *Brain, Behaviour and Evolution*, **78**, 286–301.
19. Weger, M. and Wagner, H. 2016. Morphology variations of leading-edge serrations in owls (strigiformes). *PLoS ONE*, **11** (3), e0149236. doi: 10.1371/journal.pone.0149236.

Chapter 11: The Oilbird's Story

1. Humboldt, A. and Bonpland, A. 1822. *Narrative of the Travels to the Equinoctial Regions of the New Continent, During the Years 1799–1804.* (trans. H.M. Williams). Vol III, 2nd edition. London: Longman, Hurst, Rees, Orme & Brown, pp. 111–156.
2. Mayr, G. 2010. Phylogenetic relationships of the paraphyletic 'caprimulgiform' birds (nightjars and allies). *Journal of Zoological Systematics and Evolutionary Research*, **48**, 126–137.
3. Van Remsen, J. 2016. Elevate Steatornithidae and Nyctibiidae to rank of Order. Proposal (703) to South American Classification Committee. http://www.museum.lsu.edu/~Remsen/SACCprop703.htm (accessed December 2017).
4. Jetz, W., Thomas, G.H., Joy, J.B. *et al.* 2014. Global distribution and conservation of evolutionary distinctness in birds. *Current Biology*, **24**, 919–930.
5. Isaac, N.J.B., Turvey, S.T., Collen, B. *et al.* 2007. Mammals on the EDGE: conservation priorities based on threat and phylogeny. *PLoS One*, **2** (3), e296. doi: 10.1371/journal.pone.0000296.
6. Gretz, E. 2014. Scientists rank world's most 'evolutionarily distinct' birds. *Popular Science*, 15 April 2014. www.popsci.com/article/science/scientists-rank-worlds-most-evolutionarily-distinct-birds (accessed December 2017).
7. *EDGE: Evolutionarily Distinct & Globally Endangered.* www.edgeofexistence.org/birds (accessed December 2017).
8. Rojas, L.M., Ramirez, M., McNeil, R. *et al.* 2004. Retinal morphology and electrophysiology of two Caprimulgiformes birds: the cave-living and nocturnal oilbird (*Steatornis caripensis*), and the crepuscularly and nocturnally foraging common pauraque (*Nyctidromus albicollis*). *Brain, Behavior and Evolution*, **64**, 19–33.
9. Fidler, A.E., Kuhn, S. and Gwinner, E. 2004. Convergent evolution of strigiform and caprimulgiform dark-activity is supported by phylogenetic analysis using the arylalkylamine N-acetyltransferase (*Aanat*) gene. *Molecular Phylogenetics and Evolution*, **33**, 908–921.
10. Holyoak, D.T. 2001. *Nightjars and Their Allies*. Oxford: Oxford University Press.
11. Olson, S.L. 1987. An early Eocene oilbird from the Green River Formation of Wyoming (Caprimulgiformes: Steatornithidae). *Documents des Laboratoires de Géologie de Lyon*, **99**, 57–69.
12. Holland, R.A., Wikelski, M. Kümmeth, F. *et al.* 2009. The secret life of oilbirds: new insights into the movement ecology of a unique avian frugivore. *PLoS One*, **4** (12), e8264. doi: 10.1371/journal.pone.0008264.
13. Griffin, D.R. and Galambos, R. 1941. The sensory basis of obstacle avoidance by flying bats. *Journal of Experimental Zoology*, **86**, 481–506.
14. Griffin, D.R. 1958. *Listening in the Dark: the Acoustic Orientation of Bats and Men*. New Haven, CT: Yale University Press.

15. Konishi, M. and Knudsen, E.I. 1979. The oilbirds: hearing and echolocation. *Science*, **204**, 425–427.
16. Martin, G., Rojas, L.M., Ramírez, Y. *et al.* 2004. The eyes of oilbirds (*Steatornis caripensis*): pushing at the limits of sensitivity. *Naturwissenschaften*, **91**, 26–29.
17. Brinkløv, S., Fenton, M.B. and Ratcliffe, J.M. 2013. Echolocation in oilbirds and swiftlets. *Frontiers in Physiology*, **4**, 123. doi: 10.3399/fphys.2013.00123.
18 Giannini, N.P. and Simmons, N.B. 2003. A phylogeny of megachiropteran bats (Mammalia: Chiroptera: Pteropodidae) based on direct optimization analysis of one nuclear and four mitochondrial genes. *Cladistics*, **19**, 496–511.
19. Conway Morris, S. 2003. *Life's Solution: Inevitable Humans in a Lonely Universe*. Cambridge: Cambridge University Press.

Chapter 12: The Hummingbird's Story

1. Ksepka, D.T., Clarke, J.A., Nesbitt, S.J. *et al.* 2013. Fossil evidence of wing shape in a stem relative of swifts and hummingbirds (Aves, Pan-Apodiformes). *Proceedings of the Royal Society B*, **280**, 20130580. doi: 10.1098/rspb.2013.0580.
2. Hedenström, A., Norevik, G., Warfvinge, K. *et al.* 2016. Annual 10-month aerial life phase in the common swift *Apus apus*. *Current Biology*, **26**, 3066–3070.
3. Mayr, G. 2003. A new Eocene swift-like bird with a peculiar feathering. *Ibis*, **145**, 382–391.
4. Mayr, G. 2009. *Paleogene Fossil Birds*. Berlin: Springer-Verlag, p. 135.
5. Mayr, G. 2004. Old World fossil record of modern-type hummingbirds. *Science*, **304**, 861–864.
6. Mayr, G. 2005. Fossil hummingbirds in the Old World. *Biologist*, **52**, 12–16.
7. Louchart, A., Tourment, N., Carrier, J. *et al.* 2008. Hummingbird with modern feathering: an exceptionally well-preserved Oligocene fossil from southern France. *Naturwissenschaften*, **95**, 171–175.
8. McGuire, J.A., Witt, C.C., Remsen, J.V. *et al.* 2014. Molecular phylogenetics and the diversification of hummingbirds. *Current Biology*, **24**, 910–916.
9. Baldwin, M.W., Toda, Y., Nakagita, T. *et al.* 2014. Evolution of sweet taste perception in hummingbirds by transformation of the ancestral umami receptor. *Science*, **345**, 929–933.
10. Rico-Guevara, A., Fan, T-H., Rubega, M.A. *et al.* 2015. Hummingbird tongues are elastic micropumps. *Proceedings of the Royal Society B*, **282**: 20151014. doi: org/10.1098/rspb.2015.1014.
11. Yanega, G.M. and Rubega, M.A. 2004. Feeding mechanisms: hummingbird jaw bends to aid insect capture. *Nature*, **428**, 615.
12. Estades, C.F., Vukasovic, M.A., Tomasevic, J.A. *et al.* 2008. Giant hummingbirds (*Patagona gigas*) ingest calcium-rich minerals. *The Wilson Journal of Ornithology*, **120**, 651–653.
13. Hedrick, T.L., Tobalske, B.W., Ros, I.G. *et al.* 2011. Morphological and kinematic basis of the hummingbird flight stroke: scaling of flight muscle transmission ratio. *Proceedings of the Royal Society B*, **279**, 1986–1992. doi: 10.1098/rspb.2011.2238.
14. Suarez, R.K., Lighton, J.R.B., Brown, G.S. *et al.* 1991. Mitochondrial respiration in hummingbird flight muscles. *Proceedings of the National Academy of Sciences of the United States of America*, **88**, 4870–4873.
15. Gaede, A.H., Goller, B., Lam, J.P. M., *et al.* 2017. Neurons responsive to global visual motion have unique tuning properties in hummingbirds. *Current Biology*, **27**, 279–285.
16. Ward, B.J., Day, L.B., Wilkening, S.R. *et al.* 2012. Hummingbirds have a greatly enlarged hippocampal formation. *Biology Letters*, **8**, 657–659.
17. Abrahamczyk, S., Sonto-Vilarós, D. and Renner, S.S. 2014. Escape from extreme specialization: passionflowers, bats and the sword-billed hummingbird. *Proceeding of the Royal Society B*, **281**: 20140888. doi: 10.1098/rspb.2014.0888.
18. Bradshaw, H.D. and Schemske, D.W. 2003. Allele substitution at a flower colour locus produces a pollinator shift in monkeyflowers. *Nature*, **426**, 176–178.

19. Abrahamczyk, S. and Renner, S.S. 2015. The temporal build-up of hummingbird/plant mutualisms in North America and temperate South America. *BMC Evolutionary Biology*, **15**: 104. doi: 10.1186/s12862-015-0388-z.
20. Foote, C. 2015. Hummingbirds and plants: an evolutionary love affair? https://blogs.biomedcentral.com/.../06/.../hummingbirds (accessed December 2017).

Chapter 13: The Parrot's Story

1. Widescreen Arkive. Puerto Rican Amazon (*Amazona vittata*). www.arkive.org/puerto-rican-amazon/amazona-vittata (accessed December 2017).
2. Olah, G., Butchart, S.H.M., Symes, A. *et al.* 2016. Ecological and socio-economic factors affecting extinction risk in parrots. *Biodiversity and Conservation*, **25**, 205–223.
3. Schweizer, M., Seehausen, O., Güntert, M. *et al.* 2010. The evolutionary diversification of parrots supports a taxon pulse model with multiple trans-oceanic dispersal events and local radiations. *Molecular Phylogenetics and Evolution*, **54**, 984–994.
4. Jarvis, E.D., Mirarab, S., Aberer, A.J. *et al.* 2014. Whole-genome analyses resolve early branches in the tree of life of modern birds. *Science*, **346**, 1320–1331.
5. Hackett, S.J., Kimball, R.T., Reddy, S. *et al.* 2008. A phylogenetic study of birds reveals their evolutionary history. *Science*, **320**, 1763–1768.
6. Mayr, G. 2009. Cariamae (seriemas and allies). In *Paleogene Fossil Birds*. Berlin: Springer-Verlag, pp. 139–152.
7. Mayr, G. 2002. On the osteology and phylogenetic affinities of the Pseudasturidae – Lower Eocene stem-group representatives of parrots (Aves, Psittaciformes). *Zoological Journal of the Linnean Society*, **136**, 715–729.
8. Wright, T.F., Schirtzinger, E.E., Matsumoto, T. *et al.* 2008. A multilocus molecular phylogeny of the parrots (Psittaciformes): support for a Gondwanan origin during the Cretaceous. *Molecular Biology and Evolution*, **25**, 2141–2156.
9. Schweizer, M., Seehausen, O. and Hertwig, S.T. 2011. Macroevolutionary patterns in diversification of parrots: effect of climate change, geological events and key innovations. *Journal of Biogeography*, **38**, 2176–2194.
10. Schweizer, M. 2011. The evolutionary diversification of parrots (Aves: Psittaciformes): an integrated approach. http://library.eawag.ch/EAWAG-Publications/openaccess/Eawag_06930.pdf (accessed December 2017).
11. Critically endangered kakapo on the increase. *New Zealand Herald*, 27 November 2016. http://archive.is/99jos (accessed December 2017).
12. Kirchman, J.J., Schirtzinger, E.E. and Wright, T.F. 2012. Phylogenetic relationships of the extinct Carolina Parakeet (*Conuropsis carolinensis*) inferred from DNA sequence data. *The Auk*, **129**, 197–204.
13. Urantówka, A.D., Mackiewicz, P. and Strzala, T. 2014. Phylogeny of *Amazona barbadensis* and the yellow-headed amazon complex (Aves: Psittacidae): a new look at South American parrot evolution. *PLoS ONE*, **9** (5), e97228. doi: 10.1371/journal.pone.0097228.
14. Jønsson, K.A., Bowie, R.C.K., Nylander, J.A.A. *et al.* 2010. Biogeographical history of cuckoo-shrikes (Aves: Passeriformes): transoceanic colonization of Africa from Australo-Papua. *Journal of Biogeography*, **37**, 1767–1781.
15. Ekstrom, J.M.M., Burke, T., Randrianaina, L. *et al.* 2007. Unusual sex roles in a highly promiscuous parrot: the greater vasa parrot *Caracopsis vasa*. *Ibis*, **149**, 313–320.
16. Dilger, W. 1962. The behaviour of lovebirds. *Scientific American*, **206**, 89–98.
17. Zelenkov, N.V. 2016. The first fossil parrot (Aves, Psittaciformes) from Siberia and its implications for the historical biogeography of Psittaciformes. *Biology Letters*, **12** (10), 20160717. doi: 10.1098/rsbl.2016.0717.
18. Groombridge, J.J., Jones, C.G., Nichols, R.A. *et al.* 2004. Molecular phylogeny and morpho-

logical change in the *Psittacula* parakeets. *Molecular Phylogenetics and Evolution*, **31**, 96–108.
19. Heinsohn, R. 2008. Ecology and evolution of the enigmatic electus parrot (*Eclectus roratus*). *Journal of Avian Medicine and Surgery*, **22**, 146–150.
20. Schweizer, M., Güntert, M., Seehausen, O. *et al.* 2014. Parallel adaptations to nectarivory in parrots, key innovations and the diversification of the Loriinae. *Ecology and Evolution*, **4**, 2867–2883.
21. White, N.E., Phillips, M.J., Gilbert, M.T.P. *et al.* 2011. The evolutionary history of cockatoos (Aves: Psittaciformes: Cacatuidae). *Molecular Phylogenetics and Evolution*, **59**, 615–622.
22. Ericson, P.G.P., Anderson, C.L., Britton, T. *et al.* 2006. Diversification of Neoaves: integration of molecular sequence data and fossils. *Biological Letters*, **2**, 543–547.
23. Suh, A., Paus, M., Kiefmann, M. *et al.* 2011. Mesozoic retroposons reveal parrots as the closest living relatives of passerine birds. *Nature Communications*, **2**, 443. doi: 10.1038/ncomms1448.
24. Fuchs, J., Johnson, J.A. and Mindell, D.P. 2015. Rapid diversification of falcons (Aves: Falconidae) due to the expansion of open habitats in the Late Miocene. *Molecular Phylogenetics and Evolution*, **82**, 166–182.

Chapter 14: The New Zealand Wren's Story

1. Gill, F. and Donsker, D. (eds). 2017. *IOC World Bird List* (v 7.3). doi: 10.14344/IOC.ML.7.3.
2. Robert J. Raikow described 18 distinctive characters shared by passerines, but concluded that only five were unique: the aegithognathous palate (a special variation on the neognathan palate), the 'passerine' tensor propatagialis brevis (a forearm, or wing muscle), bundled spermatozoa with coiled heads and large acrosomes, an enlarged hallux, and type VII deep plantar tendons. See Raikow, R.J. 1982. Monophyly of the Passeriformes: test of a phylogenetic hypothesis. *The Auk*, **99**, 431–455.
3. Müller, J.P. 1847. Über die bisher unbekannten typischen Verschiedenheiten der Stimmorgane der Passerinen. Abhandlungen Königlich (Preussische) Akademische Wissenschraft, Berlin, 1–71; see also Müller, J.P. 1878. *On Certain Variations in the Vocal Organs of the Passeres that Have Hitherto Escaped Notice*. Oxford: Clarendon Press.
4. Ames, P. 1971. The morphology of the syrinx in passerine birds. *Bulletin of the Peabody Museum of Natural History*, **37**, 1–194.
5. Feduccia, A. 1974. Morphology of the bony stapes in new and Old World suboscines: new evidence for common ancestry. *The Auk*, **91**, 427–429.
6. This account is taken from Birkhead, T., Wimpenny, J. and Montgomerie, B. 2014. *Ten Thousand Birds: Ornithology Since Darwin*. Princeton, NJ: Princeton University Press, p. 108. For primary publication see Sibley, C.G. and Ahlquist, J.E. 1990. *Phylogeny and Classification of Birds: a Study in Molecular Evolution*. New Haven, CT: Yale University Press.
7. Sibley, C.G. and Ahlquist, J.E. 1990. *Phylogeny and Classification of Birds: a Study in Molecular Evolution*. New Haven, CT: Yale University Press.
8. Low, T. 2014. *Where Song Began: Australia's Birds and How They Changed the World*. Melbourne: Penguin Random House, p. 65.
9. Galbreath, R. and Brown, D. 2004. The tale of the lighthouse-keeper's cat: discovery and extinction of the Stephens Island wren (*Traversia lyalli*). *Notornis*, **51**, 193–200.
10. Lionel Walter Rothschild (1868–1937), scion of the Rothschild family and inveterate species collector, formed the largest zoological collection ever amassed by a private individual. As well as vast numbers of butterflies and beetles, it included 300,000 bird skins and 200,000 birds' eggs. In all, Rothschild acquired the skins of nine Stephens Island Wrens, which, after various sales, gifts and exchanges, left three remaining at Tring (Natural History Museum).
11. Mearns, B. and Mearns, R. 1998. *The Bird Collectors*. London: Academic Press, p. 124.

12. Donald, P.F., Collar, N.J., Marsden, S.J. *et al.* 2010. *Facing Extinction: the World's Rarest Birds, and the Race to Save Them*. London: T. & A.D. Poyser, p. 112.
13. Feduccia, A. 1975. Morphology of the bony stapes in Menuridae and Acanthisittidae: evidence for oscine affinities. *Wilson Bulletin*, **87**, 418–420.
14. Raikow, R.J. 1987. Hindlimb myology and evolution of the Old World suboscine passerine birds (Acanthisittidae, Pittidae, Philepittidae, Eurylaimidae). *Ornithological Monograms*, **41**, 1–81.
15. Sibley, C.G. and Ahlquist, J.E. 1990. *Phylogeny and Classification of Birds: a Study in Molecular Evolution*. New Haven, CT: Yale University Press, p. 582.
16. Raikow, R.J. and Bledsoe, A.H. 2000. Phylogeny and evolution of the passerine birds. *BioScience*, **50**, 487–499.
17. Barker, F.K., Barrowclough, G.F., Groth, J.G. 2002. A phylogenetic hypothesis for passerine birds: taxonomic and biogeographical implications of an analysis of nuclear DNA sequence data. *Proceedings of the Royal Society B*, **269**, 295–308.
18. Ericson, P.G.P., Christidis, L., Cooper, A. *et al.* 2002. A Gondwanan origin of passerine birds supported by DNA sequences of the endemic New Zealand wrens. *Proceedings of the Royal Society B*, **269**, 235–241. doi: 10.1098/rspb.2001.1877.
19. Cooper, A. and Penny, D. 1997. Mass survival of birds across the Cretaceous–Tertiary boundary: molecular evidence. *Science*, **275**, 1109–1113; see also van Tuinen, M. and Hedges, S.B. 2001. Calibration of avian clocks. *Molecular Biology and Evolution*, **18**, 206–213.
20. Ericson, P.G.P., Klopfstein, S., Irestedt, M. *et al.* 2014. Dating the diversification of the major lineages of Passeriformes (Aves). *BMC Evolutionary Biology*, **14**, 8. doi: 10.1186/1471-2148-14-8; Barker, F.K., Cibois, A., Schikler, P. *et al.* 2004. Phylogeny and diversification of the largest avian radiation. *Proceedings of the National Academy of Sciences of the United States of America*, **101**, 11040–11045.
21. The four New Zealand wrens that have become extinct over the last 800 years are Long-billed Wren (*Dendroscansor decurvirostris*), endemic to the South Island, with a long and curved bill unlike that of the rest of the family; Stout-billed Wren (*Pachyplichus yaldwyni*), the largest of the family; Stephens Island or Lyall's Wren (*Traversia lyalli*); and Bushwren (*Xenicus longipes*).
22. Mayr, G. 2013. The age of the crown group of passerine birds and its evolutionary significance: molecular calibrations versus the fossil record. *Systematics and Biodiversity*, **11**, 1–6.
23. Worthy, T.H., Hand, S.J., Nguyen, J.M.T. *et al.* 2010. Biogeographical and phylogenetic implications of an early Miocene wren (Aves: Passeriformes: Acanthisittidae) from New Zealand. *Journal of Vertebrate Paleontology*, **30**, 479–498.
24. Ericson, P.G.P., Klopfstein, S., Irestedt, M. *et al.* 2014. Dating the diversification of the major lineages of Passeriformes (Aves). *BMC Evolutionary Biology*, **14**, 8. doi: 10.1186/1471-2148-14-8.
25. de Queiroz, A. 2014. *The Monkey's Voyage: How Improbable Journeys shaped the History of the World*. New York: Basic Books, p. 109. This fascinating book gives an in-depth analysis of the dispersal versus vicariance debate.
26. Landis, C.A., Campbell, H.J. and Begg, J.G. 2008. The Waipounamu Erosion Surface: questioning the antiquity of the New Zealand land surface and terrestrial fauna and flora. *Geological Magazine*, **145**, 173–197.
27. Granscolas, P., Robillard, M.J., Desutter-Grandcolas, L. *et al.* 2008. New Caledonia: a very old Darwinian island? *Philosophical Transactions of the Royal Society B*, **363**, 3309–3317. For further discussion of the implications of marine ingressions of Zealandia see de Queiroz, A. 2014.
28. For a readable analysis of the dispersal versus vicariance debate, see de Queiroz, A. 2014. *The Monkey's Voyage: How Improbable Journeys shaped the History of the World*. New York: Basic Books.

29. Hand, S.J., Weisbecker, V., Beck, R.M. *et al.* 2009. Bats that walk: a new evolutionary hypothesis for the terrestrial behaviour of New Zealand's endemic mystacinids. *BMC Evolutionary Biology*, **9**, 169. doi: 10,1186/1471-2148-9-169.
30. Cooper, A. and Cooper, R.A. 1995. The Oligocene bottleneck and New Zealand biota: genetic record of a past environmental crisis. *Proceedings of the Royal Society B*, **261**, 293–302.
31. Mitchell, K.T., Wood, J.R., Llamas, B. *et al.* 2016. Ancient mitochondrial genomes clarify the evolutionary history of New Zealand's enigmatic acanthisittid wrens. *Molecular Phylogenetics and Evolution*, **102**, 295–304.
32. Recent geological data support Cooper's biogeographical interpretation. Oligocene non-marine deposits have been found overlying the Taranaki Basin (sediments derived from the hinterland), a feature inconsistent with a total Oligocene drowning of Zealandia. See Strogen, D.P., Bland, K.J., Nicol, A. *et al.* 2014. Paleogeography of the Taranaki basin region during the latest Eocene–early Miocene and implications for the 'total drowning' of Zealandia. *New Zealand Journal of Geology and Geophysics*, **57**, 110–127. doi: 10.1080/00288306.2014.901231.
33. Jarvis, E.D., Mirarab, S., Aberer, A.J. *et al.* 2014. Whole-genome analyses resolve early branches in the tree of life of modern birds. *Science*, **346**, 1320–1331.
34. E.D. Jarvis, personal communication.
35. Prum, R.O., Berv, J.S., Dornburg, A. *et al.* 2015. A comprehensive phylogeny of birds (Aves) using targeted next-generation DNA sequencing. *Nature*, **526**, 569–573.
36. For further information, see Bird 10,000 Genomes (B10K) Project, http://b10k.genomics.cn.
37. Mayr, G. 2015. A reassessment of Eocene parrotlike fossils indicates a previously undetected radiation of zygodactyl stem group representatives of passerines (Passeriformes). *Zoologica Scripta*, **44**, 587–602.
38. Botelho, J.F., Smith-Paredes, D., Nuñez-Leon, D. *et al.* 2014. The developmental origin of zygodactyl feet and its possible loss in the evolution of Passeriformes. *Proceedings of the Royal Society B*, **281**, 20140765. doi: 10.1098/ rspb.2014.0765.

Chapter 15: The Manakin's Story

1. Ericson, P.G.P., Christidis, L., Cooper, A. *et al.* 2002. A Gondwanan origin of passerine birds supported by DNA sequences of the endemic New Zealand wrens. *Proceedings of the Royal Society B*, **269**, 235–241. doi: 10.1098/rspb.2001.1877.
2. Ericson, P.G.P., Klopfstein, S., Irestedt, M. *et al.* 2014. Dating the diversification of the major lineages of Passeriformes (Aves). *BMC Evolutionary Biology*, **14**, 8. doi: 10.1186/1471-2148-14-8.
3. Wilf, P., Cūneo, N.R., Johnson, K.R. *et al.* 2003. High plant diversity in Eocene South America: evidence from Patagonia. *Science*, **300**, 122–125.
4. Svensen, H., Plunke, S., Malthe-Sórenssen, A. *et al.* 2004. Release of methane from a volcanic basin as a mechanism for initial Eocene global warming. *Nature*, **429**, 542–545.
5. Zeebe, R.E., Dickens, G.R., Ridgwell, A. *et al.* 2014. Onset of carbon isotope excursion at the Paleocene–Eocene thermal maximum took millennia, not 13 years. *Proceedings of the National Academy of Sciences of the United States of America*, **111** (12), E1062–E1063. doi: 10.1073/pnas.1321177111.
6. Intergovernmental Panel on Climate Change. 2001. *Climate Change 2001: the Scientific Basis*. www.grida.no/publications/270 (accessed December 2017).
7. Scher, H.D. and Martin, E.E. 2006. Timing and climatic consequences of the opening of Drake Passage. *Science*, **312**, 428–430.
8. Zachos, J., Pagani, M., Sloan, L. *et al.* 2001. Trends, rhythms, and aberration in global climate 65 Ma to present. *Science*, **292**, 686–693.
9. Jacobs, B.F., Kingston, J.D. and Jacobs, L.L. 1999. The origin of grass-dominated ecosystems. *Annals of the Missouri Botanical Garden*, **86**, 590–643.

10. Ricklefs, R.E. 2002. Splendid isolation: historical ecology of the South American passerine fauna. *Journal of Avian Biology*, **33**, 207–211.
11. Goldner, A., Herold, N. and Huber, M. 2014. The challenge of stimulating the warmth of the mid-Miocene climatic optimum in cesm1. *Climate of the Past*, **10**, 523–536.
12. MacLeod, N. 2013. *The Great Extinctions: What Causes Them and How They Shape Life*. London: Natural History Museum, p. 160.
13. Rheindt, F.E., Christidis, L. and Norman, J.A. 2008. Habitat shifts in the evolutionary history of a Neotropical flycatcher lineage from forest and open landscapes. *BMC Evolutionary Biology*, **8**, 193. doi: 10.1186/1471-2148-8-193.
14. Gill, F. and Donsker, D. (eds). 2017. *IOC World Bird List* (v 7.3). doi: 10.14344/IOC.ML.7.3.
15. Wallace, A.R. 1852. On the monkeys of the Amazon. *Proceedings of the Zoological Society of London*, **20**, 107–110.
16. Wallace, A.R. 1853. *A Narrative of Travels on the Amazon and Rio Negro*. London: Reeve & Co.
17. Slotten, R.A. 2004. *The Life of Alfred Russel Wallace: the Heretic in Darwin's Court*. New York: Columbia University Press, p. 94.
18. Shephard, G.E., Müller, R.D., Liu, L. *et al.* 2010. Miocene drainage reversal of the Amazon River driven by plate–mantle interaction. *Nature Geoscience*, **3**, 870–875.
19. Mayr, E. 1942. *Systematics and the Origin of Species from the Viewpoint of a Zoologist*. New York: Columbia University Press.
20. Sick, H. 1967. Rios e enchentes na Amazônia como obstáculo para a avifauna. In *Atas do Simpósio sôbre a Biota Amazônica, Vol. 5 (Zoologia)* (ed. H. Kent). Rio di Janeiro: Conselho di Pequisas, pp. 495–520.
21. Capparella, A.P. 1988. Genetic variation in Neotropical birds: implications for species process. *Acta XIX Congress of International Ornithology*, 1658–1664; Neotropical avian diversity and riverine barriers. *Acta XX Congress of International Ornithology*, 307–316.
22. Cracraft, J. 1985. Historical biogeography and patterns of differentiation within the South American avifauna: areas of endemism. *Ornithological Monographs*, **36**, 49–84.
23. Cohn-Haft, M., Pacheco, A.M.F., Bechtoldt, C.L. *et al.* 2007. Inventário ornitológico. In L.R. Py-Daniel, C.P. Deus, A.L. Henriques *et al.* (eds), *Biodiversidade do Médio Madeira: Bases Cientificas para Propostas de Conservação*. Manaus: INPA, pp. 147–178.
24. Isler, M.L., Isler, P.R. and Whitney, B.M. 2007. Species limits in antbirds (thamnophilidae): the warbling antbird (*Hypocnemmis cantator*) complex. *The Auk*, **124**, 11–28.
25. Whitney, B.M., Schunck, F., Rêgo, M.A. *et al.* 2013. A new species of *Zimmerius* tyrannulet from the Upper Madeira–Tapajós Interfluvium in central Amazonian Brazil: birds don't always occur where they 'should'. In J. del Hoyo, A. Elliott, J. Sargatal and D. Christie (eds), *Handbook of the Birds of the World*, special issue: new species and global index. Barcelona: Lynx Edicions, pp. 286–291. Whitney, B. M., Schunck, F., Rêgo, M.A. *et al.* 2013. A new species of flycatcher in the *Tolmomyias assimilis* radiation from the Lower Sucunduri–Tapajó Interfluvium in central Amazonian Brazil heralds a new chapter in Amazonian biogeography. In J. del Hoyo, A. Elliott, J. Sargatal and D. Christie (eds), *Handbook of the Birds of the World*, special issue: new species and global index. Barcelona: Lynx Edicions, pp. 297–300.
26. Sardelli, C.H. 2005. Variação geográfica e genética de *Hemitriccus minor* (Aves: Tyrannidae) na bacia do Rio Madeira am/Brasil. Masters thesis, Instituto Nacional de Pesquisas da Amazônia, Manaus, Brazil.
27. Fernandes, A.M. 2012. Phylogeography and speciation of selected passerine birds of lowland Amazonia: implications for historical biogeography and conservation. PhD thesis, University of Heidelberg.
28. Fernandes, A.M., Gonzalez, J., Wink, M. *et al.* 2013. Multilocus phylogeography of the wedge-billed woodcreeper *Glyphorynchus spirurus* (Aves, Furnariidae) in lowland Amazonia: widespread cryptic diversity and paraphyly reveal a complex diversification

pattern. *Molecular Phylogenetics and Evolution*, **66**, 270–282.
29. Hayes, F.E. and Sewlal, J-A.N. 2004. The Amazon river as a dispersal barrier to passerine birds, effects of river width, habitat and taxonomy. *Journal of Biogeography*, **31**, 1809–1818.
30. A 'bivouac' is a structure of several hundred thousand worker ants linked together in a ball-shaped mass up to a metre across, within which are buried a queen and thousands of larvae. During the migratory phase, the bivouacs dissolve into raiding columns, or ant swarms, that seek out new foraging areas.
31. Remsen, J.V. and Parker, T.A. 1983. Contribution of river created habitats to bird species richness in Amazonia. *Biotropica*, **15**, 223–231.
32. Nores, M. 1999. An alternative hypothesis for the origin of Amazonian bird diversity. *Journal of Biogeography*, **26**, 475–485.
33. Haffer, J. 1969. Speciation in Amazonian forest birds. *Science*, **165**, 131–137.
34. Colinvaux, P.A., De Oliveira, P.E., Moreno, J.E. *et al.* 1996. A long pollen record from lowland Amazonia: forest and cooling in glacial times. *Science*, **274**, 85–88. For a racy first-hand account of Colinvaux's search for lakes containing mud from the last glacial maximum, see Colinvaux, P. 2008. *Amazon Expeditions: My Quest for the Ice-Age Equator*. New Haven, CT: Yale University Press.
35. Maslin, M.A., Ettwein, V.J., Boot, C.S. *et al.* 2012. Amazon fan biomarker evidence against the Pleistocene rainforest refuge hypothesis? *Journal of Quaternary Science*, **27**, 451–460.
36. Solomon, S.E., Bacci, M., Martins, J. *et al.* 2008. Paleodistributions and comparative molecular phylogeography of leafcutter ants (*Atta* spp.) provide new insight into the origin of Amazonian diversity. *PLoS ONE*, **3** (7), e2738. doi: 10.1371/journal.pone.0002738.
37. d'Horta, F.M., Cuervo, A.M., Ribas, C.C. *et al.* 2013. Phylogeny and comparative phylogeography of *Sclerurus* (Aves: Furnariidae) reveal constant and cryptic diversification in an old radiation of rain forest understorey specialists. *Journal of Biogeography*, **40**, 37–49.
38. Smith, B.T., McCormack, J.E., Cuervo, A.M. *et al.* 2014. The drivers of tropical speciation. *Nature*, **515**, 406–409. doi: 10.1038/nature13687.
39. Chevron, Z.A., Hackett, S.J. and Capparella, A.P. 2005. Complex evolutionary history of a Neotropical lowland forest bird (*Lepidothrix coronata*) and its implications for historical hypotheses of the origin of Neotropical avian diversity. *Molecular Phylogenetics and Evolution*, **36**, 338–357.
40. Irestedt. M., Fjeldså, J. and Ericson, P.G.P. 2006. Evolution of the ovenbird–woodcreeper assemblage (Aves: Furnariidae): major shifts in nest architecture and adaptive radiation. *Journal of Avian Biology*, **37**, 260–272.
41. Irestedt, M., Fjeldså, J., Dalén, L. *et al.* 2009. Convergent evolution, habitat shifts and variable diversification rates in the ovenbird–woodcreeper family (Furnariidae). *BMC Evolutionary Biology*, **9**, 268. doi: 10.1186/1471-2148-9-268.
42. Nadkarni, N.M. and Longino, J.T. 1990. Invertebrates in canopy and ground organic matter in a Neotropical forest, Costa Rica. *Biotropica*, **22**, 286–289.
43. Sillett, T.S. 1994. Foraging ecology of epiphyte-searching insectivorous birds of Costa Rica. *The Condor*, **96**, 863–877.
44. Parker, T.A., Stotz, D.F. and Fitzpatrick, J.W. 1996. Ecological and distributional databases. In D.E. Stotz, J.W. Fitzpatrick, T.A. Parker *et al.* (eds), *Neotropical Birds: Ecology and Conservation*. Chicago, IL: University of Chicago Press, pp. 113–436.
45. Cockle, K.L. and Areta, J.I. 2013. Specialization on bamboo by Neotropical birds. *The Condor*, **115**, 217–220.
46. Kratter, A.W. 1997. Bamboo specialization by Amazonian birds. *Biotropica*, **29**, 100–110.
47. Remsen, J.V. and Parker, T.A. 1984. Arboreal dead-leaf-searching birds of the Neotropics. *The Condor*, **86**, 36–41.
48. Isler, M.L., Lacerda, D.R., Isler, P.R. *et al.* 2006. *Epinecrophylla*, a new genus of antwrens (Aves: Passeriformes: Thamnophilidae). *Proceedings of the Biological Society of Washington*, **119**, 522–527.

49. Gradwohl, J. and Greenberg, R. 1982. The effect of a single species of avian predator on the arthropods of aerial leaf litter. *Ecology*, **63**, 581–583.
50. Rosenberg, K.V. 1997. Ecology of dead-leaf foraging specialists and their contribution to Amazonian bird diversity. *Ornithological Monographs*, **48**, 673–700.
51. Wrege, P. H., Wikelski, M., Mandel, J.T. *et al.* 2005. Antbirds parasitize foraging army ants. *Ecology*, **86**, 555–559.
52. Willis, E.O. and Oniki, Y. 1978. Birds and army ants. *Annual Review of Ecological Systems*, **9**, 243–263.
53. Brumfield, R.T., Tello, J.G., Cheviron, Z.A. *et al.* 2007. Phylogenetic conservatism and antiquity of a tropical specialization: army-ant-following in the typical antbirds (Thamnophilidae). *Molecular Phylogenetics and Evolution*, **45**, 1–13.
54. Brady, S.G. 2003. Evolution of the army-ant syndrome: the origin and long-term evolutionary stasis of a complex of behavioral and reproductive adaptations. *Proceedings of the National Academy of Sciences of the United States of America*, **100**, 6575–6579.

Chapter 16: The Sapayoa's Story

1. Ericson, P.G.P., Christidis, L., Cooper, A. *et al.* 2002. A Gondwanan origin of passerine birds supported by DNA sequences of the endemic New Zealand wrens. *Proceedings of the Royal Society B*, **269**, 235–241. doi: 10.1098/rspb.2001.1877.
2. Ericson, P.G.P., Klopfstein, S., Irestedt, M. *et al.* 2014. Dating the diversification of the major lineages of Passeriformes (Aves). *BMC Evolutionary Biology*, **14**, 8. doi: 10.1186/1471-2148-14-8.
3. Moyle, R.G., Chesser, R.T., Prum, R.O. *et al.* 2006. Phylogeny and evolutionary history of Old World suboscine birds (Aves: Eurylaimides). *American Museum Novitates*, **3544**, 1–22.
4. Malaysian Birds: Pitta birds. www.malaysianbirds.com/bird-family/pitta.htm (accessed December 2017).
5. Gooddie, C. 2010. *The Jewel Hunter.* Old Basing: WILDguides Ltd.
6. Clark, C.J., Kirschel, A.N.G., Hadjioannou, L. *et al.* 2016. *Smithornis* broadbills produce loud wing song by aeroelastic flutter of medial primary wing feathers. *Journal of Experimental Biology*, **219**, 1069–1075. doi: 10.1242/jeb.131664.
7. Bostwick, K.S., Elias, D.O., Mason, A. *et al.* 2010. Resonating feathers produce courtship song. *Proceedings of the Royal Society B*, **277**, 835–841. doi: 10.1098/rspb.2009.1576.
8. Prum, R.O., Morrison, R.L. and Ten Eyck, G.R. 1994. Structural color production by constructive reflection from ordered collagen arrays in a bird (*Philepitta castanea*: Eurylaimedae). *Journal of Morphology*, **222**, 61–72.
9. Raikow, R.J. 1987. Hindlimb myology and evolution of the Old World suboscine passerine birds (Acanthisittidae, Pittidae, Philepittidae, Eurylaimidae). *Ornithological Monographs*, **41**, 1–81.
10. Prum, R.O. 1993. Phylogeny, biogeography, and evolution of the broadbills (Eurylaimedae) and asities (Philepittidae) based on morphology. *The Auk*, **110**, 304–324.
11. The Sapayoa's skin was collected by R. Miketta on behalf of W.F.H. Rosenberg on 2 November 1901 in the Esmeraldas area of northwest Ecuador. The specimen was bought by Walter Rothschild and described by Ernst Hartert before finally being sold to the American Museum of Natural History (AMMNH 493552).
12. Hartert, E. 1903. On a remarkable new oligomyodian genus and species from Ecuador. *Novitates Zoologicae*, **10**, 117–118.
13. Fjeldså, J., Zuccon, D., Irestedt, M. *et al.* 2003. *Sapayoa aenigma*: a New World representative of 'Old World suboscines'. *Proceedings of the Royal Society B*, **270**, 238–241. doi: 10.1098/rsbl.2003.0075.
14. Sibley, C.G. and Ahlquist, J.E. 1990. *Phylogeny and Classification of Birds: a Study in Molecular Evolution.* New Haven, CT: Yale University Press.

15. Irestedt, M., Ohlson, J.I., Zuccon, D. *et al.* 2006. Nuclear DNA from old collections of avian study skins reveals the evolutionary history of the Old World suboscines (Aves, Passeriformes). *Zoologica Scripta*, **35**, 567–580.
16. Dzielski, S.A., Van Doren, B.M., Hruska, J.P. *et al.* 2016. Reproductive biology of the Sapayoa (*Sapayoa aenigma*), the 'Old World suboscine' of the New World. *The Auk*, **133**, 347–363.
17. Pletsch, T., Erbacher, J., Holbourn, A.E.L. *et al.* 2001. Cretaceous separation of Africa and South America: the view from the West African margin (ODP leg 159). *Journal of South American Earth Sciences*, **2**, 147–174.
18. Gibbons, A.D., Whittaker, J.M. and Müller, R.D. 2013. The breakup of East Gondwana: assimilating constraints from Cretaceous ocean basins around India into a best-fit tectonic model. *Journal of Geophysical Research. Solid Earth*, **118**, 808–822.
19. Bossuyt, F. and Milinkovitch, M.C. 2001. Amphibians as indicators of early Tertiary 'out-of-India' dispersal of vertebrates. *Science*, **292**, 93–95; Gower, D.J., Kupfer, A., Oommen, O.V. *et al.* 2002. A molecular phylogeny of ichthyophid caecilians (Amphiba: Gymnophiona: Ichthyophiidae): out of India or out of South East Asia? *Proceedings of the Royal Society B*, **269**, 1563–1569.
20. Estes, R. and Hutchinson, J.H. 1980. Eocene lower vertebrates from Ellesmere Island, Canadian Arctic Archipelago. *Palaeogeography, Palaeoclimatology, Palaeoecology*, **30**, 325–347.
21. Mayr, G. and Manegold, A. 2006. A small suboscine-like passeriform bird from the Early Oligocene of France. *The Condor*, **108**, 717–720; Ballman, P. 1969. Die Vögel aus der altburdigalen Spaltenfüllung von Wintershof (West) bei eichstätt in bayern. *Zitteliana*, **I**, 5–60.

Chapter 17: The Scrubbird's Story

1. Serventy, V. 1966. *A Continent in Danger*. London: Andre Deutsch, p. 144.
2. Taylor, S. 2012. *John Gould's Extinct and Endangered Birds of Australia*. Canberra: NLA Publishing, p. 173.
3. Gould, J. 1840–48. *The Birds of Australia*. London: published by the author.
4. Serventy, V. 1966. *A Continent in Danger*. London: Andre Deutsch, p. 150.
5. BirdLife International. Noisy Scrub-bird *Atrichornis clamosus*. www.birdlife.org/datazone/speciesfactsheet.php?id=5164 (accessed December 2017).
6. Gould, J. 1869. *The Birds of Australia, Supplement*. London: published by the author.
7. Low, T. 2014. *Where Song Began: Australia's Birds and How They Changed the World*. Melbourne: Penguin Random House.
8. Edwards, S.V. and Boles, W.E. 2002. Out of Gondwana: the origin of passerine birds. *Trends in Ecology and Evolution*, **17**, 347–349.
9. Mayr, E. 1944. Timor and the colonization of Australia by birds. *Emu*, **44**, 113–130.
10. Michael Heads is a New Zealand botanist and anti-dispersalist and used this evidence as an argument for widespread vicariance; see Heads, M. 2013. *Biogeography of Australasia: a Molecular Analysis*. Cambridge: Cambridge University Press, p. 317.
11. Sibley, C.G. On the phylogeny and classification of living birds. www.scricciolo.com/classificazione/sequence5.htm (accessed December 2017); Sibley, C.G. and Ahlquist, J.E. 1990. *Phylogeny and Classification of Birds: a Study in Molecular Evolution*. New Haven, CT: Yale University Press.
12. Christidis, L. and Schodde, R. 1991. Relationships of the Australo-Papuan songbirds: protein evidence. *Ibis*, **133**, 277–285; Simpson, K. and Day, N. 1996. *A Field Guide to the Birds of Australia*, 5th edition. Melbourne: Penguin.
13. Low, T. 2014. *Where Song Began: Australia's Birds and How They Changed the World*. Melbourne: Penguin Random House, p. 68.

14. Kuhn, T.S. 1962. *The Structure of Scientific Revolutions*. Chicago, IL: University of Chicago Press.
15. Feduccia, A. and Olson, S.L. 1982. Morphological similarities between the Menurae and Rhinocryptidae, relict passerine birds of the southern hemisphere. *Smithsonian Contributions to Zoology*, **366**, 1–22.
16. Barker, F.K., Barrowclough, G.F. and Groth, J.G. 2002. A phylogenetic hypothesis for passerine birds: taxonomic and biogeographic implications of an analysis of nuclear dna sequence data. *Proceedings of the Royal Society B*, **269**, 295–308; Ericson, P.G.P., Christidis, L., Cooper, A. *et al.* 2002. A Gondwanan origin of passerine birds supported by DNA sequences of the endemic New Zealand wrens. *Proceedings of the Royal Society B*, **269**, 235–241. doi: 10.1098/rspb.2001.1877.
17. Chesser, R.T. and ten Have, J. 2007. On the phylogenetic position of the scrub-birds (Passeriformes: Menurae: Atrichornithidae) of Australia. *Journal of Ornithology*, **148**, 471–476.
18. Barker, K. 2011. Phylogeny and diversification of modern passerines. In G. Dyke and G. Keuf (eds), *Living Dinosaurs: the Evolutionary History of Modern Birds*. Oxford: Wiley-Blackwell, pp. 235–257.

Chapter 18: The Bowerbird's Story

1. Ericson, P.G.P., Klopfstein, S., Irestedt, M. *et al.* 2014. Dating the diversification of the major lineages of Passeriformes (Aves). *BMC Evolutionary Biology*, **14**, 8. doi: 10.1186/1471-2148-14-8.
2. Ames, P. L. 1987. The unusual syrinx morphology of Australian treecreepers *Climacteris*. *Emu*, **87**, 192–195.
3. Nguyen, J.M.J. 2015. Australo-Papuan treecreepers (Passeriformes: Climacteridae) and a new species of *Sittella* (Neosittidae: Daphoenositta) from the Miocene of Australia. *Palaeontological Electronica* 19.1.1A, 1–13.
4. Boles, W.E. 1995. A preliminary analysis of the Passeriformes from Riversleigh, northwestern Queensland, Australia, with a description of a new species of lyrebird. *Courier Forschungsinstitut Senckenberg*, **181**, 163–170.
5. Low, T. 2014. *Where Song Began: Australia's Birds and How They Changed the World*. Melbourne: Penguin Random House, p. 260.
6. Bock, W.J. 1963. Relationships between the birds of paradise and the bower birds. *The Condor*, **65**, 91–125.
7. Les Christidis, personal communication.
8. Baldauf, S.L. 2003. Phylogeny for the faint of heart: a tutorial. *Trends in Genetics*, **19**, 345–351.
9. Gilliard, E.T. 1969. *Birds of Paradise and Bower Birds*. London: Weidenfeld & Nicolson, p. xii.
10. Kusmierski, R., Borgia, G., Crozier, R.H. *et al.* 1993. Molecular information on bowerbird phylogeny and the evolution of exaggerated male characteristics. *Journal of Evolutionary Biology*, **6**, 737–752.
11. Irestedt, M., Batalha-Filho, H., Roselaar, C.S. *et al.* 2016. Contrasting phylogeographic signatures in two Australo-Papuan bowerbird species complexes (Aves: *Ailuroedus*). *Zoologica Scripta*, **45**, 365–379.
12. Diamond, J. 1986. Animal art: variation in bower decorating style among male bowerbirds *Ambylornis inornatus*. *Proceedings of the National Academy of Sciences of the United States of America*, **83**, 3042–3046.
13. Madden, J.R., Lowe, T.J., Fuller, H.V. *et al.* 2004. Local traditions of bower decoration by spotted bowerbirds in a single population. *Animal Behaviour*, **68**, 759–765.
14. Uy, J.A.C. and Borgia, G. 2000. Sexual selection drives rapid divergence in bowerbird display traits. *Evolution*, **54**, 273–278.

15. Uy, J.A.C. Say it with bowers. www.naturalhistorymag.com/htmlsite/0302/0302_feature.html (accessed December 2017).
16. Borgia, G. and Presgraves, D.C. 1998. Coevolution of elaborated male display traits in the spotted bowerbird: an experimental test of the threat reduction hypothesis. *Animal Behaviour*, **56**, 1121–1128.
17 Patricelli, G.L., Coleman, S.W. and Borgia, G. 2006. Male satin bowerbirds, *Ptilonorhynchus violaceus*, adjust their display intensity in response to female startling: an experiment with robotic females. *Animal Behaviour*, **71**, 49–59.
18. Borgia, G. 1995. Why do bowerbirds build bowers? *American Scientist*, **83**, 542–547.
19. Borgia, G. 1985. Bower quality, number of decorations and mating success of male satin bowerbirds (*Ptilonorhynchus violaceus*): an experimental analysis. *Animal Behaviour*, **33**, 266–271.
20. Borgia, G. and Keagy, J. 2015. Cognitively driven co-option and the evolution of complex sexual displays in bowerbirds. In Irschick, D.J., Briffa, M. and Podos, J. (eds), *Animal Signaling and Function: an Integrated Approach*. Hoboken, NJ: John Wiley & Sons, pp. 75–109.
21. Kelley, L.A. and Endler, J.A. 2012. Illusions promote mating success in great bowerbirds. *Science*, **335**, 335–338.
22. It should be noted that Gerald Borgia offers a more prosaic hypothesis for why decoration displays often show a stepwise pattern of size change. Smaller decorations near the bower allow the male to display near the bower entrance, while the larger decorations are less likely to impede male movements during courtship if they are placed further away. Furthermore, recent studies suggest that such male displays are influenced more by object availability than by individual 'artistic' skill. See Doerr, N.R. and Endler, J.A. 2015. Illusions vary because of the types of decorations at bowers, not male skill at arranging them, in great bowerbirds. *Animal Behaviour*, **99**, 73–82.
23. Coleman, S.W., Patricelli, G.L., Coyle, B. *et al.* 2007. Female preferences during the evolution of mimetic accuracy in male sexual displays. *Biology Letters*, **3**, 463–466.
24. Keagy, J., Savard, J-F. and Borgia, G. 2009. Male satin bowerbird problem-solving ability predicts mating success. *Animal Behaviour*, **78**, 809–817; Keagy, J., Savard, J-F., and Borgia, G. 2011. Cognitive ability and the evolution of multiple behavioural display traits. *Behavioral Ecology*, **23**, 448–456.
25. Dawkins, R. 1982. *The Extended Phenotype: the Long Reach of the Gene*. Oxford. Oxford University Press.
26. Soler, M., Martín-Vivaldi, M., Martin, J.M. *et al.* 1999. Weight lifting and health status in the black wheatear. *Behavioral Ecology*, **10**, 281–286.
27. Dawkins, R. 2015. *Brief Candle in the Dark: My Life in Science*. London: Bantam Press, p. 322.

Chapter 19: The Crow's Story

1. MacArthur, R.H. and Wilson, E.O. 1967. *The Theory of Island Biogeography*. Princeton, NJ: Princeton University Press.
2. Filardi, C.E., and Moyle, R.G. 2005. Single origin of a pan-Pacific bird group and upstream colonization of Australasia. *Nature*, **438**, 216–219.
3. Jønsson, K.A., Bowie, R.C.K., Moyle, R.G. *et al.* Historical biogeography of an Indo-Pacific passerine bird family (Pachycephalidae): different colonization patterns in the Indonesian and Melanesian archipelagos. *Journal of Biogeography*, **37**, 245–257.
4. Bellemain, E. and Ricklefs, R.E. 2008. Are islands the end of the colonization road? *Trends in Ecology and Evolution*, **23**, 461–468.
5. Jønsson, K.A. and Holt, B.G. 2015. Islands contribute disproportionately high amounts of evolutionary diversity in passerine birds. *Nature Communications*, **6**, 8538. doi: 10.1038/ncomms9538.

6. Jønsson, K.A., Fabre, P-H., Ricklefs, R.E. *et al.* 2011. Major global radiation of corvoid birds originated in the proto-Papuan archipelago. *Proceedings of the National Academy of Sciences of the United States of America*, **108**, 2328–2333.
7. Aggerbeck, M., Fjeldså, J., Christidis, L. *et al.* 2014. Resolving deep lineage divergence in core corvoid passerine birds supports a proto-Papuan island origin. *Molecular Phylogenetics and Evolution*, **70**, 272–285.
8. Moyle, R.G., Oliveros, C.H., Andersen, M.J. *et al.* 2016. Tectonic collision and uplift of Wallacea triggered the global songbird radiation. *Nature Communications*, **7**, 12709. doi: 10.1038/ncomms12709.
9. Jønsson, K.A., Fabre, P-H. and Irestedt, M. 2012. Brains, tools, innovation and biogeography in crows and ravens. *BMC Evolutionary Biology*, **12**, 72. doi: 10.1186/1471-2148-12-72.
10. Hunt, G.R. 1996. Manufacture and use of hook-tools by New Caledonian crows. *Nature*, **379**, 249–251.
11. Rutz, C., Klump, B.C., Komarczyk, L. *et al.* 2016. Discovery of species-wide tool use in the Hawaiian crow. *Nature*, **537**, 403–407.
12. Clayton, N.S. and Dickinson, A. 1998. Episodic-like memory during cache recovery by scrub jays. *Nature*, **395**, 272–274; Clayton, N.S., Bussey, T.J., Dickinson, A. 2003. Can animals recall the past and plan for the future? *Nature Reviews Neuroscience*, **4**, 685–691.
13. Watanabe, A., Grodzinski, U. and Clayton, N.S. 2014. Western scrub-jays allocate longer observation time to more valuable information. *Animal Cognition*, **17**, 859–867.
14. Veit, L. and Nieder, A. 2013. Abstract rule neurons in the endbrain support intelligent behaviour in corvid songbirds. *Nature Communications*, **4**, 2878. doi: 10.1038/ncomms3878.
15. Güntürkün, O. 2005. The avian 'prefrontal cortex' and cognition. *Current Opinion in Neurobiology*, **15**, 686–693.
16. Olkowicz, S., Kocourek, M., Lučan, R.K. *et al.* 2016. Birds have primate-like numbers of neurons in the forebrain. *Proceedings of the National Academy of Sciences of the United States of America*, **113**, 7255–7260.
17. Kapusta, A., Suh, A. and Feschotte, C. 2017. Dynamics of genome size evolution in birds and mammals. *Proceedings of the National Academy of Sciences of the United States of America*, **114**, E1460–E1469. doi/10.1073/pnas.1616702114.

Chapter 20: The Bird-of-Paradise's Story

1. Mužinić, J., Bogdan, J.F. and Beehler, B. 2009. Julije Klović: the first colour drawing of greater bird of paradise *Paradisaea apoda* in Europe and its model. *Journal of Ornithology*, **150**, 645–649.
2. Belon, P. 1555. *L'Histoire de la nature des oyseaux, avec leurs descriptions et naïfs portraits retirez du naturel.* Paris.
3. Frith, C. and Beehler, B. 1998. *The Birds of Paradise*. Oxford: Oxford University Press.
4. Irestedt, M., Jønsson, K.A., Fjeldså, J. *et al.* 2009. An unexpectedly long history of sexual selection in birds-of-paradise. B*MC Evolutionary Biology*, **9**, 235. doi: 10.1186/1471-2148-9-235.
5. For a detailed review of the complex subject of plate tectonics and their influence on the formation of New Guinea see Hall, R. 2002. Cenozoic geological and plate tectonic evolution of SE Asia and the SW Pacific: computer-based reconstructions, models and animations. *Journal of Asian Earth Sciences*, **20**, 253–431.
6. Laman, T. and Scholes, E. 2012. *Birds of Paradise: Revealing the World's Most Extraordinary Birds.* Washington, DC: National Geographic, p. 31.
7. Darwin, C. 1871. *The Descent of Man, and Selection in Relation to Sex*. London: John Murray.
8. Fisher, R.A. 1930. *The Genetical Theory of Natural Selection*. Oxford: Clarendon Press.
9. Dawkins, R. 2006. *The Blind Watchmaker*. London: Longmans, Penguin. p. 215. For an in-depth but very readable account of the various theories of sexual selection see Cronin,

H. 1991. *The Ant and the Peacock*. Cambridge: Cambridge University Press.
10. Prum, R.O. 2017. *The Evolution of Beauty: How Darwin's Forgotten Theory of Mate Choice Shapes the Animal World – and Us*. New York: Doubleday.
11. Zahavi, A. 1975. Mate selection: a selection for a handicap. *Journal of Theoretical Biology*, **53**, 205–214.
12. Andersson, M. 1982. Female choice selects for extreme tail length in a widowbird. *Nature*, **299**, 818–820.
13. Grafen, A. 1990. Sexual selection unhandicapped by the Fisher process. *Journal of Theoretical Biology*, **144**, 473–516.
14. Frith, C.B. and Frith, D.W. 1997. Courtship and mating of the King of Saxony bird of paradise *Pteridophora alberti* in New Guinea with comment on their taxonomic significance. *Emu*, **97**, 185–193.
15. Taken from film commentary to Attenborough's *Paradise Birds*.

Chapter 21: The Starling's Story

1. Jønsson, K.A., Fjeldså, J., Ericson, P.G.P. *et al.* 2007. Systematic placement of an enigmatic southeast Asian taxon *Eupetes macrocerus* and implications for the biogeography of a main songbird radiation, the Passerida. Biology Letters, 3, 323–326.
2. Johansson, U.S., Fjeldså, J. and Bowie, R.C.K. 2008. Phylogenetic relationships with Passerida (Aves: Passeriformes): a review and a new molecular phylogeny based on three nuclear intron markers. *Molecular Phylogenetics and Evolution*, **48**, 858–876.
3. There are two types of polytomy, a term meaning many temporal-based branches in phylogenetics. A 'soft' polytomy exists when scientists cannot be certain of the resolution, while a 'hard' polytomy implies that simultaneous divergences have occurred and that all the daughter lineages are equally closely related to one another.
4. Barker, F.K., Cibois, A., Schikler, P. *et al.* 2004. Phylogeny and diversification of the largest avian radiation. *Proceedings of the National Academy of Sciences of the United States of America*, **101**, 11040–11045.
5. The first of the two dispersal hypotheses builds on the present geographical proximity between Australasia and Asia, whereas the second hypothesis is an inference from the observations that several basal Passerida lineages are represented in Africa. For further discussion see Fuchs, J., Fjeldså, J., Bowie, R.C. *et al.* 2006. The African warbler genus *Hyliota* as a lost lineage in the oscine songbird tree: molecular support for an African origin of Passerida. *Molecular Phylogenetics and Evolution*, **39**, 186–197.
6. Zuccon, D. Cibois, A., Pasquet, E. *et al.* 2006. Nuclear and mitochondrial sequence data reveal the major lineages of starlings, mynas and related taxa. *Molecular Phylogenetics and Evolution*, **41**, 333–344.
7. Charles Darwin noted that the Galápagos mockingbirds from the various islands were morphologically different. He correctly concluded that they all probably derived from related species on the South American mainland. Interestingly, Darwin carefully tagged his mockingbird skins, collected from four different islands, in contrast to the desultory manner in which he labelled his famous finch specimens. It was the mockingbirds, or 'mocking-thrushes' as Darwin called them, that led him to question the 'stability of species' and gave him the idea of evolution by natural selection.
8. Curry, R.L. and Anderson, D.J. 1987. Interisland variation in blood drinking by Galápagos mockingbirds. *The Auk*, **104**, 517–521.
9. It has been stressed that because the relationships among the Eurasian and various African clades are not well resolved, this 'out-of-Africa' hypothesis for the Eurasian group is equally parsimonious with a number of alternative scenarios. See Lovette, I.J. and Rubenstein, D.R. 2007. A comprehensive molecular phylogeny of the starlings (Aves: Sturnidae) and mockingbirds (Aves: Mimidae): congruent mtDNA and nuclear trees for

a cosmopolitan avian radiation. *Molecular Phylogenetics and Evolution*, **44**, 1031–1056.
10. Zuccon, D., Pasquet, E. and Ericson, P.G.P. 2008. Phylogenetic relationships among Palearctic-Oriental starlings and mynas (genera *Sturnus* and Acridotheres: Sturnidae). *Zoologica Scripta*, **37**, 469–481.
11. Lovette, I.J., McCleery, B.V., Talaba, A.L. *et al.* 2008. A complete species-level molecular phylogeny for the 'Eurasian' starlings (Sturnidae: *Sturnus*, *Acridotheres*, and allies): recent diversification in a high social and dispersive avian group. *Molecular Phylogenetics and Evolution*, **47**, 251–260.
12. Unnatural mechanisms can overcome this constraint. In 2004, after 13 years of collaborative research, an Australian company, Florigene, and a Japanese company, Suntory, created a blue rose by genetically expressing the blue pigment delphinidin in white roses. See Stoddard, M.C. and Prum, R.O. 2011. How colorful are birds? Evolution of the avian plumage color gamut. *Behavioral Ecology*, **22**, 1042–1052. doi: 10.1093/beheco/arr088.
13. Pickrell, J. 2014. *Flying Dinosaurs: How Fearsome Reptiles Became Birds*. Sydney: NewSouth Publishing, p. 136.
14. Prum, R.O., Dufresne, E.R., Quinn, T. *et al.* 2009. Development of colour-producing β-keratin nanostructures in avian feather barbs. *Journal of the Royal Society. Interface* **6**, S253–S265.
15. Stavenga, D. G., Leertouwer, H. L., Marshall, N. J. *et al.* 2010. Dramatic colour changes in a bird of paradise caused by uniquely structured breast feather barbules. *Proceedings of the Royal Society B*, **278**, 2098–2104. doi: 10.1098/rspb.2010.2293.
16. Maia, R., Rubenstein, D.R. and Shawkey, M.D. 2013. Key ornamental innovations facilitate diversification in an avian radiation. *Proceedings of the National Academy of Sciences of the United States of America*, **110**, 10687–10692.
17. For further discussion see Eliason, C.M., Bitton, P-B. and Shawkey, M.D. 2013. How hollow melanosomes affect iridescent colour production in birds. *Proceedings of the Royal Society B*, **280**, 20131505. doi: 10.1098/rspb.2013.1505.
18. Yong, E. 2013. On the origin of really shiny species. http://phenomena.nationalgeographic.com/2013/06/10/on-the-origin-of-really-shiny-species (accessed December 2017).
19. Anderson, N. 2013. Study sheds light on evolution of plumage coloration in African starlings. http://www.sci-news.com/biology/article01148-african-starlings.html (accessed December 2017).

Chapter 22: The Thrush's Story

1. Ames, P. L. 1971. The morphology of the syrinx in passerine birds. *Bulletin of the Peabody Museum of Natural History, Yale University*, **37**, 1–194.
2. Voelker, G., Rohwer, S., Bowie, R.C.K. *et al.* 2007. Molecular systematics of a speciose, cosmopolitan songbird genus: defining the limits of, and relationships among, the *Turdus* thrushes. *Molecular Phylogenetics and Evolution*, **42**, 422–434.
3. Voelker, G., Rohwer, S., Outlaw, D.C. *et al.* 2009. Repeated trans-Atlantic dispersal catalysed a global songbird radiation. *Global Ecology and Biogeography*, **18**, 41–49.
4. George Gaylord Simpson (1902–1984) was Professor of Zoology at Columbia University. He was one of the most influential palaeontologists of the twentieth century and a major architect of the modern evolutionary synthesis. His major work was entitled *Tempo and Mode in Evolution* (1944). However, he stubbornly remained opposed to Wegener's theory of continental drift.
5. Humphries, C.J. and Parenti, L.R. 1999. *Cladistic Biogeography*, 2nd edition. Oxford: Oxford University Press.
6. Salomonsen, F. 1950. The immigration and breeding of the fieldfare (*Turdus pilaris* L.)

in Greenland. *Proceedings of the International Ornithological Congress*, **10**, 515–526.
7. Simpson, G.G. 1983. *Splendid Isolation: the Curious History of South American Mammals*, new edition. New Haven, CT: Yale University Press.
8. Wier, J.T., Bermingham, E. and Schluter, D. 2009. The great American biotic interchange in birds. *Proceedings of the National Academy of Sciences of the United States of America*, **106**, 21737–21742.
9. Haug, G.H., Tiedemann, R., Zahn., R. *et al.* 2001. Role of Panama uplift on oceanic freshwater balance. *Geology*, **29**, 207–210.
10. Garcia-Moreno, J. 2004. Is there a universal mtDNA clock for birds? *Journal of Avian Biology*, **35**, 465–468.

Chapter 23: The Sparrow's Story

1. Ericson, P.G.P., Klopfstein, S., Irestedt, M. *et al.* 2014. Dating the diversification of the major lineages of Passeriformes (Aves). *BMC Evolutionary Biology*, **14**, 8. doi: 10.1186/1471-2148-14-8; Barker, F.K., Cibois, A., Schikler, P. *et al.* 2004. Phylogeny and diversification of the largest avian radiation. *Proceedings of the National Academy of Sciences of the United States of America*, **101**, 11040–11045.
2. Summers-Smith, J.D. 1988. *The Sparrows*. London: T. & A.D. Poyser. pp. 276–296.
3. Tchernov, E. 1962. Paleolithic avifauna in Palestine. *Bulletin of the Research Council of Israel, Section B, Zoology*, **11**, 95–131.
4. Harari, Y.N. 2014. *Sapiens: a Brief History of Humankind*. London: Harvill Secker, pp. 84–85.
5. Riyahi, S., Hammer, Ø., Arbabi, T. *et al.* 2013. Beak and skull shapes of human commensal and non-commensal house sparrows *Passer domesticus*. *BMC Evolutionary Biology*, **13**, 200. doi: 10.1186/1476-2148-13-200.
6. Sætre, G.P., Riyahi, S., Aliabadian, M *et al.* 2012. Single origin of human commensalism in the house sparrow. *Journal of Evolutionary Biology*, **25**, 788–796.
7. Meise, W. 1936. Zur systematik und Verbreitungsgeschichte der Haus- und Weidensperlinge, *Passer domesticus* (L.) und *hispaniolensis* (T.). *Journal of Ornithology*, **84**, 631–672.
8. Fisher, R.A. 1930. *The Genetical Theory of Natural Selection*. Oxford: Clarendon Press, p. 130.
9. Weiner, J. 1995. *The Beak of the Finch: a Story of Evolution in Our Time*. New York: Vintage, p. 199.
10. For a review of the role of hybridisation in plant evolution, see Mallet, J. 2007. Hybrid speciation. *Nature*, **446**, 279–283. doi: 10.1038/nature05706.
11. Mavárez, J. and Linares, M. 2008. Homoploid speciation in animals. *Molecular Ecology*, **17**, 4181–4185.
12. Mayr, E. 1963. *Animal Species and Evolution*. Cambridge, MA: Belknap Press, p. 114.
13. Meise, W. (1975). Natürliche Bastardpopulationen und Speziationsprobleme bei Vögeln. *Abhandlungen des Naturwissenschaftlichen Vereins in Hamburg*, **18/19**, 187–254.
14. Grant, P.R. and Grant, B.R. 1992. Hybridization of bird species. *Science*, **256**, 193–197.
15. Mallet, J. 2005. Hybridization as an invasion of the genome. *Trends in Ecology and Evolution*, **20**, 229–237.
16. Price, T. 2008. *Speciation in Birds*. Greenwood Village, CO: Roberts & Company, p. 304.
17. Price, T. 2008. *Speciation in Birds*. Greenwood Village, CO: Roberts & Company, p. 390.
18. For a further discussion of hybrid breakdown, see Lane, N. 2015. *The Vital Question: Why is Life the Way It Is?* London: Profile Books, pp. 237–279.
19. The sex chromosomes in birds are designated Z and W. Males possess two copies of chromosome Z (ZZ) while females have one of each (ZW). This contrasts with the mammalian arrangement (males XY, females XX), so that in birds the ovum determines

the sex, rather than the sperm. For variations in fitness determined by sex chromosomes (in fruit flies), see Gibson, J.R., Chippindale, A.K. and Rice, W.R. 2002. The X chromosome is a hot spot for sexually antagonistic fitness variation. *Proceedings of the Royal Society B*, **269**, 499–505. doi: 10.1098/rspb.2001.1863.

20. For further discussion on genetic introgression and references to its role in human evolution, see Rheindt, F.E. and Edwards, S.V. 2011. Genetic introgression: an integral but neglected component of speciation in birds. *The Auk*, **128**, 620–632.
21. Rieseberg, L.H. 1997. Hybrid origins of plant species. *Annual Reviews of Ecology and Systematics*, **28**, 359–389.
22. Hermansen, J.S., Sæther, S.A., Elgvin, T.O. *et al.* 2011. Hybrid speciation in sparrows 1: phenotypic intermediacy, genetic admixture and barriers to gene flow. *Journal of Ecology*, **20**, 3812–3822.
23. Zeder, M.A. 2008. Domestication and early agriculture in the Mediterranean basin: origins, diffusion, and impact. *Proceedings of the National Academy of Sciences of the United States of America*, **105**, 11597–11604.
24. Brelsford, A. 2011. Hybrid speciation in birds: allopatry more important than ecology? *Molecular Ecology*, **20**, 3705–3707.
25. Trier, C.N., Hermansen, J.S., Sætre, G-P. 2014. Evidence for mito-nuclear and sex-linked reproductive barriers between the hybrid Italian sparrow and its parent species. *PLoS Genetics*, **10** (1), e1004075. doi: 10.1371/journal.pgen.1004075.
26. In most multicellular organisms, mitochondrial DNA (mtDNA) is inherited from the female and not the male. Explanations include dilution (a healthy sperm contains only a fraction of the mtDNA possessed by an ovum) and the fact that the ovum targets and destroys male mtDNA.
27. Brelsford, A., Milá, B. and Irwin, D.E. 2011. Hybrid origin of Audubon's warbler. *Molecular Ecology*, **20**, 2380–2389; Toews, D.P.L., Brelsford, A., Grossen, C. *et al.* 2016. Genomic variation across the yellow-rumped warbler species complex. *The Auk*, **133**, 698–717. doi: 10.1642/auk-16-61.
28. Hermansen, J.S., Haas, F. Trier, C.N. *et al.* 2014. Hybrid speciation through sorting of parental incompatibilities in Italian sparrows. *Molecular Ecology*, **23**, 5831–5842.

Chapter 24: The Zebra Finch's Story

1. Arnaiz-Villena, A., Ruiz-del-Valle, V., Gomez-Prieto, P. *et al.* 2009. Estrildinae finches (Aves, Passeriformes) from Africa, South Asia and Australia: a molecular phylogeographic study. *Open Ornithological Journal*, **2**, 29–36.
2. Forshaw, J.M. and Shephard, M. 2012. *Grassfinches in Australia*. Collingwood: CSIRO Publishing, pp. v–vi.
3. Cocker, M. 2005. *Birds Britannica*. London: Chatto and Windus, p. 341. Original data from *Birds of the Western Palearctic*, edited S. Cramp, volume 5, p. 634.
4. Cocker, M. 2014. *Claxton: Field Notes From a Small Planet*. London: Jonathan Cape, p. 50.
5. Dowsett-Lemaire, F. 1979. The imitative range of the song of the marsh warbler *Acrocephalus palustris*, with special reference to imitations of African birds. *Ibis*, **121**, 453–468.
6. Coleman, S.W., Patricelli, G.L., Coyle, B. *et al.* 2007. Female preferences drive the evolution of mimetic accuracy in male sexual displays. *Biology Letters*, **3**, 463–466.
7. Boogert, N.J., Giraldeau, L-A. and Lefebvre, L. 2008. Song complexity correlates with learning ability in zebra finches. *Animal Behaviour*, **76**, 1735–1741.
8. Laje, R., Sciamarella, D., Zanella, J. *et al.* 2008. Bilateral source acoustic interaction in a syrinx model of an oscine bird. *Physical Review E*, **77**, 011912. doi: 10.1103/PhysRevE.77.011912.
9. Reide, T. and Goller, F. 2013. Morphological basis for the evolution of acoustic diversity in oscine songbirds. *Proceedings of the Royal Society B*, **281**, 20132306. doi: 10.1098/rspb.2013.2306.

10. Brenowitz, E.A. 1997. Comparative approaches to the avian song system. *Journal of Neurobiology*, **33**, 517–531.
11. Liu, W-C., Wada, K., Jarvis, E. *et al.* 2013. Rudimentary substrates for vocal learning in a suboscine. *Nature Communications*, **4**, 2082. doi: 10.1038/ncomms3082.
12. Dittrich, F., Maat, A.T., Jansen, R.F. *et al.* 2013. Maximised song learning of juvenile male zebra finches following BDNF expression in the HVC. *European Journal of Neuroscience*, **38**, 3338–3344.
13. Maat, A.T., Trost, L., Sagunsky, H. *et al.* 2014. Zebra Finch mates use their forebrain song system in unlearned call communication. *PLoS ONE*, **9** (10), e109334. doi: 10.1371/journal.pone.0109334.
14. Chakraborty, M., Walløe, S., Nedergaard, S. *et al.* 2015. Core and shell song systems unique to the parrot brain. *PLoS ONE*, **10** (6), e0118496. doi: 10.1371/journal.pone.0118496.
15. Feenders, G., Liedvogel, M., Rivas, M. *et al.* 2008. Molecular mapping of movement-associated areas in the avian brain: a motor theory for vocal learning origins. *PLoS ONE*, **3** (3), e1768. doi: 10.1371/journal.pone.0001768.
16. Liu, W-C., Wada, K., Jarvis, E. *et al.* 2013. Rudimentary substrates for vocal learning in a suboscine. *Nature Communications*, **4**, 2082. doi: 10.1038/ncomms3082; Kroodsma, D., Hamilton, D., Sánchez, J.E. *et al.* 2013. Behavioral evidence for song learning in the suboscine bellbirds (*Procnias* spp.; Cotingidae). *Wilson Journal of Ornithology*, 125, 1–14.
17. Lai, C.S.L., Fisher, S.E., Hurst, J.A. *et al.* 2001. A Forkhead-domain gene is mutated in a severe speech and language disorder. *Nature*, **413**, 519–523.
18. Wang, R., Chen, C-C., Hara, E. *et al.* 2014. Convergent differential regulation of slit-robo axon guidance genes in the brain of vocal learners. *Journal of Comparative Neurology*, 523, 892–906.
19. Pfenning, A.R., Hara, E., Whitney, O. *et al.* 2014. Convergent transcriptional specializations in the brains of humans and song-learning birds. *Science*, **346** (6215), 1256846. doi: 10.1126/science.1256846.
20. Nottebohm, F. 2005. The neural basis of birdsong. *PLoS Biology*, **3** (5), e164. doi: 10.1371/journal.pbio.0030164.

Chapter 25: The White-eye's Story

1. Chandler, D. and Couzens, D. 2011. *100 Birds to See in Your Lifetime*. London: Carlton Books.
2. Mayr, E. and Diamond, J. 2001. *The Birds of Melanesia: Speciation, Ecology, and Biogeography*. Oxford: Oxford University Press, p. vii.
3. Mayr, E. 1942. *Systematics and the Origin of Species*. New York: Columbia University Press.
4. Diamond, J. 1974. Colonization of exploded volcanic islands by birds: the supertramp strategy. *Science*, new series, **184**, 803–806.
5. Diamond, J, Gilpin, M.E. and Mayr, E. 1976. Species-distance relation for birds of the Solomon archipelago, and the paradox of the great speciators. *Proceedings of the National Academy of Sciences of the United States of America*, **73**, 2160–2164.
6. Moyle, R.G., Filardi, C.E., Smith, C.E. *et al.* 2009. Explosive Pleistocene diversification and hemispheric expansion of a 'great speciator'. *Proceedings of the National Academy of Sciences of the United States of America*, **106**, 1863–1868.
7. GrrlScientist. 2009. Meet the great speciators: the white-eyes. *The Guardian*, 27 January 2009. www.theguardian.com/science/punctuated-equilibrium/2009/jan/27/evolution-birds (accessed December 2017).
8. Diamond, J. 1998. Geographic variation in vocalisations of the white-eye superspecies *Zosterops* [*griseotinctus*] in the New Georgia group. *Emu*, **98**, 70–74.

9. Hooker, J.D. 1849. On the vegetation of the Galapagos Archipelago. *Transactions of the Linnean Society of London*, **20**, 235–262.
10. A detailed account is given in Carlquist, S. 1965. *Island Life: a Natural History of the Islands of the World*. Garden City, NY: Natural History Press, pp. 215–246.
11. For dandelions on the islands of British Columbia, see Cody, M. L. 2006. *Plants on Islands: Diversity and Dynamics on a Continental Archipelago*. Berkeley, CA: University of California Press. For flowering plants in pavements, see Cheptou, P-O., Carrue, O., Rouifed, S. *et al.* 2008. Rapid evolution of seed dispersal in an urban environment in the weed *Crepis sancta*. *Proceedings of the National Academy of Sciences of the United States of America*, **105**, 3796–3799.
12. Diamond, J. 1981. Flightlessness and fear of flying in island species. *Nature*, **293**, 507–508.
13. Komdeur, J., Piersma, T., Kraaijeveld, K. *et al.* 2004. Why Seychelles warblers fail to recolonize nearby islands: willing or unable to fly there? *Ibis*, **146**, 298–302.

Chapter 26: The Crossbill's Story

1. Dawkins, R. 1986. *The Blind Watchmaker*. London: Longman.
2. Johnson, P.E. 1991. *Darwin on Trial*. Washington, DC: Regnery Gateway Publishing Co.
3. Benkman, C.W. and Lindholm, A.K. 1991. The advantages and evolution of a morphological novelty. *Nature*, **349**, 519–520.
4. Arnaiz-Villena, A., Guillén, J., Ruiz-del-Valle, V. *et al.* 2001. Phylogeography of crossbills, bullfinches, grosbeaks, and rosefinches. *Cellular and Molecular Life Sciences*, **58**, 1159–1166.
5. Marthinsen, G., Wennerberg, L. and Lifjeld, J.T. 2008. Low support for separate species within the redpoll complex (*Carduelis flammea–hornemanni–cabaret*) from analyses of mtDNA and microsatellite markers. *Molecular Phylogenetics and Evolution*, **47**, 1005–1017.
6. Carl Georg Bergmann (1814–1865) was a German anatomist, physiologist and biologist who emphasised the relationship between latitude and body size in 1847. The rule is most often applied to mammals and birds and has not been proven for plants. Gloger's rule, named after the German ornithologist Constantin Wilhelm Gloger (1803–1863), applies not only to many bird species but also to some mammals, including humans, and plants.
7. Newton, I. 1972. *Finches*. New Naturalist Library. London: Collins.
8. Krebs, J.R. 1991. The case of the curious bill. *Nature*, **349**, 465.
9. Smith, J.W., Sjoberg, S.M., Mueller, M.C. *et al.* 2012. Assortative flocking in crossbills and implications for ecological speciation. *Proceedings of the Royal Society B*, **279**, 4223–4229. doi: 10.1098/rspb.2012.1500.
10. Parchman, T.L., Benkman, C.W. and Britch, S.C. 2006. Patterns of genetic variation in the adaptive radiation of New World crossbills (Aves: *Loxia*). *Molecular Ecology*, **15**, 1873–1887.
11. Gill, F. and Donsker, D. (eds). 2017. *IOC World Bird List* (v 7.3). doi: 10.14344/IOC.ML.7.3.
12. Parchman, T.L., Buerkle, C.A., Soria-Carrasco, V. *et al.* 2016. Genome divergence and diversification within a geographic mosaic of coevolution. *Molecular Ecology*, **25**, 5705–5718.
13. Benkman, C.W. and Smith, J.W. 2007. A coevolutionary arms race causes ecological speciation in crossbills. *American Naturalist*, **169**, 455–465.
14. New bird species found in Idaho. https://phys.org/news/2007-03-bird-species-idaho.html (accessed December 2017).

15. Parchman, T.L., Benkman, C.W. and Mezquida, E.T. 2007. Coevolution between hispaniolan crossbills and pine: does more time allow for greater phenotypic escalation at lower altitudes? *Evolution*, **61**, 2142–2153.
16. Benkman, C.W. 2016. The natural history of the South Hills Crossbill in relation to its impending extinction. *The American Naturalist*, **188**, 589–601.

Chapter 27: The Tanager's Story

1. Fjeldså, J. 2013. The global diversification of songbirds (oscines) and the build-up of the Sino-Himalayan diversity hotspot. *Chinese Birds*, **4**, 132–143.
2. Hall, K.S.S. 2005. Do nine-primaried passerines have nine or ten primary feathers? The evolution of a concept. *Journal of Ornithology*, **146**, 121–126.
3. Barker, F.K., Burns, K.J., Klicka, J. *et al.* 2015. New insights into New World biogeography: an integrated view from the phylogeny of blackbirds, cardinals, sparrows, tanagers, warblers, and allies. *The Auk*, **132**, 333–348.
4. Weir, J.T., Bermingham, E. and Schluter, D. 2009. The great American biotic interchange in birds. *Proceedings of the National Academy of Sciences of the United States of America*, **106**, 21737–21742.
5. Ricklefs, R.E. 2002. Splendid isolation: historical ecology of the South American passerine fauna. *Journal of Avian Biology*, **33**, 207–211.
6. Swanson, D.L. and Bozinovic, F. 2011. Metabolic capacity and the evolution of biogeographic patterns in oscine and suboscine passerine birds. *Physiological and Biological Zoology*, **84**, 185–194.
7. Burns, K.J., Shultz, A.J., Title, P.O. *et al.* 2014. Phylogenetics and diversification of tanagers (Passeriformes: Thraupidae), the largest radiation of Neotropical songbirds. *Molecular Phylogenetics and Evolution*, **75**, 41–77.
8. Hilty, S.L., Parker, T.A. and Silliman, J. 1979. Observations on plush-capped finches in the Andes with a description of the juvenile and immature plumages. *Wilson Bulletin*, **91**, 145–148.
9. Chaves, J.A., Hidalgo, J.R. and Klicka, J. 2013. Biogeography and evolutionary history of the Neotropical genus *Saltator* (Aves: Thraupini). *Journal of Biogeography*, **40**, 2180–2190.
10. Campagna, L., Benites, P., Lougheed, S.C. *et al.* 2005. Rapid phenotypic evolution during incipient speciation in a continental avian radiation. *Proceedings of the Royal Society B*, **279**, 1847–1856.
11. Benites, P., Eaton, M.D., Lijtmaer, D.A. *et al.* 2010. Analysis from avian visual perspective reveals plumage colour differences among females of capuchino seedeaters (*Sporophila*). *Journal of Avian Biology*, **41**, 597–602.
12. Campagna, L., Repenning, M., Silveira, L.F. *et al.* 2017. Repeated divergent selection on pigmentation genes in a rapid finch radiation. *Science Advances*, **3** (5), e1602404. doi: 10.1126/sciadv.1602404.
13. Uy, J.A.C., Cooper, E.A., Cutie, S. *et al.* 2016. Mutations in different pigmentation genes are associated with parallel melanism in island flycatchers. *Proceedings of the Royal Society B*, **283**, 20160731. doi: 10:1098/rspb:2016.0731.
14. Vijay, N., Bossu, C.M., Poelstra, J.W. *et al.* 2016. Evolution of heterogeneous genome differentiation across multiple contact zones in a crow species complex. *Nature Communications*, **7**, 13195. doi: 10.1038/ncomms13195; Toews, D.P.L., Taylor, S.A., Vallender, R. *et al.* 2016. Plumage genes and little else distinguish the genomes of hybridising warblers. *Current Biology*, **26**, 2313–2318.
15. Fjeldså, J. and Rahbek, C. 2006. Diversification of tanagers, a species rich bird group, from lowlands to montane regions of South America. *Integrative Comparative Biology*, **46**, 72–81.
16. Dawkins, R. 2004. *The Ancestor's Tale: a Pilgrimage to the Dawn of Life.* London: Weidenfeld & Nicolson, pp. 220–222.

17. Lamichhaney, S., Berglund, J., Almén, M.S. *et al.* 2015. Evolution of Darwin's finches and their beaks revealed by genome sequencing. *Nature*, **518**, 371–375.
18. Abzhanov, A., Protas, M., Grant, R.B. *et al.* 2004. Bmp4 and morphological variation of beaks in Darwin's finches. *Science*, **305**, 1462–1465.
19. Lamichhaney, S., Han, F., Berglund, J. *et al.* 2016. A beak size locus in Darwin's finches facilitated character displacement during a drought. *Science*, **352**, 470–474.
20. Skinner, M.K., Gurerrero-Bosagna, C., Haque, M.M. *et al.* 2014. Epigenetics and the evolution of Darwin's finches. *Genome Biology and Evolution*, **6**, 1972–1989.
21. Schondube, J.E. and Martinez del Rio, C. 2003. The flowerpiercers' hook: an experimental test of an evolutionary trade-off. *Proceedings of the Royal Society B*, **270**, 195–198.
22. Mauck, W.M. and Burns, K.J. 2009. Phylogeny, biogeography, and recurrent evolution of divergent bill types in the nectar-stealing flowerpiercers (Thraupini: *Diglossa* and *Diglossopsis*). *Biological Journal of the Linnean Society*, **98**. 14–24.
23. Doyle, A.C. 1912. *The Lost World*. London: Hodder & Stoughton.

Postscript: The Sixth Extinction

1. Avery, M. 2014. *A Message from Martha: the Extinction of the Passenger Pigeon and its Relevance Today*. London: Bloomsbury.
2. Goodall, J. 2010. *Hope for Animals and Their Worlds: How Endangered Species are Being Rescued From the Brink*. London: Icon Books.
3. von Euler, F. 2001. Selective extinction and rapid loss of evolutionary history in the bird fauna. *Proceedings of the Royal Society B*, **268**, 127–130; Lockwood, J.L., Brooks, T.M. and McKinney, M.L. 2000. Taxonomic homogenization of the global avifauna. *Animal Conservation*, **3**, 27–35.
4. Plumber, B. 2014. There have been five mass extinctions in Earth's history. Now we're facing a sixth. *Washington Post*, 11 February 2014. www.washingtonpost.com/news/wonk/wp/2014/02/11/there-have-been-five-mass-extinctions-in-earths-history-now-were-facing-a-sixth (accessed December 2017). See also Kolbert, E. 2014. *The Sixth Extinction: an Unnatural History*. London: Bloomsbury.
5. De Vos, J.M., Joppa, L.N., Gittleman, J.L. *et al.* 2015. Estimating the normal background rate of species extinction. *Conservation Biology*, **29**, 452–462.
6. Mora, C., Tittensor, D.P., Adl, S. *et al.* 2011. How many species are there on Earth and in the ocean? *PLoS Biology*, **9** (8), e1001127. doi: 10.1371/journal.pbio.1001127.

Bibliography

Al-Khalili, J. and McFadden, J. 2014. *Life on the Edge: the Coming Age of Quantum Biology*. London: Bantam Press.

Attenborough, D. and Fuller, E. 2012. *Drawn from Paradise: The Discovery, Art and Natural History of the Birds of Paradise*. London: HarperCollins.

Avery, M. 2014. *A Message from Martha: the Extinction of the Passenger Pigeon and its Relevance Today*. London: Bloomsbury.

Beebe, W. 1918. *Jungle Peace*. New York: Henry Holt & Company.

Belon, P. 1555. *L'Histoire de la nature des oyseaux, avec leurs descriptions et naïfs portraits retirez du naturel.* Paris.

Birkhead, T., Wimpenny, J. and Montgomerie, B. 2014. *Ten Thousand Birds: Ornithology Since Darwin*. Princeton, NJ: Princeton University Press.

Brooke, M. 2004. *Albatrosses and Petrels Across the World*. Oxford: Oxford University Press.

Burton, J.A. (ed.). 1992. *Owls of the World: Their Evolution, Structure and Ecology*, 3rd edition. London: Eurobooks.

Carlquist, S. 1965. *Island Life: a Natural History of the Islands of the World*. Garden City, NY: Natural History Press.

Chatterjee, S. 2015. *The Rise of Birds: 225 Million Years of Evolution.* Baltimore, MD: Johns Hopkins University Press.

Cherry-Garrard, A. 1922. *The Worst Journey in the World*. London: Constable.

Christidis, L. and Boles, W. 2008. *Systematics and Taxonomy of Australian Birds*. Melbourne: CSIRO Publishing.

Cocker, M. 2005. *Birds Britannica*. London: Chatto and Windus.

Cocker, M and Tipling, D. 2013. *Birds and People*. London: Jonathan Cape.

Cody, M. L. 2006. *Plants on Islands: Diversity and Dynamics on a Continental Archipelago*. Berkeley, CA: University of California Press.

Colinvaux, P. 2008. *Amazon Expeditions: My Quest for the Ice-Age Equator*. New Haven, CT: Yale University Press.

Conway Morris, S. 2003. *Life's Solution: Inevitable Humans in a Lonely Universe*. Cambridge: Cambridge University Press.

Conway Morris, S. 2014. *The Runes of Evolution: How the Universe Became Self-Aware*. West Conshohocken: Templeton Press.

Coyne, J.A and Orr, H.A. 2004. *Speciation*. Sunderland, MA: Sinauer Associates.

Cronin, H. 1991. *The Ant and the Peacock*. Cambridge: Cambridge University Press.

Darwin, C. 1859. *On the Origin of Species*. London: John Murray.

Darwin, C. 1871. *The Descent of Man, and Selection in Relation to Sex*. London: John Murray.

Dawkins, R. 1982. *The Extended Phenotype: the Long Reach of the Gene*. Oxford: Oxford University Press.

Dawkins, R. 1986. *The Blind Watchmaker*. London: Longman.

Dawkins, R. 2004. *The Ancestor's Tale: a Pilgrimage to the Dawn of Life*. London: Weidenfeld & Nicolson.

Dawkins, R. 2015. *Brief Candle in the Dark: My Life in Science*. London: Bantam Press.

de Queiroz, A. 2014. *The Monkey's Voyage: How Improbable Journeys shaped the History of the World.* New York: Basic Books.

Desmond, A. 1994. *Huxley: the Devil's Disciple.* London: Michael Joseph.

Desmond, A. and Moore, A. 1991. *Darwin.* London: Michael Joseph.

Donald, P.F., Collar, N.J., Marsden, S.J. *et al.* 2010. *Facing Extinction: the World's Rarest Birds, and the Race to Save Them.* London: T. & A.D. Poyser.

Doyle, A.C. 1912. *The Lost World.* London: Hodder & Stoughton.

Dyke, G. and Kaiser, G. (eds). 2011. *Living Dinosaurs: the Evolutionary History of Modern Birds.* Oxford: Wiley-Blackwell.

Feduccia, A. 1996. *The Origin and Evolution of Birds.* New Haven, CT, and London: Yale University Press.

Fisher, R.A. 1930. *The Genetical Theory of Natural Selection.* Oxford: Clarendon Press.

Forshaw, J.M. and Shephard, M. 2012. *Grassfinches in Australia.* Collingwood: CSIRO Publishing.

Frith, C.B. and Frith, D.W. 2004. *The Bowerbirds.* Oxford: Oxford University Press.

Fraas, A.P. 1989. *Heat Exchanger Design*, 2nd edition. New York: John Wiley & Sons.

Gavrilets, S. 2004. *Fitness, Landscapes and the Origin of Species.* Princeton, NJ: Princeton University Press.

Gilliard, E.T. 1969. *Birds of Paradise and Bower Birds.* London: Weidenfeld & Nicolson.

Goodall, J. 2010. *Hope for Animals and their Worlds: How Endangered Species are Being Rescued From the Brink.* London: Icon Books.

Gooddie, C. 2010. *The Jewel Hunter.* Old Basing: WILDguides Ltd.

Gould, J. 1840–48. *The Birds of Australia.* London: published by the author.

Gould, S.J. 1990. *Wonderful Life: the Burgess Shale and the Nature of History.* London: Hutchinson Radius.

Grant, P.R. and Grant, B.R. 2008. *How and Why Species Multiply: the Radiation of Darwin's Finches.* Princeton, NJ: Princeton University Press.

Grant, P.R. and Grant, B.R. 2014. *40 Years of Evolution: Darwin's Finches on Daphne Major Island.* Princeton, NJ: Princeton University Press.

Greenway, J.C. 1967. *Extinct and Vanishing Birds of the World.* New York: Dover Publications.

Griffin, D.R. 1958. *Listening in the Dark: the Acoustic Orientation of Bats and Men.* New Haven, CT: Yale University Press.

Harari, Y.N. 2014. *Sapiens: a Brief History of Humankind.* London: Harvill Secker.

Heads, M. 2013. *Biogeography of Australasia: a Molecular Analysis.* Cambridge: Cambridge University Press.

Hilty, S. 2005. *Birds of Tropical America: a Watcher's Introduction to Behaviour, Breeding, and Diversity.* Austin, TX: University of Texas Press.

Holyoak, D.T. 2001. *Nightjars and Their Allies.* Oxford: Oxford University Press.

Humboldt, A. and Bonpland, A. 1822. *Narrative of the Travels to the Equinoctial Regions of the New Continent, During the Years 1799–1804.* (trans. H.M. Williams).Vol III, 2nd edition. London: Longman, Hurst, Rees, Orme & Brown.

Humphries, C.J. and Parenti, L.R. 1999. *Cladistic Biogeography*, 2nd edition. Oxford: Oxford University Press.

Irschick, D.J., Briffa, M. and Podos, J. (eds). 2015. *Animal Signaling and Function: an Integrated Approach.* Hoboken, NJ: John Wiley & Sons.

Johnson, P. E. 1991. *Darwin on Trial.* Washington, DC: Regnery Gateway Publishing Co.

Kolbert, E. 2014. *The Sixth Extinction: an Unnatural History.* London: Bloomsbury.

König, C. and Weick, F. 2008. *Owls of the World*, 2nd edition. London: Helm.

Kuhn, T.S. 1962. *The Structure of Scientific Revolutions.* Chicago, IL: University of Chicago Press.

Laman, T. and Scholes, E. 2012. *Birds of Paradise: Revealing the World's Most Extraordinary Birds.* Washington, DC: National Geographic.

Lane, N. 2015. *The Vital Question: Why is Life the Way It is?* London: Profile Books.

López-Larrea, C. (ed.). 2012. *Sensing in Nature.* New York: Springer.

Losos, J.B. 2017. *Improbable Destinies: How Predictable is Evolution?* London: Allen Lane.

Low, T. 2014. *Where Song Began: Australia's Birds and How They Changed the World*. Melbourne: Penguin Random House.

MacArthur, R.H. and Wilson, E.O. 1967. *The Theory of Island Biogeography*. Princeton, NJ: Princeton University Press.

MacLeod, N. 2013. *The Great Extinctions: What Causes Them and How They Shape Life*. London: Natural History Museum.

Mayr, E. 1942. *Systematics and the Origin of Species from the Viewpoint of a Zoologist*. New York: Columbia University Press.

Mayr, E. 1963. *Animal Species and Evolution*. Cambridge, MA: Belknap Press.

Mayr, E. and Diamond, J. 2001. *The Birds of Melanesia: Speciation, Ecology, and Biogeography*. Oxford: Oxford University Press.

Mayr, G. 2009. *Paleogene Fossil Birds*. Berlin: Springer Verlag.

Mearns, B. and Mearns, R. 1998. *The Bird Collectors*. London: Academic Press.

Moffett, M.W. 2010. *Adventures Among Ants: a Global Safari with a Cast of Trillions*. Berkeley, CA: University of California Press.

Newton, I. 1972. *Finches*. New Naturalist Library. London: Collins.

Newton, I. 2003. *The Speciation and Biogeography of Birds*. London: Elsevier.

Newton, I., Kavanagh, R., Olsen, J. *et al.* (eds). 2002. *Ecology and Conservation of Owls*. Melbourne; CSIRO Publishing.

Ogilvie, M.A. 1978. *Wild Geese*. Berkhamsted: T. & A.D. Poyser.

Ohno, S. 1970. *Evolution by Gene Duplication*. Berlin: Springer-Verlag.

Pickrell, J. 2014. *Flying Dinosaurs: How Fearsome Reptiles Became Birds*. Sydney: NewSouth Publishing.

Pratt, T.K. and Beehler, B.M. 2015. *Birds of New Guinea*. Princeton, NJ: Princeton University Press.

Price, T. 2008. *Speciation in Birds*. Greenwood Village, CO: Roberts & Company.

Prins, H.H.T. and Namgail, T. (eds). 2017. *Bird Migration Across the Himalayas: Wetland Functioning Amidst Mountains and Glaciers*. Cambridge: Cambridge University Press.

Prum, R.O. 2017. *The Evolution of Beauty: How Darwin's Forgotten Theory of Mate Choice Shapes the Animal World – and Us*. New York: Doubleday.

Randall, L. 2015. *Dark Matter and the Dinosaurs: the Astounding Interconnectedness of the Universe*. London: Bodley Head.

Serventy, V. 1966. *A Continent in Danger*. London: Andre Deutsch.

Sibley, C.G. and Ahlquist, J.E. 1990. *Phylogeny and Classification of Birds: a Study in Molecular Evolution*. New Haven, CT: Yale University Press.

Simpson, G.G. 1944. *Tempo and Mode in Evolution*. New York: Columbia University Press.

Simpson, G.G. 1983. *Splendid Isolation: Curious History of South American Mammals*, new edition. New Haven, CT: Yale University Press.

Simpson, K. and Day, N. 1996. *A Field Guide to the Birds of Australia*, 5th edition. Melbourne: Penguin Books Australia.

Summers-Smith, J.D. 1988. *The Sparrows*. London: T. & A.D. Poyser.

Taylor, S. 2012. *John Gould's Extinct and Endangered Birds of Australia*. Canberra: NLA Publishing.

Tennyson, A.J.D. and Martison, P. 2006. *Extinct Birds of New Zealand*. Wellington: Te Papa Press.

Tudge, C. 2008. *Consider the Birds: Who They Are and What They Do*. London: Allen Lane.

van Grouw, K. 2012. *The Unfeathered Bird*. Princeton, NJ: Princeton University Press.

Wagner, A. 2014. *Arrival of the Fittest. Solving Evolution's Greatest Puzzle*. London: Oneworld.

Wallace, A.R. 1853. *A Narrative of Travels on the Amazon and Rio Negro*. London: Reeve & Co.

Wallace, A.R. 1889. *Darwinism*. London: Macmillan.

Weiner, J. 1994. *The Beak of the Finch: a Story of Evolution in Our Time*. New York: Knopf.

Wetmore, A. and Swales, B.H. 1931. *The Birds of Haiti and the Dominican Republic*. Bulletin 155. Washington, DC: Smithsonian Institute.

Woodley, K. 2012. *Shorebirds of New Zealand: Sharing the Margins*. London: Penguin.

Worthy, T.H. and Holdaway, R.N. 2002. *The Lost World of the Moa: Prehistoric Life of New Zealand*. Bloomington, IN: Indiana University Press.

Dramatis Personae

Below is a list of species mentioned in the text. The anglicised names are given as they appear in the book, together with their scientific names (see the IOC World Bird List – www.worldbirdnames.org). The names of families, such as tinamous and birds-of-paradise, are not included. Species are arranged in alphabetical order for convenience.

Abbott's Booby *Papasula abbotti*
Abyssinian Thrush *Turdus abyssinicus*
African Broadbill *Smithornis capensis*
African Grey Parrot *Psittacus erithacus*
African Penguin *Spheniscus demersus*
African Thrush *Turdus pelios*
Albert's Lyrebird *Menura alberti*
Alder Flycatcher *Empidonax alnorum*
American Barn Owl *Tyto furcata*
American Black Duck *Anas ruprides*
American Robin *Turdus migratorius*
Amsterdam Albatross *Diomedea amsterdamensis*
Andean Condor *Vultur gryphus*
Andean Goose *Chloephaga melanoptera*
Andean Hillstar *Oreotrochilus estella*
Anna's Hummingbird *Calypte anna*
Antipodean Albatross *Diomedea antipodensis*
Archbold's Bowerbird *Archboldia papuensis*
Arctic Redpoll *Acanthis hornemanni*
Ascension Frigatebird *Fregata aquila*
Atlantic Canary *Serinus canaria*
Atoll Starling *Aplonis feadensis*
Austral Thrush *Turdus falcklandii*
Band-rumped Storm Petrel *Oceanodroma castro*
Bar-headed Goose *Anser indicus*

Bar-tailed Godwit *Limosa lapponica*
Barnacle Goose *Branta leucopsis*
Bateleur *Terathopius ecaudatus*
Bee Hummingbird *Mellisuga helenae*
Black Robin *Petroica traversi*
Black Sicklebill *Epimachus fastosus*
Black Swan *Cygnus atratus*
Black Wheatear *Oenanthe leucura*
Black-browed Albatross *Thalassarche melanophris*
Black-footed Albatross *Phoebastria nigripes*
Black-legged Seriema *Chunga burmeisteri*
Black-necked Swan *Cygnus melanocoryphus*
Black-shouldered Kite *Elanus axillaris*
Black-tailed Godwit *Limosa limosa*
Black-tailed Leaftosser *Sclerurus caudacutus*
Black-winged Kite *Elanus caeruleus*
Blue Bird-of-Paradise *Paradisaea rudolphi*
Blue Jay *Cyanocitta cristata*
Blue-and-yellow Macaw *Ara ararauna*
Blue-backed Manakin *Chiroxiphia pareola*
Blue-crowned Manakin *Lepidothrix coronata*
Blue-grey Tanager *Thraupis episcopus*
Blue-winged Warbler *Vermivora cyanoptera*
Booted Eagle *Hieraaetus pennatus*
Boreal Owl *Aegolius funereus*
Brazilian Merganser *Mergus octosetaceus*
Brent Goose *Branta bernicla*
Bronze Parotia *Parotia berlepschi*
Brown Sicklebill *Epimachus meyeri*
Brünnich's Guillemot *Uria lomvia*
Budgerigar *Melopsittacus undulatus*
Buff-tailed Sicklebill *Eutoxeres condamini*
Buffy Tuftedcheek *Pseudocolaptes lawrencii*
Burrowing Owl *Athene cunicularia*
Bushwren *Xenicus longipes*
Cackling Goose *Branta hutchinsii*
California Condor *Gymnogyps californianus*
California Scrub Jay *Aphelocoma californica*
Cape Barren Goose *Cereopsis novaehollandiae*
Canada Goose *Branta canadensis*

Canary Islands Chiffchaff *Phylloscopus canariensis*
Cape Parrot *Poicephalus robustus*
Cape Sparrow *Passer melanurus*
Carolina Parakeet *Conuropsis carolinensis*
Carrion Crow *Corvus cornone*
Cassia Crossbill *Loxia sinesciuris*
Cebu Flowerpecker *Dicaeum quadricolor*
Chatham Island Robin *Petroica traversi*
Chatham Kaka *Nestor chathamensis*
Checker-throated Antwren *Epinecrophylla fulviventris*
Chestnut-bellied Monarch *Monarcha castaneiventris*
Chestnut-tailed Antbird *Myrmeciza hemimalaena*
Chico's Tyrannulet *Zimmerius chicomendesi*
Chiloe Wigeon *Anas sibilatrix*
Chimney Swift *Chaetura pelagica*
Chubut Steamer Duck *Tachyeres leucocephalus*
Cinnamon-bellied Flowerpiercer *Diglossa baritula*
Clark's Nutcracker *Nucifraga columbiana*
Club-winged Manakin *Machaeropterus deliciosus*
Coal-crested Finch *Charitospiza eucosma*
Cockatiel *Nymphicus hollandicus*
Cocos Finch *Pinaroloxias inornata*
Collared Flycatcher *Ficedula albicolis*
Common Blackbird *Turdus merula*
Common Buzzard *Buteo buteo*
Common Cactus Finch *Geospiza scandens*
Common Crossbill *Loxia curvirostra*
Common Nightingale *Luscinia megarhynchos*
Common Quail *Coturnix coturnix*
Common Raven *Corvus corax*
Common Redpoll *Acanthis flammea*
Common Starling *Sturnus vulgaris*
Common Swift *Apus apus*
Coscoroba Swan *Coscoroba coscoroba*
Crab Plover *Dromas ardeola*
Crested Eagle *Morphnus guianensis*
Crimson Fruitcrow *Haematoderus militaris*
Cuckoo Roller *Leptosomus discolor*
Curl-crested Manucode *Manucodia comrii*
Dodo *Raphus cucullatus*

Domestic Chicken *Gallus gallus*
Dunlin *Calidris alpina*
Dusky Broadbill *Corydon sumatranus*
Eastern Barn Owl *Tyto javanica*
Eastern Bluebird *Sialia sialis*
Eastern Buzzard *Buteo japonicus*
Eastern Meadowlark *Sturnella magna*
Eastern Osprey *Pandion cristatus*
Eastern Phoebe *Sayornis phoebe*
Eclectus Parrot *Eclectus roratus*
Egyptian Plover *Pluvianus aegyptius*
Elegant Pitta *Pitta elegans*
Emperor Bird-of-Paradise *Paradisaea guilielmi*
Emperor Penguin *Aptenodytes forsteri*
Emu *Dromaius novaehollandiae*
Eurasian Eagle-Owl *Bubo bubo*
Eurasian Skylark *Alauda arvensis*
Eurasian Wren *Troglodytes troglodytes*
European Robin *Erithacus rubecula*
Fairy Pitta *Piita nympha*
Falkland Steamer Duck *Tachyeres brachypterus*
Fan-tailed Berrypecker *Melanocharis versteri*
Fawn-breasted Bowerbird *Chlamydera cerviniventris*
Fieldfare *Turdus pilaris*
Fischer's Lovebird *Agapornis fischeri*
Flammulated Owl *Psiloscops flammeolus*
Florida Scrub Jay *Aphelocoma coerulescens*
Fuegian Steamer Duck *Tachyeres pteneres*
Galápagos Mockingbird *Mimus parvulus*
Galápagos Penguin *Spheniscus meniculus*
Galah *Eolophus roseicapilla*
Gang-gang Cockatoo *Callocephalon fimbriatum*
Garnet Pitta *Erythropitta granatina*
Genovesa Cactus Finch *Geospiza propinqua*
Giant Hummingbird *Patagona gigas*
Giant Ibis *Thaumatibus gigantia*
Gibson's Albatross *Diomedea gibsoni*
Glossy-mantled Manucode *Manucodia ater*
Goldcrest *Regulus regulus*
Golden Eagle *Aquila chrysaetos*

Golden-breasted Starling *Lamprotornis regius*
Golden-capped Parakeet *Aratinga auricapillus*
Golden-fronted Bowerbird *Amblyornis flavifrons*
Golden-headed Manakin *Pipra erythrocephala*
Golden-winged Warbler *Vermivora chrysoptera*
Grauer's Broadbill *Pseudocalyptomena graueri*
Great Auk *Pinguinus impennis*
Great Bowerbird *Chlamydera nuchalis*
Great Grey Owl *Strix nebulosa*
Great Horned Owl *Bubo virginianus*
Great Knot *Calidris tenuirostris*
Great Sapphirewing *Pterophanes cyanopterus*
Greater Bird-of-Paradise *Paradisaea apoda*
Greater Blue-eared Starling *Lamprotornis chalybaeus*
Greater Flowerpiercer *Diglossa major*
Greater Sooty Owl *Tyto tenebricosa*
Greater Vasa Parrot *Caracoptis vasa*
Green-winged Saltator *Saltator similis*
Grey Tinamou *Tinamus tao*
Grey Warbler-Finch *Certhidea fusca*
Grey-bellied Flowerpiercer *Diglossa carbonaria*
Grey-headed Broadbill *Smithornis sharpei*
Grey-headed Lovebird *Agapornis canus*
Grey-headed Picathartes *Picathartes orea*s
Greylag Goose *Anser anser*
Guaiabero *Bolbopsittacus lunulatus*
Haast's Eagle *Harpagornis moorei*
Harpy Eagle *Harpia harpyja*
Hawaiian Crow *Corvus hawaiiensis*
Hawaiian Duck *Anas wyvilliana*
Hawaiian Petrel *Pterodroma sandwichensis*
Heath Hen *Tympanuchus cupido cupido*
Hepatic Tanager *Piranga hepatica*
Himalayan Buzzard *Buteo burmanius*
Hispaniolan Crossbill *Loxia megaplaga*
Hoatzin *Opisthocomus hoazin*
Hooded Crow *Corvus cornix*
Hose's Broadbill *Calyptomena hosii*
House Sparrow *Passer domesticus*
Hyacinth Macaw *Anodorhynchus hyacinthinus*

Italian Sparrow *Passer italiae*
Jerdon's Babbler *Chrysomma altirostre*
Junin Grebe *Podiceps taczanowski*
Kagu *Rhynochetos jubatus*
Kakapo *Strigops habroptilus*
Kea *Nestor notabilis*
King of Saxony Bird-of-Paradise *Pteridophora alberti*
King Penguin *Aptenodytes patagonicus*
Klages's Antwren *Myrmotherula klagesi*
Lake Duck *Oxyura vittata*
Large Ground Finch *Geospiza magnirostris*
Lawes's Parotia *Parotia lawesii*
Laysan Albatross *Phoebastria immutabilis*
Laysan Duck *Anas laysanensis*
Lear's Macaw *Anodorhynchus leari*
Lesser Antillean Saltator *Saltator albicollis*
Lesser Bird-of-Paradise *Paradisaea minor*
Lesser Redpoll *Acanthis cabaret*
Letter-winged Kite *Elanus scriptus*
Little Eagle *Hieraaetus morphnoides*
Long-tailed Widowbird *Euplectes progne*
Louisiade White-eye *Zosterops griseotinctus*
MacGregor's Bowerbird *Amblyornis macgregoriae*
Madagascan Buzzard *Buteo brachypterus*
Madeiran Storm Petrel *Oceanodroma castro*
Magellanic Plover *Pluvianellus socialis*
Magnificent Bird-of-Paradise *Diphyllodes magnificus*
Magnificent Riflebird *Ptiloris magnificus*
Magpie Goose *Anseranas semipalmata*
Mallard *Anas platyrhynchos*
Marquesan Ground Dove *Gallicolumba rubescens*.
Marsh Warbler *Acrocephalus palustris*
Marsh Wren *Cistothorus palustris*
Masked Duck *Nomonyx dominicus*
Masked Flowerpiercer *Diglossa cyanea*
Medium Ground Finch *Geospiza fortis*
Mexican Duck *Anas diazi*
Mexican Jay *Aphelocoma wollweberi*
Mistle Thrush *Turdus viscivorus*
Monteiro's Storm Petrel *Oceanodroma monteiro*

Mottled Duck *Anas fulvigula*
Muscovy Duck *Cairina moschata*
Nanday Parakeet *Nandayus nenday*
Nēnē *Branta sandvicensis*
New Caledonian Crow *Corvus moneduloides*
New Caledonian Lorikeet *Charmosyna diadema*
New Caledonian Owlet-nightjar *Aegotheles savesi*
New Zealand Kaka *Nestor meridionalis*
New Zealand Merganser *Mergus australis*
New Zealand Rockwren *Xenicus gilviventris*
New Zealand Storm Petrel *Fregetta maoriana*
Night Parrot *Pezoporus occidentalis*
Noisy Scrubbird *Atrichornis clamosus*
Norfolk Kaka *Nestor productus*
Northern Hawk Owl *Surnia ulula*
Northern Pygmy Owl *Glaucidium californicum*
Northern Saw-whet Owl *Aegolius acadicus*
Oilbird *Steatornis caripensis*
Palm Cockatoo *Probosciger aterrimus*
Palmchat *Dulus dominicus*
Papuan Eagle *Harpyopsis novaeguinae*
Papuan Treecreeper *Cormobates placens*
Paradise Parrot *Psephotus pulcherrimus*
Paradise-crow *Lycocorax pyrrhopterus*
Parrot Crossbill *Loxia pytyopsittacus*
Passenger Pigeon *Ectopistes migratorius*
Philippine Eagle *Pithecophaga jefferyi*
Pied Flycatcher *Ficedula hypoleuca*
Pink-footed Goose *Anser brachyrhynchus*
Pink-headed Duck *Rhodonessa caryophyllacea*
Plains-wanderer *Pedionomus torquatus*
Plushcap *Catamblyrhynchus diadema*
Pohnpei Starling *Aplonis pelzelni*
Point-tailed Palmcreeper *Berlepschia rikeri*
Poo-uli *Melamprosops phaeosoma*
Przevalski's Finch *Urocynchramus pylzowi*
Puerto Rican Amazon *Amazona vittata*
Puerto Rican Screech Owl *Megascops nudipes*
Rail-babbler *Eupetes macrocerus*
Rainbow Pitta *Pitta iris*

Raja Shelduck *Tadorna radjah*
Rarotonga Starling *Aplonis cinerascens*
Razorbill *Alca torda*
Red Crossbill *Loxia curvirostra*
Red Knot *Calidris canutus*
Red-breasted Goose *Branta ruficollis*
Red-headed Manakin *Ceratopipra rubracapilla*
Red-legged Seriema *Cariama cristata*
Red-necked Buzzard *Buteo auguralis*
Red-winged Blackbird *Agelaius phoeniceus*
Redwing *Turdus iliacus*
Regent Bowerbird *Sericulus chrysocephalus*
Rifleman *Acanthisitta chloris*
Ring Ouzel *Turdus torquatus*
Rinjani Scops Owl *Otus jolandae*
River Tyrannulet *Serpophaga hypoleuca*
Rodrigues Solitaire *Pezophaps solitaria*
Rondonia Warbling Antbird *Hypocnemis ochrogyna*
Rook *Corvus frugilegus*
Rose-ringed Parakeet *Psittacula krameri*
Rosy-faced Lovebird *Agapornis roseicollis*
Ruby-throated Hummingbird *Archilochus colubris*
Ruddy Duck *Oxyura australis*
Ruddy Treerunner *Margarornis rubiginsus*
Ruddy Turnstone *Arenaria interpres*
Rufous Scrubbird *Atrichornis rufescens*
Rufous-sided Broadbill *Smithornis rufolateralis*
Sapayoa *Sapayoa aenigma*
Satin Bowerbird *Ptilonorhynchus violaceus*
Scaled Flowerpiercer *Diglossa duidae*
Scaled Spinetail *Craniolenea muelleri*
Scaly-throated Leaftosser *Sclerurus guatemalensis*
Scarlet Tanager *Piranga olivacea*
Secretarybird *Sagittarius serpentarius*
Senegal Parrot *Poicephalus senegalus*
Seychelles Parakeet *Psittacula wardi*
Seychelles Warbler *Acrocephalus sechellensis*
Sharp-beaked Ground Finch *Geospiza difficilis*
Sharp-tailed Streamcreeper *Lochmias nematura*
Short-tailed Albatross *Phoebastria albatrus*

Short-tailed Pygmy Tyrant *Myiornis ecaudatus*
Silvereye *Zosterops lateralis*
Silvery Grebe *Podiceps occipitalis*
Slaty Flowerpiercer *Diglossa plumbea*
Slender-billed Curlew *Numenius tenuirostris*
Snethlage's Tody-Tyrant *Hemitriccus minor*
Solomons Frogmouth *Rigidipenna inexpectata*
Song Thrush *Turdus philomelos*
Sooty Shearwater *Ardenna grisea*
South Georgia Pipit *Anthus antarcticus*
South Island Piopio *Turnagra capensis*
Spanish Sparrow *Passer hispaniolensis*
Spectacled Warbler *Sylvia conspicillata orbitalis*
Spix's Macaw *Cyanopsitta spixii*
Spot-backed Antbird *Hylophylax naevius*
Spot-backed Antwren *Herpsilochmus dorsimaculata*
Spot-crowned Woodcreeper *Lepidocolaptes affinis*
Spotless Starling *Sturnus unicolor*
Spotted Bowerbird *Chlamydera maculata*
Spotted Owl *Strix occidentalis*
Stephens Island Wren *Traversia lyalli*
Steppe Buzzard *Buteo buteo vulpinus*
Stout-legged Wren *Pachyplichas yaldwyni*
Streaked Berrypecker *Melanocharis striativentris*
Streaked Bowerbird *Amblyornis subalaris*
Sucunduri Flycatcher *Tolmomyias sucunduri*
Sulphury Flycatcher *Tyrannopsis sulphurea*
Sun Parakeet *Aratinga solstitialis*
Superb Lyrebird *Menura novaehollandiae*
Superb Pitta *Pitta superba*
Superb Starling *Lamprotornis superbus*
Sword-billed Hummingbird *Ensifera ensifera*
Taiga Bean Goose *Anser fabalis*
Taita Thrush *Turdus helleri*
Tawny Owl *Strix aluco*
Tooth-billed Bowerbird *Scenopoeetes dentirostris*
Tristan Albatross *Diomedea dabbenena*
Trumpet Manucode *Phonygammus keraudrenii*
Two-barred Crossbill *Loxia leucoptera*
Upland Buzzard *Buteo hemilasius*

Ural Owl *Strix uralensis*
Vampire Ground Finch *Geospiza septentrionalis*
Vegetarian Finch *Platyspiza crassirostris*
Violet-backed Starling *Cinnyricinclus leucogaster*
Vogelkop Bowerbird *Amblyornis inornatus*
Wandering Albatross *Diomedea exulans*
Wedge-billed Woodcreeper *Glyphorynchus spirurus*
Wedge-tailed Eagle *Aquila audax*
Western Barn Owl *Tyto alba*
Western Meadowlark *Sturnella neglecta*
Western Osprey *Pandion haliaetus*
White-breasted Antbird *Rhegmatorhina hoffmannsi*
White-cheeked Starling *Poliopsar cineraceus*
White-tailed Kite *Elanus leucurus*
White-throated Dipper *Cinclus cinclus*
Whitehead's Broadbill *Calyptomena whiteheadi*
Whooping Crane *Grus americana*
Wilson's Bird-of-Paradise *Diphyllodes respublica*
Woodpecker Finch *Camarhynchus pallidus*
Yellow-billed Teal *Anas flavirostris*
Yellow-breasted Bowerbird *Chlamydera lauterbachi*
Yellow-headed Picathartes *Picathartes gymnocephalus*
Zebra Finch *Taeniopygia guttata*

Index

Page numbers in *italics* refer to figures

Also from Pelagic Publishing

Ornithology

The Barn Owl Conservation Handbook – Barn Owl Trust
Fighting for Birds – Mark Avery
Urban Peregrines – Ed Drewitt
Bird Conservation – David Williams *et al*
Behind the Binoculars – Mark Avery & Keith Betton
Lapwings, Loons and Lousy Jacks – Ray Reedman
Behind More Binoculars – Keith Betton & Mark Avery

Popular Wildlife

Shrewdunnit – Conor Mark Jameson
Call of Nature: The Secret Life of Dung – Richard Jones

Data in the Wild Series

Remote Sensing and GIS for Ecologists – Wegmann, Leutner & Dech
Community Ecology – Mark Gardener
Camera Trapping for Wildlife Research – Rovero & Zimmermann
CCTV for Wildlife Monitoring – Susan Young
Statistics for Ecologists Using R and Excel – Mark Gardener

Naturalists' Handbooks Series

Bumblebees – Oliver Prŷs-Jones & Sarah Corbet
Ladybirds – Roy *et al*
Solitary bees – Ted Benton
Ponds and small lakes – Brian Moss
Leaf beetles – Dave Hubble